Electromagnetism and Special Methods for Electric Circuits Analysis

Online at: https://doi.org/10.1088/978-0-7503-5854-5

Electromagnetism and Special Methods for Electric Circuits Analysis

Ştefan Antohe
Faculty of Physics, University of Bucharest, Bucharest, Romania

Vlad-Andrei Antohe
Faculty of Physics, University of Bucharest, Bucharest, Romania

IOP Publishing, Bristol, UK

ISBN 978-0-7503-5854-5 (ebook)
ISBN 978-0-7503-5852-1 (print)
ISBN 978-0-7503-5855-2 (myPrint)
ISBN 978-0-7503-5853-8 (mobi)

DOI 10.1088/978-0-7503-5854-5

Version: 20250101

IOP ebooks

British Library Cataloguing-in-Publication Data: A catalogue record for this book is available from the British Library.

Published by IOP Publishing, wholly owned by The Institute of Physics, London

IOP Publishing, No.2 The Distillery, Glassfields, Avon Street, Bristol, BS2 0GR, UK

US Office: IOP Publishing, Inc., 190 North Independence Mall West, Suite 601, Philadelphia, PA 19106, USA

The authors dedicate this book to their families.

Contents

Preface x

Acknowledgements xi

Author biographies xii

Introduction xiv

1 The electrokinetic state: direct electric current **1-1**

1.1 Electric current 1-1
1.2 Electric current densities 1-3
1.3 Fundamental laws of electrokinetics 1-8
1.4 Microscopic approach to electric conduction in metals 1-27
1.5 Potential energy of electrons in solids 1-36
References and further reading 1-42

2 Electric charge transport in media other than metals **2-1**

2.1 Electric conduction of free electrons in vacuum 2-1
2.2 Electrical conduction in ionized gases 2-16
2.3 Electric conduction in electrolytic solutions 2-18
2.4 Thermoelectric effects 2-23
References and further reading 2-34

3 Direct current circuits **3-1**

3.1 Electromotive force and the induced electric field 3-1
3.2 Ohm's law for inhomogeneous conducting media 3-8
3.3 Joule–Lenz law for inhomogeneous conducting media 3-10
3.4 Linear and filiform DC circuits 3-12
References and further reading 3-31

4 Magnetic field of the electrokinetic state **4-1**

4.1 Magnetic fields. Definition and characterization 4-1
4.2 Magnetic fields of DC circuits in vacuum 4-13
4.3 Lorentz force 4-49
4.4 Force of interaction between two circuits carrying electric currents 4-56
4.5 Magnetic dipole moment 4-57
4.6 Force and torque exerted by a magnetic field on an electric circuit 4-61

4.7 Energy of a magnetic moment placed in a magnetic field 4-65
References and further reading 4-66

5 Electromagnetic induction and self-induction 5-1
5.1 Electromagnetic induction 5-1
5.2 Self-induction 5-16
Further reading 5-24

6 Electromagnetic energy produced by electric current 6-1
6.1 Energy stored in the magnetic fields of coils 6-1
6.2 Electromagnetic energy of a system of electrical circuits 6-5
6.3 Location of magnetic energy. Magnetic energy density 6-7
6.4 The theorems of generalized forces in magnetic fields 6-9
Further reading 6-13

7 Magnetization state 7-1
7.1 Magnetic properties of substances 7-1
7.2 The relationship between the vectors $\vec{B}$, $\vec{H}$, and $\vec{M}$ 7-5
7.3 Microscopic approach to the magnetization of substances 7-11
7.4 The precession of a magnetic moment in magnetic field 7-25
7.5 Diamagnetism 7-27
7.6 Paramagnetism 7-30
7.7 Ferromagnetism 7-36
Further reading 7-50

8 Transient phenomena in electrical circuits 8-1
8.1 DC and AC electrical circuits in the transient regime 8-1
Further reading 8-20

9 Electrical circuits in the alternating current regime 9-1
9.1 Physical quantities describing alternating current 9-1
9.2 Analysis of AC electrical circuits 9-6
9.3 Power in alternating current circuits 9-18
9.4 Resonance in AC circuits 9-24
9.5 Complex representation for alternating current 9-34
9.6 Special methods for the analysis of AC electric circuits 9-47

9.7 Complex representation of power in AC circuits 9-54
Further reading 9-57

10 Electromagnetic fields and electromagnetic waves 10-1

10.1 Maxwell's equations 10-1
10.2 The electromagnetic wave equation 10-3
10.3 Electromagnetic energy theorem and the Poynting vector 10-9
Further reading 10-11

Preface

This textbook is the second in a series of two, covering the ***Electricity and Magnetism*** course unit offered by the Faculty of Physics at the University of Bucharest. The training of a physicist cannot be considered complete without a deep knowledge of the complex mechanisms and phenomena related to classical electromagnetism. Therefore, the main goal of this course is to provide a thorough understanding of electric and magnetic phenomena, starting from a phenomenological analysis of the main laws governing the electromagnetism. Essentially, this important part of physics can be summarized by the complete system of Maxwell's equations, whose establishment is pursued as a leitmotif throughout a full academic year dedicated to electricity and magnetism at the Faculty of Physics.

While the first part of the series, already published [1], is exclusively focused on the study of electrostatic fields in both vacuum and condensed matter, this second part, entitled '***Electromagnetism and Special Methods for Electric Circuits Analysis***', describes phenomena related to moving electric charges (i.e. electrokinetics and magnetism). Like the previous part, it represents another set of 14 lectures (of two hours each) held for students enrolled in different physics programs. Its content was developed over the last 30 years, including consistent and regular updates required by various changes in curricula within the Faculty of Physics and other science faculties. In particular, this textbook is dedicated to studying the electric and magnetic fields in both vacuum and matter, being organized into a brief introduction and ten chapters. Each chapter includes a list of a few references, which collectively contributed to shaping the entire lecture as it is captured in this second textbook. Notably, like the first preceding textbook, which is completed by two appendices detailing specific elements of algebra and vectorial analysis, this second textbook also introduces math operators (i.e. gradient, divergence, curl, and Laplace operators) starting from the physical premises that require their use.

The authors consider that the way the content is provided is a very efficient approach to gaining new knowledge. Moreover, this textbook contains practical examples and exercises, written in *italic* characters, that help the reader to assimilate the presented theoretical concepts. The authors strongly believe that this work will attract the attention of readers dealing with the theory and applications of classical electromagnetism, since it is particularly intended to serve as an academic course support for international students of various degrees (i.e. bachelor, engineer, master, and PhD) working in the domains of physics and the exact sciences in general.

Acknowledgements

During the elaboration of this work, the authors respectfully remembered their professors, who laid the foundations of this discipline at the Faculty of Physics at the University of Bucharest, Romania. The authors are also deeply grateful to their colleagues from the Department of Electricity, Solid State Physics and Biophysics, as well as to the members of the Research and Development Center for Materials and Electronic & Optoelectronic Devices (MDEO), who sincerely engaged in various discussions and an exchange of ideas on the theory of classical electromagnetism and its best teaching practices.

Professor Ştefan Antohe, PhD
Professor Vlad-Andrei Antohe, PhD Eng. Habil.
Măgurele—Bucharest, 2024

Author biographies

Mr Ştefan Antohe, PhD
Professor Emeritus
University of Bucharest, Faculty of Physics,
Bucharest - Măgurele, Romania
Academy of Romanian Scientists (AOSR),
Ilfov Street 3, 050045 Bucharest, Romania

Mr Ştefan Antohe graduated from the Faculty of Physics, the University of Bucharest, Romania, in 1977, and in 1994, he completed a PhD program in physics at the same institution. He is currently **professor emeritus** at the University of Bucharest, **head** of the Research and Development Center for Materials and Electronic and Optoelectronic Devices (MDEO) of the Faculty of Physics, and a **full member** of the Academy of Romanian Scientists (AOSR).

His main research activities are dedicated to the physics of organic and inorganic semiconducting thin films and nanomaterials, with a special emphasis on the investigation of charge transport mechanisms within materials and interfaces. He also possesses great knowledge of fabrication technologies and the characterization of various electronic and optoelectronic devices, especially photovoltaic cells and transparent field-effect transistors. He has coauthored more than 250 scientific publications in International Scientific Indexing (ISI)-indexed journals, acquiring more than 3000 independent citations, and 25 books or book chapters with national and international publishing houses. He has supervised more than 50 PhD and postdoctoral fellows and has also coordinated numerous research projects.

Mr Vlad-Andrei Antohe, PhD Eng. Habil.
Full Professor
University of Bucharest, Faculty of Physics,
Bucharest - Măgurele, Romania

In 2002, Mr Vlad-Andrei Antohe received his bachelor's degree (BSc) in physics education, and in 2005, he received his master's diploma (MSc) in physical electronics from the Faculty of Physics, University of Bucharest, Romania. In 2003, he also obtained an engineering degree (MEng) in electronics from the Faculty of Electronics and Telecommunications, Polytechnic University of Bucharest, Romania. In 2012, he was awarded a PhD diploma in applied engineering sciences by the Catholic University of Louvain, Belgium, where he continued his postdoctoral studies until 2016. He was then an associate professor at the University of Bucharest, and since February 2024, he has been a **full professor** at the same institution.

In addition to his teaching activities with undergraduate and graduate students, his main research interests lie in the areas of materials science and nanotechnology, with a particular focus on the development and investigation of nanostructured materials and low-dimensional architectures, with the aim of generating novel structural arrangements tailored to specific desired properties. He also has expertise in the fabrication and characterization of electronic and optoelectronic devices based on inorganic, organic, or hybrid organic/inorganic nanostructured materials, such as photovoltaic cells, sensors and biosensors, magnetic media for various applications, as well as microbatteries, microsupercapacitors, and other electrochemical systems. He has coauthored around 65 scientific publications, of which more than 15 ISI-indexed research papers were published in journals with high impact factors ranging approximately from 5 to 32.

Introduction

The electromagnetic force, like the gravitational force, is a fundamental force of nature that acts on all things that possess electric charge, just as gravity acts on all things that have mass. The electromagnetism is thus the dominant force in interactions between atoms and molecules that occur via electromagnetic fields. Effectively, the electromagnetism can be interpreted as a combination of electricity and magnetism, which are different but closely related phenomena.

The electromagnetism has been studied since ancient times, although electricity and magnetism were originally considered to be two distinct and unrelated forces. Their study began about 50 centuries ago when ancient Chinese, Mayan, and Egyptian civilizations exploited the attractive properties of naturally occurring magnetite, incorporating it within their works of art and architecture. Later on, **the magnetism** was actually discovered when people noticed that the mineral magnetite could attract iron. The word 'magnet' comes from the Greek term *magnetis lithos,* which is translated as 'magnesian stone'. At that time, ancient people were also aware of lightning and static electricity, although they possessed no effective knowledge about the mechanisms behind these phenomena. In the 6th century BCE, the Greek philosopher *Thales of Miletus*[1] discovered that amber (*elektron* in Greek) could acquire an electric charge when it was rubbed with cloth, which allowed it to pick up light objects such as pieces of straw. This is considered the first observation leading to the discovery of **the electricity**. The first scientific discussion of magnetism is considered to have occurred between *Aristotle*[2] and Thales of Miletus. However, the first scientific publication documenting electric and magnetic phenomena was written much later, i.e. in 1600 CE, by the English physicist *William Gilbert*. The work was entitled *De Magnete*, and it is mainly of historical importance, as the correlation between electric and magnetic phenomena was not yet clearly established.

The laws of electricity and magnetism were practically discovered by experimentalists who had little or no knowledge of the modern theory of the atomic and molecular structure of matter. For instance, the law describing electrostatic interactions between differently charged bodies was discovered in 1785 by the French physicist *Charles A Coulomb*, who established for the first time that the force of interaction between two point charges placed in vacuum and separated by a distance d between their centers is proportional to the product of their charges, inversely proportional to the square of d, and always oriented along the line joining their centers. In 1799, the Italian physicist *Alessandro Volta* built the first voltaic pile (electric battery), thus creating the possibility of extending electricity and magnetism experiments to many laboratories all over the world. Consequently, the forthcoming

[1] Thales (624–546 BC) was a Greek mathematician, astronomer, and philosopher from Miletus (an ancient Greek city on the western coast of Asia Minor). He was one of the Seven Sages of Greece, recognized as the first in Western civilization to engage in scientific philosophy.

[2] Aristotle (384–322 BC) was an ancient Greek philosopher and polymath, as well as the founder of the Peripatetic school of philosophy in the Lyceum in Athens, which laid the groundwork for the development of modern science.

19th century was characterized by substantial progress in our fundamental understanding of electric and magnetic phenomena, as well as their interaction. In this context, the first experimental investigation of the interaction between coils carrying electric currents was performed by the French physicist and mathematician *André-Marie Ampère* during 1820–25; his work was followed by that of the scientists *Hans C Oersted*, *Jean-Baptiste Biot,* and *Félix Savart.* Ampère found that two long parallel wires carrying electric currents in opposite directions repel each other, whereas when carrying electric currents in the same direction, they attract each other. Furthermore, it was also found that if a coil carrying an electric current is placed close to a compass needle, it experiences both a force and a couple. Under these circumstances, the group of scientists concluded that both the magnet and the coil carrying the current produce a magnetic field, taking into account the relation between electric and magnetic phenomena. Later on, in 1826, *Georg S Ohm* described the law of electric conduction that carries his name. In 1831–33, *Michael Faraday* and *Heinrich F E Lenz* discovered the law of electromagnetic induction, but their names are commonly used for other discoveries, such as Faraday's law of electrolysis and the Joule–Lenz law of electro-caloric effects.

The application of Faraday's ideas to electromagnetism, including his writing of the mathematical equations and his development of the field concept, is due to the English physicist *James C Maxwell.* In his work, *'Treatise about Electricity and Magnetism',* written in 1873, he established the basis of the macroscopic theory of electromagnetic fields, effectively challenging the view of electricity and magnetism as separate phenomena and strengthening the idea that the interactions between positive and negative charges are mediated by one force. In concrete terms, Maxwell theoretically foresaw the existence of electromagnetic waves and the 'displacement current' in 1862. Using these concepts, he created the electromagnetic theory of light in 1865. The experimental confirmation of Maxwell's theory was made by *Heinrich Hertz* in 1889 through his experiments on the propagation of electromagnetic waves. Essentially, the physicists of the 19th century (such as Ampère, Faraday, and Maxwell) shaped the basis of classical electromagnetism, while the physicists and chemists of the 20th century contributed further to our understanding of electromagnetism by introducing the atomic and molecular structure of matter.

Classical electromagnetism deals with the study of both stationary and moving charges. The study of stationary charges is the aim of **electrostatics** [1], while the study of moving charges is the aim of **electrodynamics**. Maxwell showed that the laws of **electromagnetism** can be expressed in the form of four differential equations, referring to the components of electric and magnetic fields. The table of Maxwell's equations, as written in 1876, is appended below:

$$\nabla \times \vec{E} = -\frac{\partial \vec{B}}{\partial t} \tag{I.1}$$

$$\nabla \times \vec{H} = \vec{j} + \frac{\partial \vec{D}}{\partial t} \tag{I.2}$$

$$\nabla \vec{D} = \rho \tag{I.3}$$

$$\nabla \vec{B} = 0 \tag{I.4}$$

Equation (I.1) is derived from Faraday's and Lenz's laws of electromagnetic induction and shows that a time-dependent magnetic field is the source of a time-dependent electric field whose lines enclose those of the magnetic field. Equation (I.2) is obtained from the law of magnetic circuits as a special case of Ampère's law (that takes into account only two sources for the magnetic field, namely, conduction current and displacement current). In the absence of the conduction current, it similarly shows that a time-dependent electric field is the source of a time-dependent magnetic field whose lines enclose those of the electric field. Equation (I.3) is a local form of Gauss's law (the theorem of electric flux) as applied to the electric field, and it describes the direct relationship between an electric field and its own sources. Equation (I.4) comes from Gauss's law (the theorem of magnetic flux) as applied to the magnetic field; it shows that the magnetic field has no polar sources, being a field with closed lines. Notably, equations (I.1) and (I.2) reveal that an electromagnetic field is an ensemble of electric and magnetic fields that can be generated reciprocally; such an electromagnetic field can leave its source, giving rise to a propagating electromagnetic wave.

Maxwell's equations were written long before the development of quantum mechanics and relativity theory. Nevertheless, they have not undergone any changes to date. The importance of the classical theory of electromagnetism in modern physics can be explained by the fact that Maxwell's equations are perfectly compatible with relativity theory. Eventually, changes arise from the impact of quantum theory on electromagnetic forces, which are important only at very small distances (10^{-10} cm), in other words, at distances a hundred times smaller than the atomic diameter. Therefore, the same laws of classical electromagnetism can also be used to study interactions at the atomic and molecular level (or scale). However, for distances smaller than 10^{-10} cm, quantum electrodynamics must also be taken into account to understand interactions at that level.

This book, entitled '*Electromagnetism and Special Methods for Electric Circuits Analysis*', was produced to complete the first part, entitled '*Electrostatics. Formalism of the Electrostatic Field in Vacuum and Matter*' [1]. Together, these books are entirely dedicated to studying the classical laws of electricity and magnetism by progressively demonstrating and constructing the entire system of Maxwell's equations for the electromagnetic field presented above.

References and further reading

[1] Antohe Ş and Antohe V-A 2023 *Electrostatics. Formalism of the Electrostatic Field in Vacuum and Matter* 1st edn (Bristol: IOP Publishing) 266 pages

[2] Feynman R P, Leighton R B and Sands M 2011 *Feynman Lectures on Physics, Vol. II: The New Millennium Edition: Mainly Electromagnetism and Matter* (Feynman Lectures on Physics) (New York: Basic Books) 589 pages

[3] Jackson J D 1998 *Classical Electrodynamics* 3rd edn (New York: Wiley) 832 pages
[4] Purcell E M and Morin D J 2013 *Electricity and Magnetism* 3rd edn (Cambridge: Cambridge University Press) 853 pages
[5] Von Laue M 2018 *History of Physics* ed V Petkov (Montréal, QC: Minkowski Institute Press) 184 pages
[6] Maxwell J C 1954 *A Treatise on Electricity and Magnetism* 3rd edn vol 1 (Mineola, NY: Dover) 560 pages

IOP Publishing

Electromagnetism and Special Methods for Electric Circuits Analysis

Ştefan Antohe and Vlad-Andrei Antohe

Chapter 1

The electrokinetic state: direct electric current

Stationary electric charges are capable of producing electric fields in vacuum or in different materials. This state is typically referred to as *static electricity*, the study of which is **electrostatics** [1]. In contrast, when electric charges are free to move in space, they produce electric currents; this state, referred to as *dynamic electricity*, is studied within the field of **electrodynamics**. This latter branch of physics is particularly important in classical field theory, as it allows us to understand classical ***electromagnetism*** by studying the interactions between electric charges and currents. In this chapter, specific notions of electrodynamics are introduced, with the main aim of defining the laws governing ***electrokinetics***, i.e. the part related to electric current and circuits.

1.1 Electric current

If a free charge is placed in an electric field, it is acted upon by a force and it moves in a direction determined by the line of action of that force. Thus, if an initial potential difference exists in a conductor, charge moves until it reaches an equilibrium position and the whole conductor becomes an equipotential surface.

Let us now consider two conducting bodies, A and B, charged with electric charge, having potentials V_A and V_B (with $V_A > V_B$) versus a common reference potential (assumed to be null for simplicity). By connecting a battery between the two conductors (A and B), a permanent potential difference may be maintained between them. Subsequently, connecting the two conducting bodies through a conducting wire (see figure 1.1) causes a movement of electric charge through the metallic wire from conductor A (with a higher potential) toward conductor B (with a lower potential); the movement takes place until the two potentials become equal. During this time, when the electric charge moves through the conducting wire, the system formed by the two conductors and the metallic wire is no longer in

doi:10.1088/978-0-7503-5854-5ch1

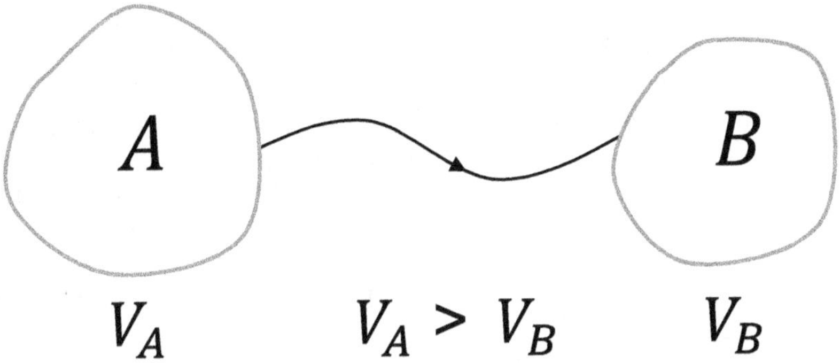

Figure 1.1. Two conducting bodies initially at different potentials, connected by a conducting wire.

electrostatic equilibrium. Under these conditions, the system behaves like a single conducting body whose surface is no longer an equipotential surface, since the electric potential takes different values at different points inside that conducting body.

Experimentally, it has been observed that the continuous movement of charge through a conductor is accompanied by different effects, such as heating, magnetic, chemical, and mechanical effects. Furthermore, all these effects can be evidenced at the same time as the movement of charge, and they vanish as soon as the electric charge ceases to move. The electric state associated with the movement of electric charge is called the ***electrokinetic state.*** The oriented movement of electric charge is called an ***electric current***. Hence, the electrokinetic state is the state associated with the electric current. To characterize the strength of the effects produced by the electric current, we must introduce a physical quantity that characterizes the strength of the electric current, i.e. the ***current intensity***, defined at the macroscopic scale as the rate at which charge passes through a given area of a conductor, such as the cross-section of a metallic wire:

$$i = \frac{dQ}{dt} \tag{1.1}$$

where dQ is the charge flowing in a time dt through the cross-section of the conductor. If we consider the relation (1.1) to be valid at the macroscopic scale, *the intensity of the electric current can be generally defined as the scalar physical quantity numerically equal to the speed at which the electric charge moves through the cross-section of the conductor:*

$$i = \lim_{\Delta t \to 0} \frac{\Delta q}{dt} \tag{1.2}$$

In the particular case of metals, a *unipolar* charge transport mechanism exists, namely the movement of quasi-free electrons with respect to the nodes of the crystal lattice. In contrast, conduction in electrolytes and semiconductors is *ambipolar*, the charge carriers being either the positive and negative ions within electrolytes or the electrons and holes within semiconductors, respectively. By convention, the positive direction of an electric current is defined as the direction of movement of positive electric charges through the conducting medium, i.e. by the direction of the electric field present in the conducting medium. Therefore, it can be stated that in a conducting medium, the current flows from a higher potential to a lower one.

The current intensity is measured in *amperes* (A). The definition of the ampere is formulated on the basis of the electrodynamic force; i.e. *the ampere represents the intensity of the electric current that, passing through two infinitely long parallel conductors placed in vacuum at a distance of* 1 m *apart, creates an interaction force between the two conductors of* 2×10^{-7} N *per meter of length.* Given the ampere, a fundamental unit in the International System (SI) of units,[1] we can define the *coulomb* (C), i.e. *the unit of electric charge that represents the charge carried by a section of a conductor in which a steady current of 1* A *flows for 1* s:

$$Q = \int_{0}^{t_0} i dt \tag{1.3}$$

Regarding the variation over time of the quantities describing the electrokinetic state, several cases can be distinguished, such as: (i) *the stationary electrokinetic state* (associated with a continuous electric current) described by quantities independent of time; (ii) *the non-stationary electrokinetic state* (associated with the transient regime) where the quantities involved are time dependent; and (iii) *the quasi-stationary electrokinetic state* (associated with alternating electric current) where the quantities involved vary over time according to a certain law (usually a periodic function in time), their duration being unlimited. As a generally accepted rule, the current intensity is denoted by a lowercase i for non-stationary and time-variable currents, whereas for stationary currents, an uppercase I is used.

1.2 Electric current densities

Any moving charge constitutes an electric current. Since charge is known to be discrete, its flow is discontinuous. Nevertheless, when there is a great number of very small charges with sufficient velocities, the current can be seen as a fluid that flows. Generally, in the study of electric current flowing through metallic conductors, the charge carriers are quasi-free electrons that undergo oriented movement relative to the nodes of the crystalline lattice, representing the so-called ***conduction current***. Sometimes, when a macroscopic charged body moves, its oriented movement is associated with an electric current called a ***convection current.***

In the most general case, the total current is given by the rate of passage of charge across a specified area of a conductor. If we are dealing with an electric current

[1] The *International System* (*Système International d'Unités,* SI) of measuring units is preferred by most physicists and engineers to solve electromagnetism problems involving large-scale objects. A different system of measuring units, called the *Gaussian System* (GS), is almost universally used in atomic physics and solid-state physics.

extended in space, we may define the current density ($\vec{j}$) as the quantity of charge passing in 1 s through the unit area of a plane normal to the direction of flow.

1.2.1 Conduction current density

The electric **conduction current** at each point of a conductor can be characterized by a vector quantity called the ***conduction current density***, denoted by $\vec{j}$. Let us consider a certain surface S_Γ, bounded by a closed curve Γ, which is part of a conductor carrying an electric current i (see figure 1.2). In these terms, the conduction current

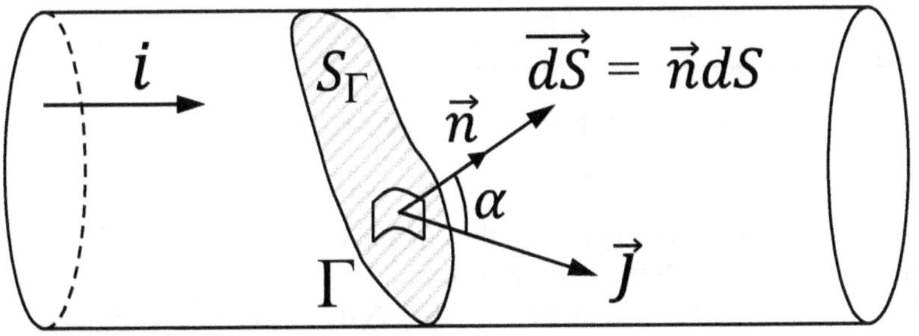

Figure 1.2. Surface (S_Γ) inside a conducting medium used to define the conduction current density vector ($\vec{j}$).

density is defined as the vector quantity $\vec{j}$ whose flux through the considered surface is equal to the intensity of the conduction current passing through that surface:

$$i = \int_{S_\Gamma} \vec{j}\,\overrightarrow{dS} = \int_{S_\Gamma} \vec{j}\,\vec{n}dS \tag{1.4}$$

In relation (1.4), the direction of the current (i) is defined by the direction of the normal ($\vec{n}$) to the respective surface, while the direction of the current density vector ($\vec{j}$) is given by the local direction of the displacement of the positive charges at the considered point. As the value of the current density is generally not the same at every point of the conductor's section, the magnitude of the current density vector is defined by the relation:

$$j = \lim_{\Delta S \to 0} \frac{\Delta i}{\Delta S} = \frac{di}{dS} \tag{1.5}$$

Consequently, the intensity of the current passing through the surface element (dS), whatever its orientation, is given by the relation:

$$di = \vec{j}\,\overrightarrow{dS} = \vec{j}\,\vec{n}dS = j_n\,dS \tag{1.6}$$

The validity of relation (1.6) is due to the fact that on the surface, the tangential component of the current density characterizes the movement of charge (the passage of current) along the dS surface and not through it; in other words, the conduction current density characterizes the movement of charge carriers through that surface

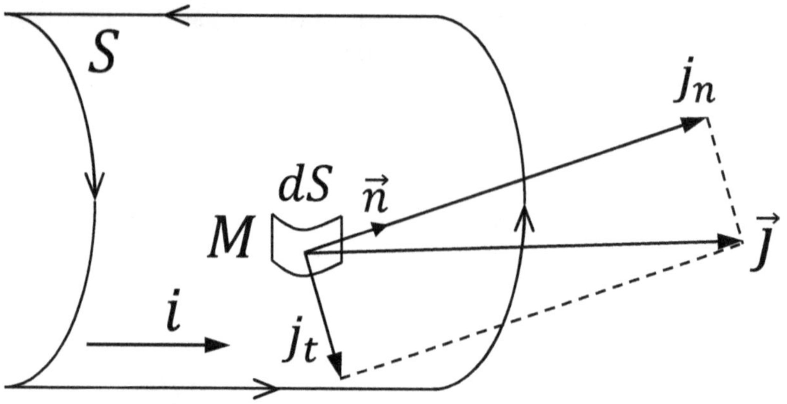

Figure 1.3. Components of the conduction current density.

(see figure 1.3). Equation (1.5) gives us the current density in SI units: $\langle j \rangle_{IS} = \text{A m}^{-2}$, and since this is too large for practical applications, its submultiples (i.e. $\text{A cm}^{-2} = 10^{-4}\ \text{A m}^{-2}$ and $\text{A mm}^{-2} = 10^{-6}\ \text{A m}^{-2}$) are often used. In practice, the conduction current density is used to select adequate dimensions of copper (Cu), aluminum (Al), or other conductors used in the construction of various electrical and electrotechnical devices and circuits, i.e. to avoid their excessive heating by the Joule effect. For example, for underground cables, a current density of 2–4 A mm^{-2} is typically recommended, while for the wire windings of electric machines, a current density of about 2–3 A mm^{-2} is usually prescribed.

The current density vector ($\vec{j}$) can be defined at any point of a conducting medium. Hence, all these vectors form a field of vectors called a current field defined by field lines, i.e. named current lines, representing the tangent curves at any point of the current density vector. An assembly of current lines forms a tube of current lines, as represented in figure 1.4. In the case of stationary current, the current density

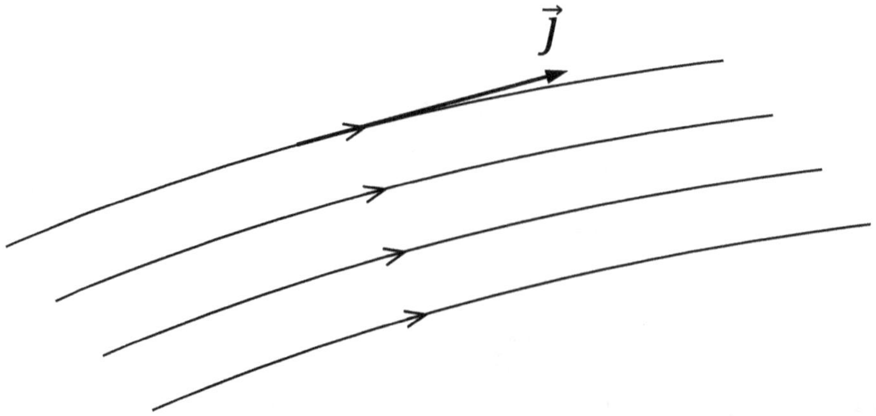

Figure 1.4. Tube of current lines.

vector ($\vec{j}$) is constant over time at any point of the conductor. The intensity of the electric current through the section of a conductor is expressed as the flux of the normal component (j_n) of the current density through that surface and has the same value regardless of the value of the tangential component of the current density (j_t), see figure 1.3.

As for the electric field produced by the stationary electric conduction currents (the electric field associated with the stationary electrokinetic state), this is a potential vector field, satisfying the conditions:

$$\begin{cases} \nabla \times \vec{E} = 0 \\ \vec{E} = -\text{grad } V = -\nabla V \end{cases} \tag{1.7}$$

Indeed, in the case of the stationary electrokinetic state, the spatial distribution of the electric charges must remain stationary, that is, unchanged over time, because if any redistribution of these charges takes place, the intensity of the electric current has to vary and therefore the current does not remain constant. At a certain point in space, some charges are replaced by others due to the flow of the electric current, but their number and therefore the charge density at each point in space remain constant, without causing the electric field to change. Under these conditions, the intensity of the electric field produced by these electric charges must be identical to that of the field that would be produced by fixed charges in space having a constant spatial distribution over time. Consequently, the stationary electric field of continuous (direct) currents is a potential field like the electrostatic field.

In metallic conductors, the charge carriers are quasi-free electrons, which is why the current flowing through these conductors is an **electronic current**. In contrast, within electrolytes, the positive and negative ions in solution are responsible for carrying the current, and a similar situation occurs in ionized gases; in both cases, we thus have an **ionic current**.

1.2.2 Convection current density

An electric current is an oriented movement of electric charges. In the case of conduction current, the movement of electric charges occurs due to the application of an electric field to the conducting medium, i.e. an electric field that acts on each charge carrier with an electric force described by:

$$\vec{F} = q\vec{E} \tag{1.8}$$

The oriented movement of the charge carriers takes place inside the conductor, and this is a movement relative to the crystal lattice of the respective conductor. In addition, an electric current also occurs when a charged body (carrying localized charges) moves as a whole with a velocity $\vec{v}$. The current generated by the overall motion of electric charges fixed on a charged body is called **convection current**. Similarly, the ***convection current density*** is related to the physical quantities describing its moving state (velocity) and its charging state (electric charge density), respectively. Indeed, if we consider a uniformly charged body that has a volume charge density of ρ, moving with a velocity vector of $\vec{v}$, the convection current density may be obtained by considering a rectangular volume element whose sides are parallel to the velocity vector and whose base is an element of area dS from a fixed surface Σ (see figure 1.5). If the base dS is in the plane Σ, which is considered

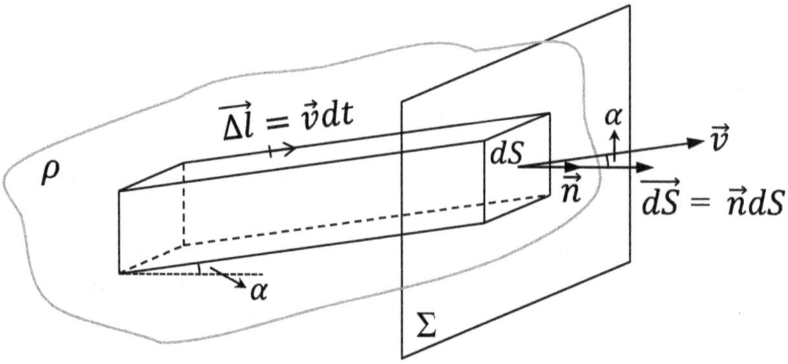

Figure 1.5. Convection current density.

vertical, the volume of the rectangular element i is:

$$dS \cdot \Delta l \cos\alpha = dSv\Delta t \cos\alpha = \overrightarrow{dS} \cdot \vec{v}\Delta t \tag{1.9}$$

During a time Δt, the area element dS carries only charged particles from the body contained in the considered volume element, since they only move in the direction of their velocity. Consequently, the elementary electric charge that passes through the surface element dS in the time Δt is the electric charge found in this rectangular volume element:

$$\Delta q = \rho\overrightarrow{dS} \cdot \vec{v}\Delta t \tag{1.10}$$

and then the convection current passing through the surface element dS is:

$$di_c = \frac{\Delta q}{\Delta t} = \rho\overrightarrow{dS} \cdot \vec{v} \tag{1.11}$$

Taking into account the definition of current density and letting $\vec{j_c}$ denote the convection current density, we obtain:

$$di_c = \vec{j_c}\,\overrightarrow{dS} = \rho\overrightarrow{dS}\vec{v} \Rightarrow \vec{j_c} = \rho\vec{v} \tag{1.12}$$

Using relation (1.12), the intensity of the convection current can be calculated (as well as the intensity of the conduction current) by taking the flux of $\vec{j_c}$ passing through the surface Σ:

$$i_c = \int_\Sigma \vec{j_c}\,\overrightarrow{dS} = \int_\Sigma \vec{j_c} \cdot \vec{n}dS = \int_\Sigma \rho\vec{v}\vec{n}dS \tag{1.13}$$

1.3 Fundamental laws of electrokinetics

Analysis of the electrokinetic state requires the study of the fundamental laws and theorems governing this state, such as: *(i) the theorem of the stationary electric potential, (ii) the law of electric charge conservation, (iii) the law of electric charge transport in conducting media*, and *(iv) the law of energy transformations in conducting media*.

1.3.1 Theorem of the electric potential

In electrostatics, the theorem of electrostatic potential was developed. This states that the line integral of the electrostatic field between two points does not depend on the line joining the two points (path of integration) and that the circulation of the electrostatic field on a closed curve (path) is null. Based on this theorem, the potential at a point in the electrostatic field is defined by the relation [1]:

$$V(P) = V(P_0) - \int_{P_0}^{P} \vec{E}\,\vec{dl} \tag{1.14}$$

In the case of the stationary electrokinetic regime, a similar theorem can be stated. In general, the stationary regime means a regime characterized by parameters that do not change over time. In addition, if no energy transformations take place in the stationary regime (i.e. the production of heat, chemical energy, light energy, etc.), then the regime is electrostatic. In contrast, the stationary electrokinetic regime may be accompanied by heat exchanges or, commonly, by certain energy transformations.

The ***stationary electrokinetic regime*** is therefore different from the electrostatic regime studied in [1]. As discussed in section 1.2, it is characterized by a nonzero current density that is invariable over time in the case of direct (continuous) current. In the case of the electric continuous current, this current density at each point of a conductor is associated with a resulting nonzero electric field that has a constant value over time. This field consists of a coulombic field and an induced electric field (a field that is non-electrical in nature) that ensures the continuity of charge movement through the conductor. Despite the nonuniform character of the stationary electrokinetic regime (of continuous electric current), analysis of the experimental data shows that in the stationary electrokinetic regime of immobile conductors, the potential character of the electric field is maintained. The theorem of the stationary electric potential makes the same statement as the electrostatic potential theorem. However, there are a few differences between the electric field of conductors in electrostatic equilibrium and the electric field of a system of conductors in the stationary electrokinetic state.

In the stationary electrokinetic regime, the conductors' surfaces are no longer equipotential surfaces, unlike in the electrostatic regime (see figure 1.6(a)). Within conductors carrying a direct current (DC), the intensity of the electric field inside is nonzero, and the boundary equations for the electric field (more precisely, the continuity equation of the tangential components of the electric field $\vec{E}$) must be

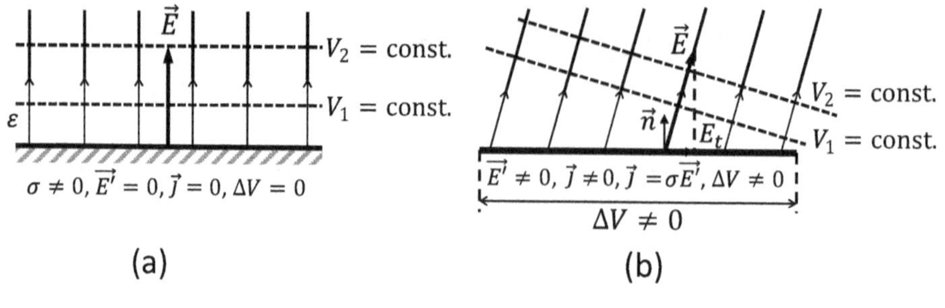

Figure 1.6. Electric field at the surface and equipotential surfaces of a conductor (a) at electrostatic equilibrium and (b) in the stationary electrokinetic state.

satisfied. Taking this into account, at the surface of the conductor, the field lines outside the conductor must make an angle with respect to the normal at the conductor's surface, so that the relation $E' = E_t \neq 0$ must be true (see figure 1.6(b)). The reason for this is that inside the conductor in the electrokinetic state, the electric field is nonzero ($E' \neq 0$).

1.3.2 Electric charge conservation law: continuity equation

1.3.2.1 Integral form of the electric charge conservation law

Consider a closed surface Σ, defined to lie within an insulating medium such that no electric current passes through it. It has been experimentally found that the total charge located inside the surface remains constant ($q_\Sigma = \text{const}$), regardless of the phenomena that occur inside the surface. If, however, the surface Σ also passes through a conducting medium carrying an electric conduction current, then the charge inside the surface varies over time in accordance with the microscopic variation of the conduction current (due to the ordered movement of the charge carriers). Let us consider, for example, a charged capacitor whose plates are connected through a conducting wire (the capacitor is discharged). After closing the switch K (see figure 1.7), the potential inside the connecting conductor cannot

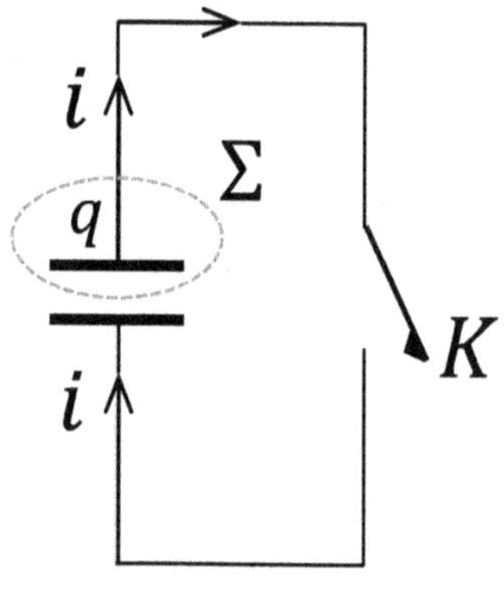

Figure 1.7. Discharge of a capacitor.

remain constant (the capacitor armatures having different potentials); hence, electrostatic equilibrium cannot be maintained. During the transient regime of discharging the capacitor, an electric current passes through the conducting wire that is equal to the decrease in charge on the capacitor plate over time. Thus, an electric current passes through the closed surface Σ, as shown in figure 1.7; this current is equal to the rate of decrease of the charge on the armature contained within the surface Σ.

This result can be generalized to all charged bodies contained in a closed surface Σ (see figure 1.8) passing through an insulating or conducting medium and which is

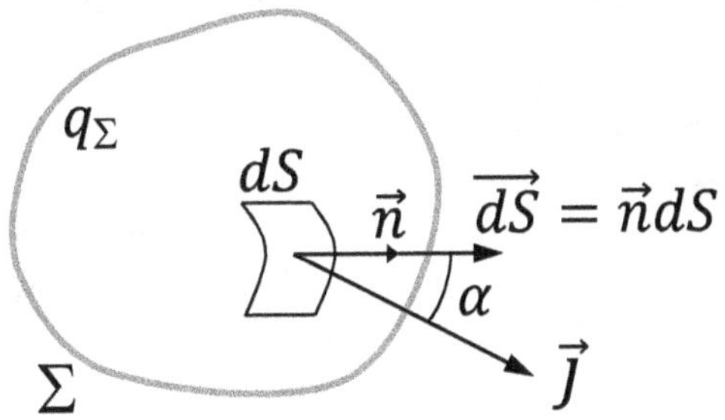

Figure 1.8. Closed surface containing the charge q_Σ, which carries a conduction current of density $\vec{j}$.

considered attached to the medium, leading to the integral form of the electric charge conservation law: *the intensity of the electric conduction current (i_Σ) leaving a closed surface (Σ) attached to a charged body is, at all times, equal to the rate of decrease of the electric charge (q_Σ) located inside the surface:*

$$i_\Sigma = -\frac{dq_\Sigma}{dt} \tag{1.15}$$

Expressing the intensity of the current (i_Σ) with the help of the conduction current density ($\vec{j}$) and the charge (q_Σ) by means of the volume charge density (ρ) in the case of a continuous charge distribution inside the surface Σ, the law of conservation of electric charge can also be written as:

$$\int_\Sigma \vec{j}\,\overrightarrow{dS} = -\frac{d}{dt}\int_{V_\Sigma} \rho dV \tag{1.16}$$

In agreement with figure 1.8, it can be seen that $i_\Sigma = \int_\Sigma \vec{j}\,\overrightarrow{dS}$ is the algebraic sum of the currents that cross the surface toward the outside (the '+' sign denotes those that exit, and the '−' sign denotes those that enter). The normal $\vec{n}$ at dS is considered positive when oriented from the inside to the outside of the closed surface. Also, when the volume integral is derived from the second part of equation (1.16), the surface Σ is considered mobile together with the bodies to which it is rigidly attached. Such a derivate is called a *substantial derivate*. Another formulation can be given to this law by considering the fixed surface, which allows the derivation of the area under the integral. In this case, however, the variation of the charge inside a fixed surface is due not only to the conduction current but also to the departure of the electrically charged bodies through that surface. Thus, the variation of the charge inside the surface is also caused by the convection current, described by:

$$i_{c_\Sigma} = \int_\Sigma \rho \vec{v} \overrightarrow{dS} \tag{1.17}$$

With this observation, *the developed integral form of the electric charge conservation law* can be deduced:

$$\int_\Sigma (\vec{j} + \rho\vec{v})\overrightarrow{dS} = -\frac{d}{dt}\int_{V_\Sigma} \rho dV = -\int_{V_\Sigma} \frac{\partial \rho}{\partial t} dV \tag{1.18}$$

which can be explained as follows: *the rate of decrease of the electric charge inside a closed fixed surface* (Σ) *is equal to the sum of the intensity of the conduction current* (i_Σ) *and the intensity of the convection current* (i_{c_Σ}) *leaving the surface* Σ.

Observation 1. This form, i.e. relation (1.18), would have been reached by calculating the derivative of the volume integral from the right-hand side of equation (1.16) in the scenario of the movable surface attached to the body. From mathematics, it is known that the time derivative of the volume integral of a scalar function (having as variables the vector $\vec{r}$ and the scalar t), such as ρ, in the case where the integration domain is 'mobile' (time dependent) or the so-called 'substantial derivative', is:

$$\frac{d}{dt}\int_{V_\Sigma} \rho dV = \int_{V_\Sigma} \left[\frac{\partial \rho}{\partial t} + \nabla\,(\vec{v}\rho)\right] dV = \int_{V_\Sigma} \frac{\partial \rho}{\partial t} dV + \int_\Sigma \rho\vec{v}\overrightarrow{dS} \tag{1.19}$$

Given the above, equation (1.16) can be rewritten as:

$$\int_\Sigma \vec{j}\,\overrightarrow{dS} = -\int_{V_\Sigma} \frac{\partial \rho}{\partial t} dV - \int_\Sigma \rho\vec{v}\overrightarrow{dS}$$

or:

$$\int_\Sigma (\vec{j} + \rho\vec{v})\overrightarrow{dS} = -\int_{V_\Sigma} \frac{\partial \rho}{\partial t} dV \tag{1.20}$$

which is the developed integral form of the electric charge conservation law, i.e. equation (1.18) written for the electric charge obtained using the convection current associated with the motion of the charged bodies through the fixed surface Σ.

1.3.2.2 Local form of the electric charge conservation law

If, on the left-hand side of equation (1.20), the surface integral is transformed to a volume integral using the Gauss–Ostrogradsky theorem [1], we obtain:

$$\int_{V_\Sigma} \nabla(\vec{j} + \rho\vec{v})dV = -\int_{V_\Sigma} \frac{\partial \rho}{\partial t} dV \tag{1.21}$$

an equality that is valid regardless of the volume V_Σ. Equation (1.21) can only hold if its integrands are equal; in this case, the ***local form of the electric charge conservation law*** is:

$$\nabla(\vec{j} + \rho\vec{v}) = -\frac{\partial\rho}{\partial t} \tag{1.22}$$

or:

$$\nabla(\vec{j} + \rho\vec{v}) + \frac{\partial\rho}{\partial t} = 0 \tag{1.23}$$

Equations (1.22) or (1.23) both express the local form of the electric charge conservation law, which is called ***the continuity equation***; this states that *the rate of decrease of the charge density at a given point is equal to the divergence of the sum of the density of the conduction current and the density of the convection current.*

In the particular case where the surface Σ lies inside a conductor carrying only the conduction current, equation (1.23) becomes:

$$\nabla\vec{j} + \frac{\partial\rho}{\partial t} = 0 \tag{1.24}$$

which shows that at any point in the conductor where there is a source of the density of the conduction current, the charge density decreases over time, and:

$$\nabla\vec{j} = -\frac{\partial\rho}{\partial t} \tag{1.25}$$

In the case of stationary currents, which are characterized by the constancy over time of the electric charge density at each point of the conductor, the continuity equation is reduced to:

$$\nabla\vec{j} = 0 \Rightarrow \vec{j} = \text{constant} \tag{1.26}$$

Relation (1.26) shows that inside a conductor where there is no change of charge over time, there are no points from which the current density starts. The current entering the volume V_Σ bounded by the closed area Σ per unit of time is equal to the current leaving the volume V_Σ in the same unit of time.

1.3.2.3 Current-line continuity theorem

This theorem is the main consequence of the law of charge conservation in the stationary electrokinetic regime (i.e. in DC). In this regime, the electric charge (q_Σ) is constant, so that the equation of charge conservation becomes:

$$i_\Sigma = -\frac{dq_\Sigma}{dt} = 0 \tag{1.27}$$

or:

$$i_\Sigma = \int_\Sigma \vec{j}\,\overrightarrow{dS} \Rightarrow \nabla\vec{j} = 0 \tag{1.28}$$

which states that *the total DC passing through a closed surface* (Σ) *is null (counting as positive the currents that leave the surface* Σ *and negative those that enter the surface* Σ*). The current lines have neither a beginning nor an end; in other words, direct electric current circulates only on closed paths.*

This theorem has the following consequences:

(a) At the surface of a conductor carrying DC, the current density is tangential, and there is no normal component (see figure 1.9):

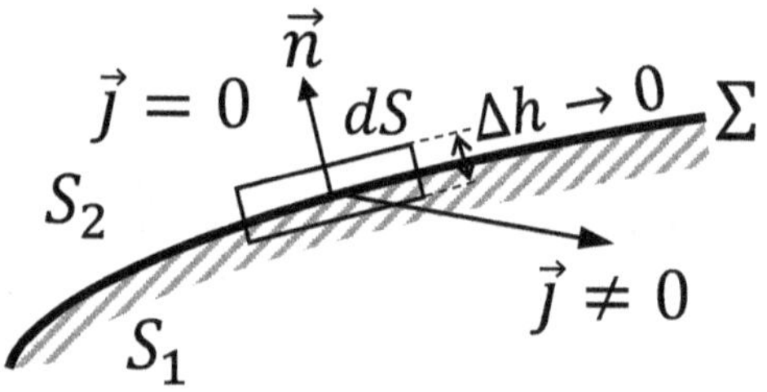

Figure 1.9. Conduction current density does not have a normal component at the conductor's surface.

$$j_n = \vec{j} \cdot \vec{n} = 0 \tag{1.29}$$

Indeed, considering a surface Σ that is a flat cylinder of height $\Delta h \to 0$, with bases S_1 and S_2, and writing the current intensity of the i_Σ that crosses surface Σ as the flux of the current density through it, gives:

$$i_\Sigma = \int_{S_1} \vec{j}\, \overrightarrow{dS} + \int_{S_2} \vec{j}\, \overrightarrow{dS} = \int_{S_1} \vec{j}\, \overrightarrow{dS} = \int_{S_1} \vec{j}\, \vec{n} dS = \vec{j}\, \vec{n} S_1 = 0 \tag{1.30}$$

It follows that $\vec{j} \cdot \vec{n} = 0$. Therefore, $\vec{j}$ has only a tangential component to the surface of the conductor. Basically, the surface of a conductor carrying DC is a tube of current lines.

(b) Consider a tube of current lines bounded by a surface Σ with bases S_1 and S_2 (see figure 1.10(a)). Writing the continuity equation for the current lines yields:

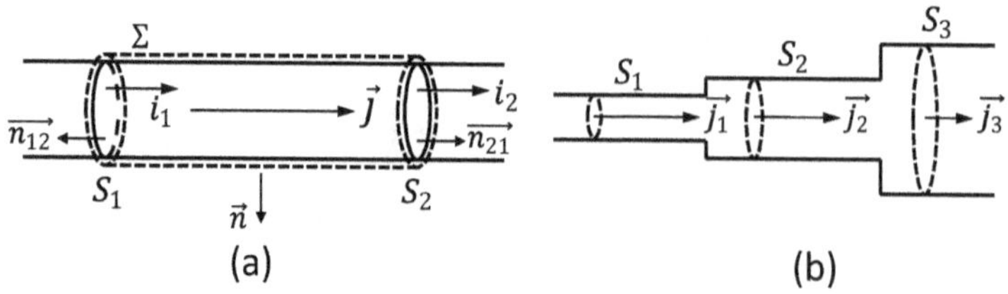

Figure 1.10. Constant current intensity along: (a) a tube of current lines, and (b) a conductor with different cross-sections.

$$i_\Sigma = \int_\Sigma \vec{j}\,\overrightarrow{dS} = \int_{S_1} \vec{j}\,\overrightarrow{dS} + \int_{S_{lat}} \vec{j}\,\vec{n}dS + \int_{S_2} \vec{j}\,\overrightarrow{dS} = 0$$

$$\int_{S_1} \vec{j}\,\overrightarrow{n_{12}}dS + \int_{S_2} \vec{j}\,\overrightarrow{n_{21}}dS = 0 \tag{1.31}$$

$$-i_1 + i_2 = 0 \Rightarrow i_1 = i_2$$

In other words, the DC current intensity is the same along a current tube, and in particular along an electric conductor. In the particular case of an electrical conductor that has different cross-sections (see figure 1.10(b)), the equation for constant current intensity (1.31) becomes:

$$j_1 S_1 = j_2 S_2 = j_3 S_3 \tag{1.32}$$

(c) The normal component of the current density is continuous when passing through the interface between two different conducting media. Let us consider a closed surface Σ of cylindrical shape; the area of its bases is dS and its height is h. It is located at the interface between two conducting media (see figure 1.11). Writing the continuity equation of the electric current lines gives:

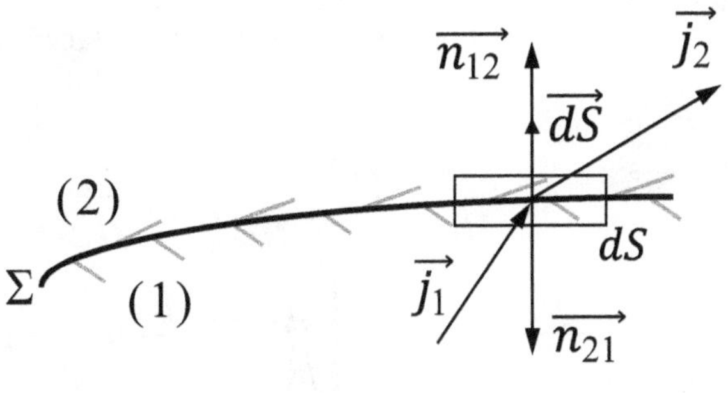

Figure 1.11. Continuity of the normal component of the conduction current density at the interface between two conducting media.

$$i_\Sigma = \int_\Sigma \vec{j}\,\overrightarrow{dS} = \overrightarrow{j_2}\,\overrightarrow{n_{12}}dS + \overrightarrow{j_1}\,\overrightarrow{n_{21}}dS = 0 \tag{1.33}$$

or, noting that $\overrightarrow{n_{12}} = -\overrightarrow{n_{21}}$, the result is that:

$$\overrightarrow{n_{12}}(\overrightarrow{j_2} - \overrightarrow{j_1})dS = 0 \Rightarrow j_{2_n} = j_{1_n} \tag{1.34}$$

(d) Kirchhoff's first theorem, as will be seen later, is also a consequence of the electric charge conservation law.

Observation 2. In the quasi-stationary regime, the rate of variation of charge is negligible everywhere (except for the capacitor plates), and therefore the continuity theorem is valid for any surface (Σ) that does not pass through the capacitor's dielectric. So, consequences (a), (b), and (c) of the continuity theorem remain valid in the quasi-stationary regime.

Observation 3. In the variable regime (where the physical quantities are time dependent), the continuity theorem is no longer valid. In particular, the electric conduction current may vary along a filiform conductor, and equation (1.34) is replaced in the case of a fixed conducting medium by:

$$\overrightarrow{n_{12}}(\vec{j_2} - \vec{j_1}) = -\frac{\partial \sigma_S}{\partial t} \tag{1.35}$$

where σ_S is the surface charge density at the interface between the two different conducting media, which is time dependent.

1.3.2.4 Displacement current

During the charging of an electric capacitor by its connection to an electric power source, there is always a displacement of electric charges through the conductors that connect the source to the capacitor's plates; therefore, a conduction current appears. In experimental circuits, this can be demonstrated by introducing an ammeter into the circuit and measuring the current. Let us consider an open surface Σ through the capacitor's dielectric, but without cutting the conductor, as shown in figure 1.12. In this case, the current intensity crossing the surface Σ will be written as:

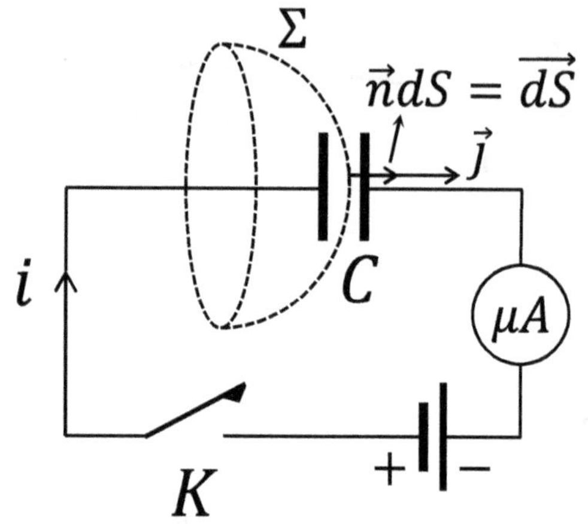

Figure 1.12. Figure illustrating the presence of the displacement current.

$$i_\Sigma = \int_\Sigma \vec{j}\,\vec{n}dS = 0 \tag{1.36}$$

But, in the dielectric of the capacitor there are no free electric charges that can move, then $\vec{j} = 0$. Thus, a contradiction (paradox) appears, namely the existence of a conduction current (because the ammeter indicates a current) in an open circuit (the dielectric is considered to have a resistance that tends to ∞). This contradiction was solved by Maxwell's introduction of the ***displacement current***. The introduction of this displacement current recalls the fact that an electric current circulates only on closed paths. In order to establish an expression for the density of the displacement current, the continuity equation was used, written for the conduction current in the connecting conductors:

$$\nabla \vec{j} = -\frac{\partial \rho}{\partial t} \tag{1.37}$$

where ρ is the volume charge density in the volume limited by the surface Σ corresponding, in this case, to the electric charges on the capacitor plate inside the surface Σ. However, $\rho = \nabla \vec{D}$ (where ρ is understood as a free charge). As a result:

$$\nabla \vec{j} = -\frac{\partial}{\partial t} \nabla \vec{D} = -\nabla \frac{\partial \vec{D}}{\partial t} \tag{1.38}$$

One can thus define a total current density $\vec{j_t}$, called the ***hertzian current density***, using the relation:

$$\vec{j_t} = \vec{j} + \frac{\partial \vec{D}}{\partial t} \tag{1.39}$$

which has the following properties:

(1) $\nabla \vec{j_t} = 0$ (this is a result of equation (1.38)). Consequently, the flux of the vector's total current density ($\vec{j_t}$) is conservative at any point of the conductors or dielectric.

(2) The normal component ($\vec{j_{t_n}}$) is continuous on the separating surface between the two different media. Indeed, $\vec{D}$ and its derivative $\left(\frac{d\vec{D}}{dt}\right)$ have normal components that are continuous at the separation surface according to the boundary conditions. It was demonstrated above that the conduction current density ($\vec{j}$) also has a continuous normal component at the interface between the two different conducting media. The current density which appears in the expression of the total current density is called the ***displacement current density***:

$$\vec{j_d} = \frac{\partial \vec{D}}{\partial t} \tag{1.40}$$

Ultimately, the intensity of the displacement current introduced by Maxwell is:

$$i_d = \int_{\Sigma} \vec{j_d}\, \vec{dS} = \int_{\Sigma} \frac{\partial \vec{D}}{\partial t} \vec{dS} \tag{1.41}$$

Relation (1.41) can be interpreted as follows: it can be agreed that the dielectric medium also carries an electric current, i.e. the displacement current, whose density is given by relation (1.40).

Recalling the fact that a current is a movement of electric charges, it can be considered that in the case of a dielectric, the displacement current is given by the movement of the elementary charges during the polarization of the dielectric. Indeed, supposing that in the applied electric field, the elementary charges from a unit of volume (nq) move across the elementary distance $\overrightarrow{dl}$ in the time dt, then we can consider that the dielectric carries a convection current that has a density of:

$$\overrightarrow{j_c} = \rho\vec{v} = nq\frac{\overrightarrow{dl}}{dt} = \frac{nd\vec{p}}{dt} = \frac{d\vec{P}}{dt} \tag{1.42}$$

As the displacement vector is $\vec{D} = \varepsilon_0\vec{E} + \vec{P}$, the result is that:

$$\overrightarrow{j_d} = \frac{\partial\vec{D}}{\partial t} = \varepsilon_0\frac{\partial\vec{E}}{\partial t} + \frac{d\vec{P}}{dt} \tag{1.43}$$

In the case of a dielectric, the displacement current may be related to the movement of the bound charges during the polarization of the dielectric, described by the second term from equation (1.43), i.e. $\left(\frac{d\vec{P}}{dt}\right)$.

In the case of a capacitor without a dielectric, the displacement current was postulated by Maxwell to be:

$$\overrightarrow{j_{d_0}} = \varepsilon_0\frac{\partial\vec{E}}{\partial t} \tag{1.44}$$

Thus, in vacuum, the displacement current is not related to the movement of charges but is given by an electric field dependent on time. This current plays a very important role in our understanding of the magnetic phenomena created by time-dependent electric fields.

Returning to the abovementioned paradox, we observe that it was solved by Maxwell, who introduced the displacement current which completes the conduction current from the conducting wires through the dielectric. The resulting current, i.e. the total current $\vec{j_t} = \vec{j} + \frac{\partial\vec{D}}{\partial t}$, has closed lines satisfying the continuity equation:

$$\nabla\left(\vec{j} + \frac{\partial\vec{D}}{\partial t}\right) = 0$$

Under these conditions, the charging circuit of the capacitor is also a closed circuit.

1.3.3 Law of electric conduction in homogeneous conducting media

1.3.3.1 Integral form of Ohm's law for homogeneous conducting media

Electric resistance. Let us consider a portion of a metallic conductor, filiform, homogeneous, and linear, connected to an electric generator that maintains a constant potential difference $U_{AB} = V_A - V_B > 0$ between points A and B (see figure 1.13).

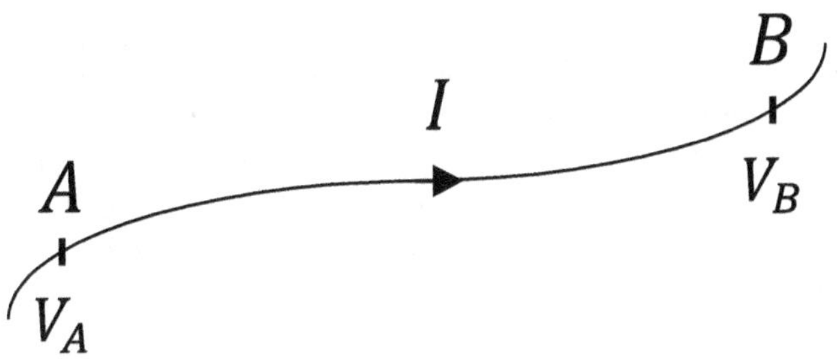

Figure 1.13. A portion of a conductor carrying the conduction current, I.

Along the conductor, there is a stationary electric field given by this potential difference, which drives the charge carriers (the quasi-free electrons) in an oriented motion that has the opposite direction to that of the field. As a result, the conductor carries a stationary electric conduction current from point A, located at a higher potential, to point B, situated at a lower potential ($V_A > V_B$). Also, it has been experimentally found that if an electric current flows through the conductor between two of its sections A and B, a potential difference of $U_{AB} = V_A - V_B$ ($V_A > V_B$) is maintained constant. Taking into account the two experimental observations mentioned above, it has been observed that the ratio between the potential difference (U_{AB}) and current intensity (I) is constant, independent of current and voltage, if the conductor temperature is maintained at a constant value. This fact can be expressed through the relationship:

$$\frac{U_{AB}}{I} = \frac{V_A - V_B}{I} = R \tag{1.45}$$

which is the ***integral form of Ohm's law*** for a homogeneous conducting medium.

The proportionality constant between voltage and current, expressed by Ohm's law, is called the ***electric resistance*** of the portion of the considered conductor. It is a scalar physical quantity that characterizes the property of a conducting medium that opposes the flow of electric current through it. The SI unit of electrical resistance is the ***ohm*** (Ω), *defined as the electric resistance between two points of a conductor when the application between them of a constant voltage of* 1 V *produces the passage through the conductor of a current with an intensity of* 1 A, *if that conductor does not host other electromotive forces.* Experimentally, it was found that the electric resistance of a portion of a conducting medium depends both on the nature of the material from which the conductor is made and on its geometry (in particular cases, on the length of the conductor, l and the area of its cross-section, S); however, it also depends on external conditions. The relationship that expresses this dependence is:

$$R = \rho \frac{l}{S} \tag{1.46}$$

in which ρ, the factor expressing the dependence of the electrical resistance on the nature and physical state of the conductor, called the ***resistivity*** or ***specific resistance*** of the conductor, is *defined as the electric resistance of a conductor with a length of*

1 m and a cross-section equal to the unit of surface (1 m^2*). The SI unit of resistivity is the* $\Omega \cdot \mathrm{m}$, *which represents the resistivity of a homogeneous and isotropic material for which a conductor with a length of* 1 m *and a cross-section of* 1 m^2 *has a resistance of* 1 Ω. In practice, since filiform conductors are often employed, featuring large lengths and very small sections, resistivity submultiples are typically used, such as: $1\,\Omega \cdot \mathrm{mm} = 10^{-3}\,\Omega \cdot \mathrm{m}$ and $1\,\Omega \cdot \mathrm{cm} = 10^{-2}\,\Omega \cdot \mathrm{m}$. The inverse of electrical resistance is called a conductor's ***conductance***:

$$G = \frac{1}{R} \tag{1.47}$$

which is measured in Ω^{-1} or S (siemens), while the inverse of electrical resistivity is called ***electrical conductivity***:

$$\sigma = \frac{1}{\rho} \tag{1.48}$$

and is measured in $\Omega^{-1} \cdot \mathrm{m}^{-1} = \mathrm{S} \cdot \mathrm{m}^{-1}$. The name 'conductance' and its definition given by relation (1.47) suggest that this parameter, unlike electrical resistance, characterizes the property of a conducting medium that allows a flow of electric current through it. Both the electric resistivity and conductivity are material-specific quantities. Table 1.1 gives the resistivity of some representative materials divided into three classes:

(a) Good conductors ($\rho \sim 10^{-8}\,\Omega \cdot \mathrm{m}$);
(b) Poor conductors and semiconductors ($\rho \sim 10^{-5} - 10^{4}\,\Omega \cdot \mathrm{m}$);
(c) Insulators ($\rho > 10^{10}\,\Omega \cdot \mathrm{m}$).

Table 1.1. Conductivity and resistivity of some representative materials.

Material	ρ ($\Omega \cdot$ m) at 20 °C	σ ($\Omega^{-1} \cdot \mathrm{m}^{-1}$) at 20 °C	Mean temperature coefficient (per K) between 0 °C and 100 °C
Good conductors			
Silver (Ag)	1.6×10^{-8}	6.2×10^{7}	0.0041
Copper (Cu)	1.7×10^{-8}	5.8×10^{7}	0.0043
Aluminum (Al)	2.7×10^{-8}	3.7×10^{7}	0.0045
Platinum (Pt)	10.6×10^{-8}	0.94×10^{7}	0.0039
Manganin[a]	42×10^{-8}	0.24×10^{7}	10^{-5}
Constantan[b]	48×10^{-8}	0.21×10^{7}	2×10^{-5}
Mercury (Hg)	96×10^{-8}	0.10×10^{7}	10^{-3}
Carbon (C) (graphite)	1.7×10^{-8}	$0.16\text{–}2.9 \times 10^{5}$	-5×10^{-4}
Poor conductors and semiconductors			
Indium antimonide (InSb)	5.7×10^{-5}	1.76×10^{4}	
Germanium (Ge)	0.47	2.1	

(Continued)

Table 1.1. (*Continued*)

Material	ρ ($\Omega \cdot$ m) at 20 °C	σ ($\Omega^{-1} \cdot$ m^{-1}) at 20 °C	Mean temperature coefficient (per K) between 0 °C and 100 °C
Silicon (Si)	2.3×10^3	4.3×10^{-4}	
Saturated sodium chloride (NaCl) solution	4.4×10^{-2}	22.6	
Insulators			
Pyrex glass	$\sim 10^{12}$	$\sim 10^{-12}$	
Paraffin wax	$\sim 10^{14}$	$\sim 10^{-14}$	
Polystyrene	$\sim 10^{15}$	$\sim 10^{-15}$	

[a] An alloy typically made of 84.2% Cu, 12.1% manganese (Mn), and 3.7% nickel (Ni), extensively used in the fabrication of resistors, in particular ammeter shunts.
[b] An alloy typically made of 55% Cu and 45% Ni, featuring a low thermal variation of its resistivity.

The specific resistances of all metals are independent of current density over an extremely large range but increase with temperature. Experimentally, it has been found that the resistivities of conductors increase with temperature, in the range of high temperatures above 0 °C, according to the relation:

$$\rho = \rho_0[1 + \alpha(t - t_0)] \tag{1.49}$$

where ρ_0 is the resistivity at temperature t_0 (°C), ρ is the resistivity at temperature t (°C), and $\alpha = \frac{d\rho}{\rho_0 dt}$ is the coefficient of variation of resistivity with temperature, expressed in (°C)$^{-1}$. Taking into account relations (1.46) and (1.49), one can also write an equation for electrical resistance that expresses its dependence on temperature:

$$R = R_0[1 + \alpha_R(t - t_0)] \tag{1.50}$$

in which $\alpha_R = \frac{dR}{R_0 dt}$ is the coefficient of variation of resistance with temperature. In relatively small ranges of temperature variation (1 °C–100 °C), it practically coincides with α (the coefficient of variation of resistivity with temperature, i.e. $\alpha_R = \alpha$). Relation (1.50) can be used practically to determine the degree of heating of an electric winding or a conductor, a measurement which cannot be obtained using a conventional thermometer. Thus, by measuring the electrical resistance before operation (when it has the ambient temperature t_0) and after a certain operational time (at an unknown temperature t), the conductor's temperature at the latter time (the heating degree) can easily be obtained from relation (1.50):

$$t - t_0 = \frac{R - R_0}{R_0 \alpha} \tag{1.51}$$

Concerning the variation of resistivity with temperature, it has been experimentally observed that:

- At very low temperatures, usually below 10 K, the resistance of many metals and alloys suddenly decreases to a negligible value, a phenomenon known as ***superconductivity***.[2] Practically, the resistivity ρ of Hg, Pb, or Sn decreases almost to zero when the temperature drops to 1–10 K.
- Semiconductors, insulators, and electrolytes all have resistivity values that decrease as the temperature rises; for example, in the case of semiconductors, the resistivity decreases with temperature according to the following equation:

$$\rho = \rho_0 e^{b/T} \tag{1.52}$$

where ρ_0 and b are constants specific to a given semiconductor.

- Bismuth (Bi) shows a remarkably high variation of resistivity when a magnetic field is applied perpendicular to the current flow, and this property has been used to measure the intensity of such fields. This phenomenon is known as ***magnetoresistance***.

1.3.3.2 Local (differential) form of Ohm's law for homogeneous conducting media

Let us consider a conductor portion of cylindrical shape, having a cross-section of dS and a length of dl; it is made from a homogeneous and isotropic material of resistivity ρ (see figure 1.14). An electric voltage $dU = V_A - V_B$, with $V_A > V_B$, is

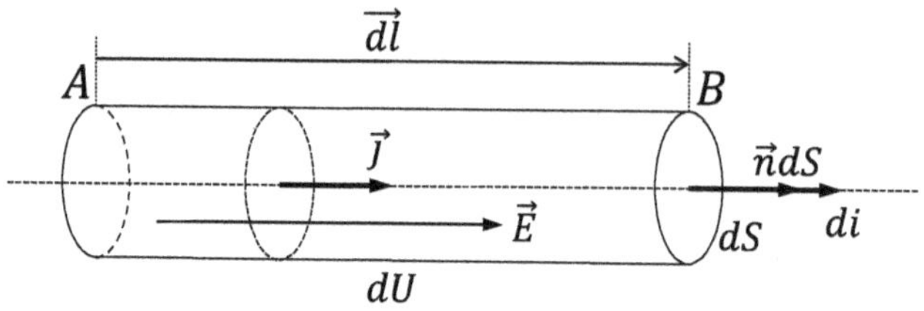

Figure 1.14. Portion of a homogeneous and isotropic conductor of length dl and cross-section dS carrying a conduction current.

applied to the ends of the conductor, which produces an electric field $\vec{E}$ along the conductor. This field gives rise to an oriented movement of charge through the conductor, i.e. to the appearance of an electric current. Taking the normal vector $\vec{n}$ at the conductor's cross-section and using the direction of the electric current through the conductor, the difference in the potential dU between the ends of the conductor can be expressed in terms of the intensity of the electric field by the relation (elementary circulation of $\vec{E}$ between A and B):

[2] Superconductivity was discovered in 1911 by **Heike K Onnes** (a Dutch physicist), who was studying the resistance of solid mercury at cryogenic temperatures using the recently produced liquid helium as a refrigerant. At a temperature of 4.2 K, he observed that the resistance abruptly disappeared [2, 3].

$$dU = \vec{E} \cdot \vec{n} dl = \vec{E} \vec{dl} \tag{1.53}$$

Moreover, the current intensity dI passing through the cross-section dS of the conductor can be expressed in terms of the density of the conduction current $\vec{j}$ using the relation (see the definition of conduction current density):

$$dI = \vec{j} \cdot \vec{n} dS = \vec{j} \vec{dS} \tag{1.54}$$

Using the integral form of Ohm's law, i.e. $R = \frac{dU}{dI}$ and the relationship that expresses the dependence of the electrical resistance on the nature and dimensions of a conductor, i.e. $R = \rho \frac{dl}{dS}$, it is possible to obtain:

$$dU = RdI \Rightarrow \vec{E} \vec{n} dl = \rho \frac{dl}{dS} \cdot \vec{j} \cdot \vec{n} \cdot dS \Rightarrow \vec{E} = \rho \vec{j} \tag{1.55}$$

or:

$$\rho = \frac{1}{\sigma} \Rightarrow \vec{E} = \frac{\vec{j}}{\sigma} \Rightarrow \vec{j} = \sigma \vec{E} \tag{1.56}$$

Relationships (1.55) and (1.56) represent ***the local (or differential) form of Ohm's law*** for a homogeneous conducting medium. These relations are valid at any point in a conducting, isotropic medium, in the absence of sources of electromotive force. In the case of anisotropic environments, the conductivity σ depends on the direction of the current, and it is no longer a scalar quantity but a tensor that has the components:

$$[\sigma] = \begin{bmatrix} \sigma_{xx} & \sigma_{xy} & \sigma_{xz} \\ \sigma_{yx} & \sigma_{yy} & \sigma_{yz} \\ \sigma_{zx} & \sigma_{zy} & \sigma_{zz} \end{bmatrix} \tag{1.57}$$

As a result, if the column matrices are used for current density and electric field intensity, respectively, i.e. $[\vec{j}\,] = \begin{bmatrix} j_x \\ j_y \\ j_z \end{bmatrix}$ and $[\vec{E}\,] = \begin{bmatrix} E_x \\ E_y \\ E_z \end{bmatrix}$, then Ohm's law can be expressed as $[\vec{j}\,] = [\sigma] \cdot [\vec{E}\,]$, showing that $\vec{j}$ and $\vec{E}$ no longer have the same orientation. In the case of an isotropic conducting medium and a stationary current, the current lines coincide with the electric field lines and are normal to the equipotential surfaces.

In addition to its theoretical importance, the local form of Ohm's law, i.e. $\vec{j} = \sigma \vec{E}$, is the fundamental law used to analyze the electric charge transport process in a conducting medium. It is also of particular practical interest because it allows the calculation of the electrical resistance of a portion of an inhomogeneous, anisotropic conductor if the law of its σ variation with position is known, or when the material is homogeneous and isotropic but the geometry of the conductor does not allow the application of the $R = \rho \frac{l}{S}$ relation. In these cases, the resistance is calculated using the integral form of Ohm's law:

$$R = \frac{U_{AB}}{I}, \quad \text{where:} \quad U_{AB} = \int_A^B \vec{E}\,d\vec{l} \tag{1.58}$$

In relation (1.58), the electric field $\vec{E}$ expressed in terms of the current intensity is introduced by means of the local form of Ohm's law, i.e. $\vec{j} = \sigma\vec{E}$, where: $j = \frac{dI}{dS}$.

Example 1. *A cube with an edge length of a is made of an inhomogeneous conducting medium whose conductivity changes linearly from σ_1 at face 1 perpendicular to the OX axis ($x = 0$) to σ_2 at face 2 ($x = a$), (see figure 1.15). Find the resistances R_{AB} and R_{CD} of the sample.*

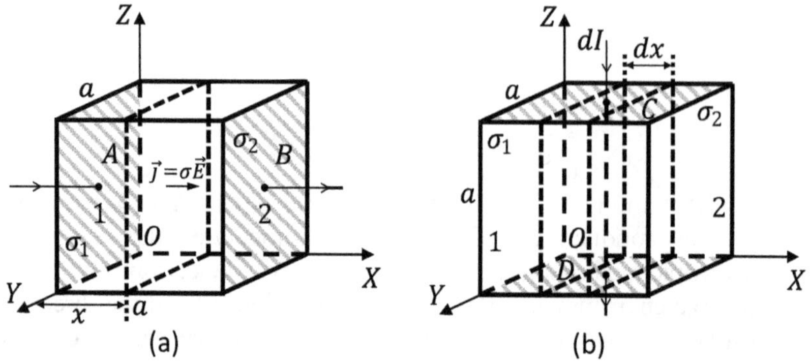

Figure 1.15. *A cube with an edge length of a, made of an inhomogeneous conducting medium.*

Solution

The conductivity at a distance x from face 1 *in* figure 1.15(a) *is:*

$$\sigma(x) = \sigma_1 + \frac{\sigma_2 - \sigma_1}{a}x \tag{1.59}$$

(a) *According to the local form of Ohm's law:*

$$\vec{j} = \sigma(x)\vec{E} \Longrightarrow \vec{E}(x) = \frac{\vec{j}}{\sigma_1 + \frac{\sigma_2 - \sigma_1}{a}x} \tag{1.60}$$

Then:

$$U_{AB} = \int_A^B \vec{E}\,d\vec{l} = \int_0^a \frac{\vec{j}}{\sigma_1 + \frac{\sigma_2 - \sigma_1}{a}x}d\vec{l} = \frac{I}{a(\sigma_2 - \sigma_1)}\ln\left(\frac{\sigma_2}{\sigma_1}\right) \tag{1.61}$$

Given the above, the resistance R_{AB} is :

$$R_{AB} = \frac{U_{AB}}{I} = \frac{1}{a(\sigma_2 - \sigma_1)}\ln\left(\frac{\sigma_2}{\sigma_1}\right) \tag{1.62}$$

(b) *Considering the surface element on the upper face of the cube (shaded side), figure 1.15(b), with an area of adx, the current intensity passing through it is:*

$$dI = \vec{j}\,(x)adx = \sigma(x)Eadx = \left(\sigma_1 + \frac{\sigma_2 - \sigma_1}{a}x\right)\frac{U_{CD}}{a}adx \tag{1.63}$$

Then:

$$\frac{I}{U_{CD}} = \int_0^a \left(\sigma_1 + \frac{\sigma_2 - \sigma_1}{a}x\right)dx = \frac{(\sigma_2 + \sigma_1)a}{2} \Rightarrow$$

$$\Rightarrow R_{CD} = \frac{U_{CD}}{I} = \frac{2}{a(\sigma_2 + \sigma_1)} \tag{1.64}$$

Example 2. *A circular disc that has a radius of a and a thickness of g is connected both at its center and its rim by a conducting wire featuring a circular cross-section of diameter d; this is very small compared to both the radius and the thickness of the disk (see figure 1.16). Supposing that the electrical conductivity of the disk (σ) is smaller*

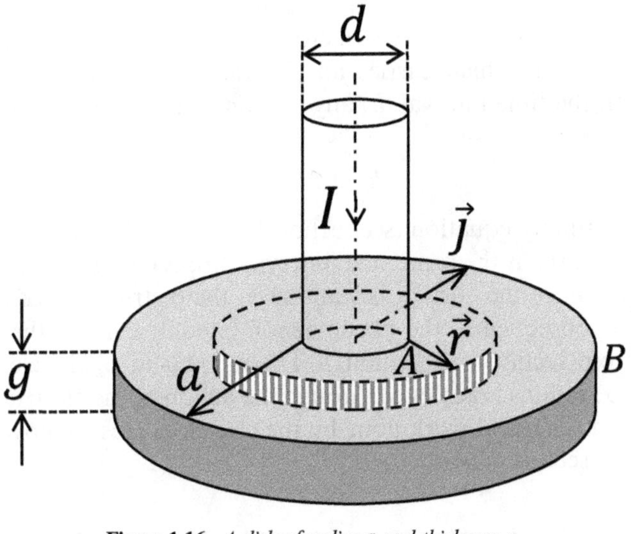

Figure 1.16. *A disk of radius a and thickness g.*

than that of the conducting wires used for connection, find the electric resistance R_{AB} between the center and rim of the disk.

Solution: *Considering a cylindrical surface of radius r, parallel with the lateral surface of the disk whose area $A = 2\pi rg$, the current density along the radius of the disk at a distance r from its center is:*

$$\vec{j} = \sigma \overrightarrow{E(\vec{r})} \Longrightarrow E(\vec{r}) = \frac{j}{\sigma} = \frac{I}{2\pi\sigma g r} \tag{1.65}$$

Given this result, the voltage U_{AB} is:

$$U_{AB} = \int_{A}^{B} \vec{E}\,\vec{dl} = \int_{\frac{d}{2}}^{a} \frac{I}{2\pi\sigma g r} dr = \frac{I}{2\pi\sigma g} \ln\left(\frac{2a}{d}\right) \tag{1.66}$$

and the resistance is:

$$R_{AB} = \frac{U_{AB}}{I} = \frac{1}{2\pi\sigma g} \ln\left(\frac{2a}{d}\right) \tag{1.67}$$

1.3.4 The Joule–Lenz law for homogeneous conducting media

1.3.4.1 Integral form of the Joule–Lenz law for homogeneous conductive media

One of the effects that accompanies the electrokinetic state is the electro-caloric effect of the electric current, which heats a conductor that carries an electric current. Experiments carried out by James P Joule[3] in 1841 and subsequently by Heinrich F E Lenz[4] in 1844 showed that the thermal energy produced when an electric current passes through a conductor is equivalent to the work done by the electric field to move electric charges through the conductor. Let us consider a filiform conductor of constant cross-section, which carries an electric conduction current that has an intensity of I. In the time interval dt, any section of the conductor carries a charge of:

$$dq = Idt \tag{1.68}$$

Based on the continuity equation (see section 1.3.2), the charge entering section 1 of the conductor departs in the same time interval through section 2 of the conductor (see figure 1.17). Since the current is stationary, the distribution of charge remains unchanged and consequently the phenomenon is equivalent to the transport of a charge of dq from section 1 to section 2. The movement of the electric charge dq between the two sections is achieved through the action of the electric field inside the conductor. The mechanical work done by the electric forces to move the charge dq from section 1 to section 2 is:

$$-dL = \int_{1}^{2} dq\vec{E} \cdot \vec{dl} = dq \cdot (V_1 - V_2) = Idt(V_1 - V_2) = IUdt = dW \tag{1.69}$$

[3] **James Prescott Joule** (1818–89) was an English physicist; his discoveries were related to the nature of heat and its relationship to mechanical work, demonstrating *the law of energy conservation* and further developing *the first law of thermodynamics*. The *joule* (J), an SI derived unit of energy, is named after him.

[4] **Heinrich Friedrich Emil Lenz** (1804–65) was a Russian physicist most noted for formulating *Lenz's law* in electrodynamics in 1834.

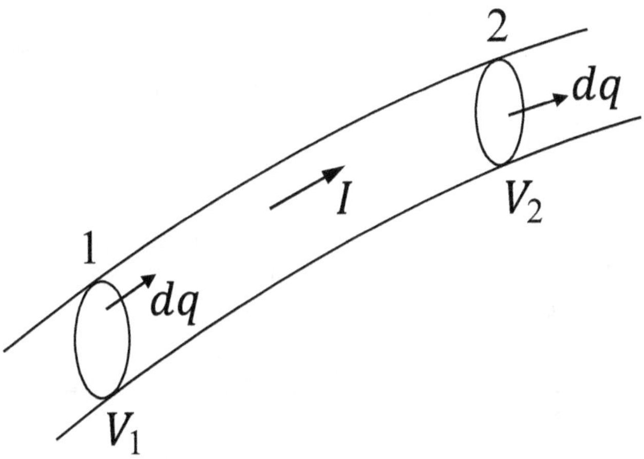

Figure 1.17. Illustration of charge transport between two points of a conductor under a potential difference, i.e. $V_1 - V_2$.

This mechanical work is equal to the energy that the system gives away to the medium around it, in this case in the form of caloric energy, which is manifested by the heating of the conductor. So, when an electric current passes through a conductor in a time interval dt, the total caloric energy released is:

$$-dL = dW = UIdt \tag{1.70}$$

Taking into account Ohm's law in its integral form, dW is also written:

$$dW = RI^2dt = \frac{U^2}{R}dt = UIdt \tag{1.71}$$

The caloric energy dissipated by an electric current is also measured in *joules* (J). The thermal phenomenon studied here is called the ***Joule–Lenz effect,*** and relations (1.71) express the *Joule–Lenz law in integral form.*

1.3.4.2 Electric power

Generally, **power** is a scalar physical quantity defined as the work done by a physical system per time unit, being mathematically defined as:

$$P = \lim_{\Delta t \to 0} \frac{\Delta W}{\Delta \mathrm{t}} = \frac{dW}{dt} \tag{1.72}$$

Taking into account the expression of electrical energy (1.71), we obtain the following for electric power:

$$P = \frac{dW}{dt} = UI = RI^2 = \frac{U^2}{R} = GU^2 \tag{1.73}$$

Power is measured in J s^{-1} = W (watts).

1.3.4.3 Local form of the Joule–Lenz law

The density of thermal power (w) is first defined as the energy dissipated per unit volume and unit time:

$$w = \lim_{\substack{\Delta V \to 0 \\ \Delta t \to 0}} \frac{\Delta W}{\Delta V \Delta t} = \lim_{\Delta V \to 0} \frac{\Delta P}{\Delta V} = \frac{dP}{dV} \tag{1.74}$$

Referring again to a volume element of a conductor that has a cross-section of dS and a length of dl, such as the one represented in figure 1.14, to which a voltage dU is applied, the power density can be expressed as follows:

$$w = \frac{\Delta W}{\Delta V \Delta t} = \frac{dUI\Delta t}{dSdl\Delta t} = \frac{RI^2}{dSdl} = \frac{\rho \frac{dl}{dS} \cdot j^2 dS^2}{dSdl} = \frac{1}{\sigma}\sigma^2 E^2 = \sigma E^2 = \rho j^2 \tag{1.75}$$

or, taking into account the local Ohm's law, i.e. $\vec{j} = \sigma \vec{E}$, the power density can also be written as follows:

$$w = \vec{j}\,\vec{E} \tag{1.76}$$

Relationships (1.74) and (1.76) are the ***local forms of the Joule–Lenz law***. In fact, these relations are general; hence, they can be applied to any conductor, regardless of its form and homogeneity, whether we are dealing with stationary or quasi-stationary currents.

1.3.5 Wiedemann–Franz law

The thermal conductivity $\varkappa$ of a conductor (the scalar physical quantity that characterizes the ability of the material to conduct thermally or allow the propagation of heat through it) and its electric conductivity σ are linked by a relationship that bears the name of the ***Wiedemann–Franz law***. This law states that the ratio between the thermal conductivity ($\varkappa$) and the electric conductivity (σ) of a conductor is directly proportional to its absolute temperature:

$$\frac{\varkappa}{\sigma} = c \cdot T \tag{1.77}$$

1.4 Microscopic approach to electric conduction in metals

The Ohm, Joule–Lenz, and Wiedemann–Franz laws, in their local forms, have only been formulated above. They will be effectively established later, using a microscopic approach to electronic conduction in metals based on the *classical theory of electric conduction in metals* or the *Drude–Lorentz theory*.

1.4.1 Free electron gas model

At the microscopic scale, a metallic conductor consists of a spatial network. At the nodes of such a network, there are positive ions among which the electrons can move 'freely'. 'Free electrons' means electrons that are not bound to any atom that is part of the crystal lattice of the metal; such electrons can move over very long distances in the metal compared to the crystal lattice constant. These electrons come from the outer layer (valence layer) of the atoms that form the crystal lattice. However, these electrons do not behave like free particles in vacuum, i.e. their energy does not have a continuous spectrum and is quantized. Their energy is quantized in wide energy intervals available to them rather than discrete energy levels in the atoms, so they are more correctly called *'quasi-free electrons'*. By releasing and collectivizing these electrons, the atoms in the metal become positively charged ions. Electrons interact strongly with these ions in the crystal lattice, providing the forces and energy necessary for the cohesion of the metal; the total energy of the electrons in the crystal remains negative, even if the interaction energy between electrons is positive. In 1901, Eduard Riecke provided experimental evidence to support the idea of the existence of free electrons in metals. For over a year, he maintained a current of $I = 0.1$ A in a circuit consisting of an Al cylinder placed between two cylinders of Cu (see figure 1.18) and found that although the charge passing through the cross-sections of the two contacts was enormous, i.e. 3.5×10^6 C, no chemical transformations occurred in the conductors, and there was no transport of substance from one cylinder to another. Thus, the charge carriers were the same in both metals, and the mass of the charge carriers was negligible. As a result, these charge carriers could only have been electrons.

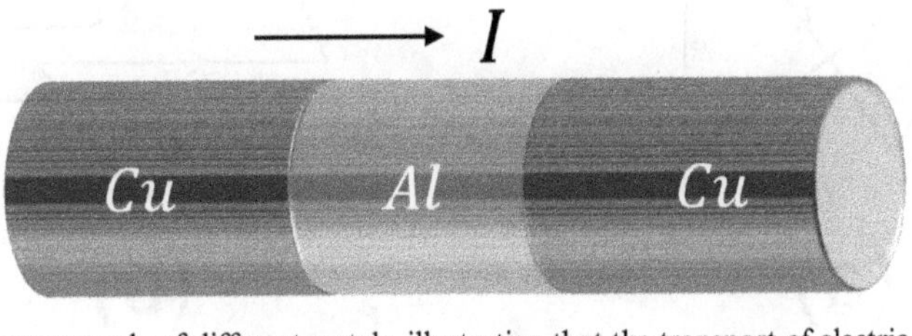

Figure 1.18. Conductors made of different metals, illustrating that the transport of electric charge is provided by quasi-free electrons.

In 1917, Thomas D Stewart and Richard C Tolman provided additional experimental evidence to support the idea of the existence of quasi-free electrons in metals by measuring the intensity of electric current occurring in a conductor in motion due to the inertia of moving electrons when the conductor suddenly stops moving. The value of the specific electric charge *q*/*m* determined in this way is in good agreement with the specific charge of an electron.

The classical theory of electronic conduction in metals was developed as early as 1900 by Paul Drude and subsequently perfected by Hendrik A Lorentz in 1907. In this model, it is assumed that the quasi-free electrons in a metal behave like an ideal gas, 'enclosed' in a container that is the volume of the metal. The volume of the electrons is negligible: $V_{\text{el}} = n\frac{4\pi r^3}{3} = 10^{28} \cdot \frac{4\pi}{3} \cdot 10^{-45} \approx 10^{-17}\ \text{m}^3$. The electronic gas only interacts with all the nodes of the network (positive ions), and this interaction is reflected primarily in the value of the *mean free path* of the electrons.

The interactions between electrons are not taken into account, even though the interaction force between them is very high $\left(F_c = \frac{e^2}{4\pi\varepsilon_0 r^2} = 2 \cdot 10^{-8} \text{ N}\right)$, causing accelerations of the order of $a = \frac{F}{m} = \frac{2 \cdot 10^{-8}}{9 \cdot 10^{-31}} = 10^{22}$ m s^{-2}; the reason for this is that the interactions with the nodes of the crystal lattice are dominant, making the total energy negative.

According to the free electron gas model, in the absence of an electric field, electrons move chaotically in all directions without achieving, over time, a net displacement through the conductor (see figure 1.19(a)). The movement due to

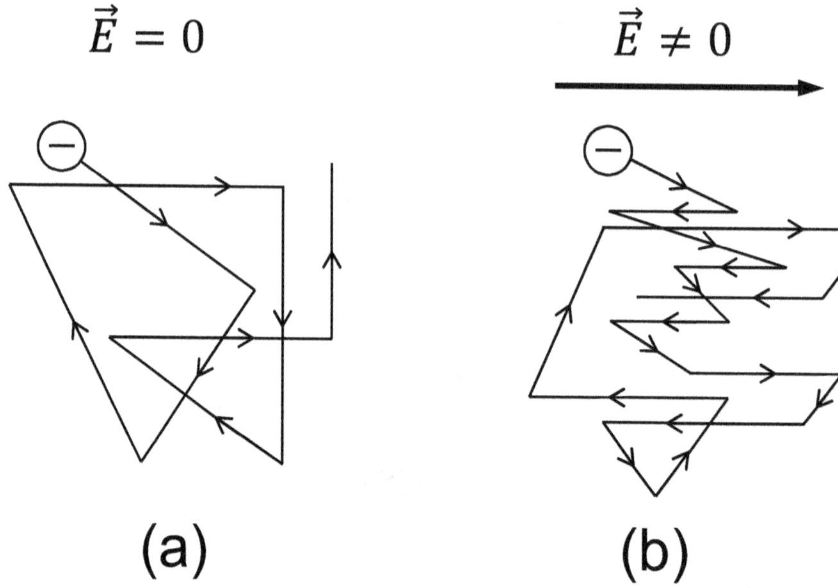

Figure 1.19. Movement of quasi-free electrons in a metal: (a) in the absence of an electric field, and (b) in the presence of an external electric field.

thermal agitation is characterized by one of the following velocities: the most probable speed, i.e. $v_p = \sqrt{\frac{2kT}{m_0}}$; the average quadratic velocity (thermal velocity), i.e. $v_t = \sqrt{\overline{v}^2} = \sqrt{\frac{3kT}{m_0}}$; or the average speed, i.e. $\overline{u} = \sqrt{\frac{8kT}{\pi m_0}}$. All these speeds are very close to each other and have an order of magnitude of $\overline{u} = \sqrt{\frac{8 \cdot 1.38 \cdot 10^{-23} \cdot 300}{3.14 \cdot 9.1 \cdot 10^{-31}}} \approx 10^5$ m s^{-2}. In the following, the average speed $\overline{u}$ is used.

When an electric field is applied to the metal, the movement due to thermal agitation overlaps with the oriented movement due to the electrons' collective movement in the opposite direction to that of the electric field (see figure 1.19(b)). This movement, called drift motion, is due to the force with which the electric field acts on the electrons. If the electrons were to move freely, i.e. without suffering any collisions, then in a constant electric field, their speed would have to increase constantly; however, as the $\vec{v_d} = \dfrac{\sigma \vec{E}}{n_0 e}$ relationship suggests, the average speed of oriented motion is constant in a constant electric field. This is explained by the fact that in their movement, electrons interact with the nodes of the crystal lattice, defects, impurities, and/or phonons of the network. In these collisions, the electrons completely lose the energy that they gain under the influence of the electric field. However, over time, they manage to move in the opposite direction to that of the electric field at a constant average speed, called *the drift speed,* denoted by $\vec{v_d}$, thus giving rise to an electric conduction current.

1.4.2 Local form of Ohm's law for homogeneous conducting media

Consider a volume element that is part of a homogeneous and isotropic conductor

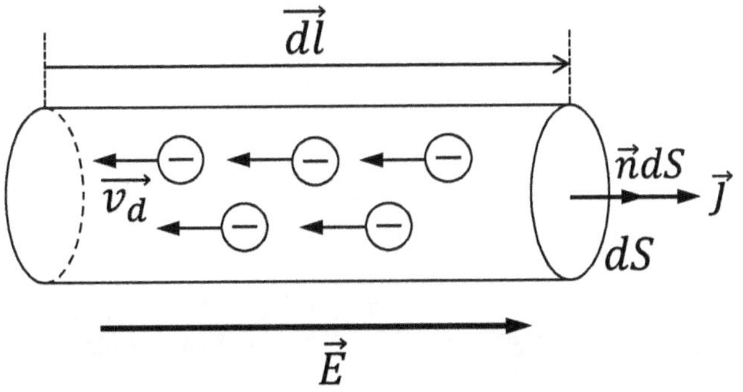

Figure 1.20. Volume element from a homogeneous and isotropic conductor carrying an electric conduction current.

(see figure 1.20) which contains n electrons per unit volume. If we apply an electric field along it, it then carries an electric conduction current whose density is:

$$\vec{j} = -ne\vec{v_d} \tag{1.78}$$

The drift velocity can be expressed as a function of the average physical dimensions characterizing the electronic gas. In the Drude–Lorentz theory, it is assumed that the collisions suffered by the electrons in their advancement through the conductor are elastic and are governed by the laws of conservation of energy and momentum. Since the mass of an electron is very small with respect to that of the ions in the network, it is considered that only the momentum and speed of the electrons undergo changes as a result of collisions, while the momentum of the ions of the crystal lattice does not change. If, at the time $t = 0$, the momentum of the electron has a well-determined orientation, after a time $t > 0$ in which it suffers one or more collisions, there is no correlation between its current momentum and its initial momentum. This minimal time that an electron needs to lose all the information about its initial state (i.e. its orientation and the value of its momentum at a given time) is called the *correlation time.* In other words, the correlation time is basically the time τ associated with the mean free path (or the average free time). In the time interval τ, the electron is accelerated in the opposite direction to that of the field. It is also assumed that the drift velocities are much smaller than v_t and that τ is unaffected by the application of $\vec{E}$:

$$\vec{a} = -\frac{e\vec{E}}{m} \tag{1.79}$$

The speed gained at the end of the time interval τ is therefore:

$$\vec{v_1} = \vec{a}\tau = -\frac{e\vec{E}}{m}\tau \tag{1.80}$$

Since it has been assumed that each electron completely loses its kinetic energy ($v_{01} = 0$) after each collision, then it moves with the following average drift speed in the time interval τ:

$$\vec{v_d} = \frac{\vec{v_{01}} + \vec{v_1}}{2} = -\frac{e\vec{E}}{2m}\tau \tag{1.81}$$

However:

$$\tau = \frac{\overline{\lambda}}{\overline{u}} \tag{1.82}$$

where $\overline{\lambda}$ is the mean free path of the electron and $\overline{u}$ is the average thermal velocity of the electron. Given these expressions, the drift speed of the electrons is:

$$\overrightarrow{\langle v_d \rangle} = -\frac{e\vec{E}\,\overline{\lambda}}{2m\,v_t} \tag{1.83}$$

Adding this to the expression for the current density, we obtain:

$$\vec{j} = (-e)n\overrightarrow{\langle v_d \rangle} = (-e)n\left(-\frac{e\vec{E}\,\overline{\lambda}}{2m\,v_t}\right) \text{ or } \vec{j} = \frac{ne^2\,\overline{\lambda}}{2m\,v_t}\vec{E} = \sigma\vec{E} \tag{1.84}$$

The local form of Ohm's law has been obtained again. This time, the electrical conductivity σ appears explicitly, being expressed according to the physical dimensions that characterize the electronic gas:

$$\sigma = \frac{ne^2\,\overline{\lambda}}{2m\,v_t} = \frac{ne^2}{2m}\tau \tag{1.85}$$

Another quantity that characterizes the oriented movement of the charge carriers is **the *mobility*** (μ), defined as the drift velocity acquired by a particle when accelerated in a unit electric field. Taking into account expressions (1.81) and (1.83), the electron mobility is:

$$\mu = \frac{v_d}{E} \Rightarrow \mu = -\frac{e\tau}{2m} = -\frac{e\,\overline{\lambda}}{2m\,v_t} \tag{1.86}$$

The SI unit of measurement for mobility is m V s^{-1}, which results from the given definition (1.86). Given this expression for mobility, the electric conductivity of the metallic conductor can also be written as:

$$\sigma = ne\mu \tag{1.87}$$

Equation (1.85) shows that σ is a function of temperature. If more than one type of carrier is present, then $\sigma = \sum ne\mu$, which includes the signs of e and μ.

Meanwhile, only the mean drift velocity of the carriers is taken into account, while neglecting the effects of ***diffusion.***[5] If the density of carriers in a conductor is greater at one point than another, diffusion takes place, and an electric current flows in addition to that resulting from the application of an electric field. This process is not important in metals, because the drift velocities are so much greater than the diffusion velocities; however, in semiconductors, ionized gases, and electrolytes, diffusion plays a significant part in determining the current.

1.4.3 Local form of the Joule–Lenz law for homogeneous conducting media

It is considered that in the process of collision between electrons and the nodes of the crystalline lattice, the electrons transfer to the crystalline lattice all their kinetic energy gained in the electric field during the mean free path. After each time interval (τ) or distance ($\overline{\lambda}$), a new process of electron acceleration starts again, and the kinetic energy gained is transferred to the crystalline lattice at the next collision, continuing this process of the dissipation of kinetic energy. If the kinetic energy gained by an electron between two collisions is:

$$w_1 = \frac{mv_1^2}{2} = \frac{me^2E^2}{2m^2}\tau^2 = \frac{e^2E^2}{2m}\tau^2 \tag{1.88}$$

then, in the amount of time in which z collisions of the electron with the nodes of the lattice can occur, the electron transfers the energy $w_1' = zw_1$ to the lattice. Multiplying the number of electrons in the unit volume (n) by w_1', we obtain the energy transferred to the crystalline lattice by the electrons in the unit volume in unit time, i.e. precisely the power density (w):

$$w = nzw_1 = n\frac{1}{\tau}\frac{e^2E^2}{2m}\tau^2 = \frac{ne^2\tau}{2m}E^2 = \sigma E^2 = \rho j^2 = \vec{j}\,\vec{E} \tag{1.89}$$

1.4.4 Applying the Wiedemann–Franz law to the Drude–Lorentz theory

In accordance with the kinetic–molecular theory of the ideal gas, the thermal conductivity of an electronic gas is given by:

[5] In general terms, **diffusion** represents the migration of atoms from regions of high concentration of mobile species to regions of low concentration.

$$\varkappa = \frac{1}{2} nkv_t \overline{\lambda} \tag{1.90}$$

where $k = 1.38 \times 10^{-23}\ \mathrm{m^2\ kg\ s^{-2}\ K^{-1}}$ is the **Boltzmann constant**. As a result, the ratio between the thermal and electric conductivity is:

$$\frac{\varkappa}{\sigma} = \frac{\frac{1}{2} nkv_t \overline{\lambda}}{\frac{1}{2} \frac{ne^2}{m} \frac{\overline{\lambda}}{v_t}} = \frac{mk}{e^2} v_t^2 \tag{1.91}$$

Considering the average thermal velocity v_t in the average thermal kinetic energy of the electron gives:

$$\frac{mv_t^2}{2} = \frac{3kT}{2} \Rightarrow mv_t^2 = 3kT \tag{1.92}$$

such that the above ratio is:

$$\frac{\varkappa}{\sigma} = 3\frac{k^2}{e^2} T \tag{1.93}$$

and the previously formulated Wiedemann–Franz relationship is found. However, this time, the constant of proportionality appears explicitly, showing that it is a universal constant. Relation (1.93) is valid for all metals, nuancing ***the Wiedemann–Franz law*** with the **Lorenz number** $L = 3\left(\frac{k}{e}\right)^2$ or $L = \frac{8}{\pi}\left(\frac{k}{e}\right)^2$. If we consider the thermal velocity to be the average velocity $\overline{v} = \sqrt{\frac{8kT}{\pi m}}$, L is about $2 \times 10^{-8}\ \mathrm{V^2\ K^{-2}}$. Many metals have Lorenz numbers close to this value (for example, at 100 °C, the Lorenz numbers for Na, Al, and Ni are 2.19, 2.23, and 1.83, respectively), but there are also many that differ (for example, W and Bi, with values of 3.20 and 2.89, respectively). Essentially, the order of magnitude remains the same (i.e. $10^{-8}\ \mathrm{V^2\ K^{-2}}$), while the significant digits change slightly, taking values around two.

1.4.5 Physical meaning of the Drude–Lorenz theory and its deficiencies

Following the Drude–Lorenz theory presented above, several conclusions can be drawn, as described below:

1. The value of the average drift velocity can be evaluated from the expression for the current density, i.e. $j = ne\langle v_d \rangle$. Taking into account, for example, $n \approx 10^{28} - 10^{29}\ \mathrm{m^{-3}}$, and the maximum permissible value for the current density, i.e. $10^7\ \mathrm{A\ m^{-2}}$, we obtain:

$$\langle v_d \rangle = \frac{j}{ne} = \frac{10^7\ \mathrm{A\ m^{-2}}}{10^{28}\ \mathrm{m^{-3}} \cdot 1.6 \cdot 10^{-19}\ \mathrm{C}} \approx 6 \cdot 10^{-3}\ \mathrm{m\ s^{-1}}$$

It is worth noting that the drift velocity is very low compared to the thermal velocity, i.e. $v_t \approx 10^5\ \mathrm{m\ s^{-1}}$. The drift speed should not be confused with the speed of propagation of electric current through conductors. Practice shows that when the circuit is closed, the electric current propagates almost instantly at any distance from the switch. This is not a paradox! The explanation of this phenomenon lies in the fact that when we connect a voltage source to the circuit, an electric field is produced in the conductor at a speed close to the speed of light in vacuum. During this time, the quasi-free electrons present at any point in the conductor are set into motion, practically instantly over the entire length of the conductor, but the drift speed does not exceed a value of $6\ \mathrm{mm\ s^{-1}}$.

2. The value of the mean free path may be calculated by measuring the resistivity ρ to which it is linked by:

$$\rho = \frac{2m\, v_t}{ne^2\, \overline{\lambda}} \Rightarrow \overline{\lambda} = \frac{2mv_t}{ne^2 \rho}$$

Considering that for iron (Fe), $\rho = 10^{-7}\ \Omega\mathrm{m}$ and $n = 10^{28}\ \mathrm{m^{-3}}$, we obtain:

$$\overline{\lambda} = \frac{2 \cdot 9.1 \cdot 10^{-31} \cdot 10^5}{10^{28} \cdot 1.6^2 \cdot 10^{-38} \cdot 10^{-7}} \approx 0.7 \cdot 10^{-8}\ \mathrm{m}$$

This value corresponds to a temperature of 300 K, at which the average thermal velocity of $v_t = 10^5\ \mathrm{m\ s^{-1}}$ was evaluated; the value thus obtained for the mean free path is seen to be about 100 times higher than the crystal lattice constant of metals.

3. The value of the correlation time τ (of the average free time) results from its definition, i.e. $\tau = \frac{\overline{\lambda}}{v_t} \Rightarrow \tau = \frac{0.7 \cdot 10^{-8}\ \mathrm{m}}{10^5\ \mathrm{m\ s^{-1}}} \approx 7 \cdot 10^{-14}\ \mathrm{s}$.

4. The value of the mobility of free electrons in metals is:

$$\mu = \frac{e\tau}{2m} = \frac{1.6 \cdot 10^{-19} \cdot 7 \cdot 10^{-14}}{2 \cdot 9.1 \cdot 10^{-31}} = 0.6 \cdot 10^{-3}\ \mathrm{m^2\ V\ s^{-1}}$$

The dependence of the mobility of the charge carriers on τ reflects the fact that the collision processes (through which the charge carriers interact with the conducting medium) are responsible for the resistance that the conductor imposes on the progress of the charge carriers. It has been shown that μ is constant, regardless of the applied electric field.

5. The temperature dependence of the electrical conductivity is obtained after the replacement of the average thermal velocity by $v_t = \sqrt{\frac{3kT}{m}}$, leading to the dependence:

$$\sigma = \frac{ne^2\, \overline{\lambda}}{2m\, v_t} = \frac{ne^2 \lambda}{2m\sqrt{\frac{3kT}{m}}} = \frac{ve^2 \overline{\lambda}}{\sqrt{12\, mk}} T^{-1/2}$$

Several deficiencies of the Drude–Lorentz theory, based on the hypothesis that τ and $\overline{\lambda}$ are constants, are explained below:

(a) As n, e, m, k are constants and $\overline{\lambda} = \dfrac{1}{\sqrt{2} \cdot n\sigma}$ (defined in the kinetic–molecular theory of gases) is considered independent of temperature, it follows that the electrical conductivity decreases with temperature, i.e. $\sigma \sim T^{-1/2}$. Experience shows that at temperatures which are not too low, the temperature dependence of conductivity is of the type $\sigma \sim T^{-1}$. This insufficiency of the classical theory of electrical conduction in metals can be attributed to the temperature dependence of the mean free path ($\overline{\lambda}$), or even to the concentration of the charge carriers. Indeed, $\overline{\lambda}$ decreases when the temperature increases, because the amplitudes of the vibrations of the lattice are increased, decreasing τ and the mean free path correspondingly ($\overline{\lambda} = v_t \tau$).

(b) The value of the electrical conductivity of the metals calculated using the relation $\sigma = \dfrac{1}{\rho} = \dfrac{ne^2}{2m}\dfrac{\overline{\lambda}}{v_t}$ is an order of magnitude lower than that measured experimentally, and the proportionality between σ and the concentration n of charge carriers is not always respected.

(c) While we do not treat the theory of the caloric capacity of metals here, we should also mention a deficiency of the classical theory of electrons in metals, namely the fact that this theory presents values of caloric capacity higher than those measured experimentally.

(d) At the same time, the electronic gas model cannot answer the question: *'Why are some materials conductors, while others are semiconductors, and yet others are insulators?'*

The relationships established by the Drude–Lorentz theory remain valid only under small applied fields, such that $\tau = \dfrac{\overline{\lambda}}{v_t}$ is constant. If the electric field intensity increases, the acceleration increases and hence the drift speed increases, so that it can become equal to v_t, i.e. $v_t \cong v_d$. In this case, $\tau = \dfrac{\overline{\lambda}}{v_t + v_d}$ no longer remains constant but decreases. Consequently, the mobility $\mu = \dfrac{e\tau}{2m}$ has the same behavior, and so the average speed of the oriented movement follows the same trend.

In order to remove the shortcomings of the classical theory of electrical conduction in metals, we must approach the electronic gas using quantum theory. The quantum theory of electronic gases in metals was elaborated in 1928 by Arnold J W Sommerfeld. In this theory, the following quantum properties of electrons are taken into account:

(1) Electrons manifest the wave–particle dualism;
(2) The movement of electrons is subject to Heisenberg's uncertainty principle;
(3) Electrons are subject to the Pauli exclusion principle;
(4) The distribution of electrons by energy follows Fermi–Dirac statistics.

Taking all these into account, a new approach must be used to understand the general process of electronic conduction in the solid state.

1.5 Potential energy of electrons in solids

1.5.1 Potential energy of electrons in a metal

The image of the electric conduction processes of metals created using the classical treatment presented so far is good but incomplete (leading to the deficiencies that have already been mentioned). This behavior is due to the fact that it was considered that electrons in metal behave like free or quasi-free particles, without taking into account the interactions between them and the ions in the nodes of the network. Therefore, an analysis of the movement of electrons in the electric field created by the ions (considered immobile and located at the nodes of the crystal lattice) is required. Let us consider a three-dimensional network of any metal. The potential energy of an electron at any point in the network is the product of the electron's charge and the electrical potential produced by all the ions of the network at that point. In order to determine the potential energy of an electron in the field of a nucleus with charge Ze, the potential created by the nucleus at a distance r from it is multiplied by the electron charge:

$$W_p = -eV = -\frac{Ze^2}{4\pi\varepsilon r} \tag{1.94}$$

By choosing the vacuum level (i.e. infinity) as the origin of potential energy, we can say that the potential energy of an electron at any point in the crystal is negative and has a hyperbolic dependence as a function of the variable r. The spatial image is difficult to represent, but an image of a vertical section according to this model is shown in figure 1.21(a), which displays the position in the crystal of the atomic

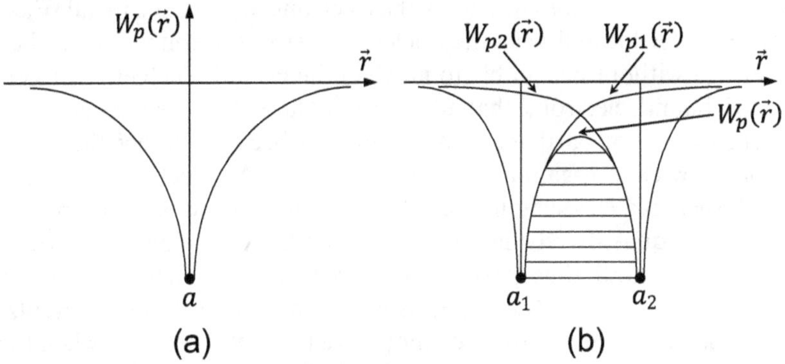

Figure 1.21. The potential energy of the electron in the electric field: (a) of a single nucleus and (b) of two nuclei placed at a distance a in the crystal.

nucleus with the electric charge Ze. We now consider two nuclei, namely the 1st and 2nd. The curves $W_{p_1}(\vec{r})$ and $W_{p_2}(\vec{r})$ are shown in figure 1.21(b), and it can be observed that the total energy given by the sum of the two nuclei has the same shape as $W_{p_1}(\vec{r})$ and $W_{p_2}(\vec{r})$ near the nuclei but flattens between the two nuclei, having lower values than the individual curves.

This scenario for two neighboring nuclei is repeated for several nuclei placed in an ionic row, so that the potential energy of the nodes of the crystal lattice naturally has a periodic character, as suggested in figure 1.22. At electrostatic equilibrium, the

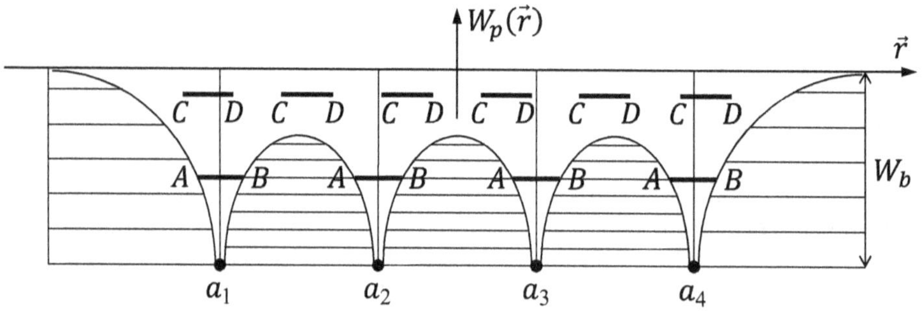

Figure 1.22. Potential energy of an electron in the periodic potential of a crystal.

interior of the metal is an equipotential region of constant potential equal to that on the surface. It is found that if one takes into account the crystalline structure of the metal, the potential energy varies significantly near the nuclei, tending even toward infinity in their strict vicinity, while between the nuclei and beyond them the potential varies very little, being almost constant. If we consider the 4th atom in figure 1.22, it is located at the surface of the crystal. Because there is no other nucleus on its right-hand side, the curve $W_p(\vec{r})$ is no longer influenced in the same way as it is between nuclei. As a result, it behaves like a barrier of height W_b in the path of the electrons—a barrier that does not allow the electrons to leave the metal. Electrons at an energy level *AB* located near the nucleus are strongly connected to the nucleus and held near it without being able to move in the crystal. Such electrons constitute the so-called bound electrons that usually populate the valence band in a solid. These electrons cannot participate in conduction because, under the action of an electric field, they cannot gain a nonzero drift speed. An electron (or several) at an energy level (such as *CD*, see figure 1.22) far from the nucleus is poorly bonded to the nucleus, having a quasi-free behavior; as such, under the action of an external field, it can easily move and thus participate in conduction. These are the *quasi-free electrons that behave like an ideal gas,* as considered in the Drude–Lorentz theory. However, these quasi-free electrons cannot leave the metal because, when they reach the potential barrier at the surface of the metal, they are reflected by the barrier and return inside the metal. An electron can leave the metal only when it acquires an energy greater than, or at least equal to, the height of the barrier (W_b) at the surface and is thus able to escape it. However, in order to evaluate the energy that needs to be supplied to an electron to leave the metal, we need to know the energetic state of the electron in the metal; that is, we need to know how the electrons are distributed inside the metal; therefore, we must know the distribution function for the energy of the electrons in the crystal.

1.5.2 The distribution function for the energy of electrons in a solid body

If we assume that inside the metal, the electrons have all the possible energies, then the number of electrons $dN(w)$ that have an energy that falls within the range from w to $w + dw$ is proportional to the density of energy states $g(w)$, to the probability $f(w)$ of occupying the energy state w, and to the width dw of the energy range:

$$dN(w) = g(w)f(w)dw \tag{1.95}$$

1.5.2.1 The effective density of energetic states in a solid

The function $g(w)$, called the effective density of states, represents the number of possible energy states per unit volume and per unit energy. A rigorous evaluation of this function $g(w)$ can be made by relating the volume of the spherical shell $V_k = 4\pi k^2 dk$ in the vector space of the $\vec{k}$ reciprocal lattice to the volume associated with the free movement of an electron, $V_n = 8\pi^3/L_x L_y L_z$. Intuitively, in a classical treatment, one could reason as follows: the state of an electron in metal is characterized by the set of coordinates (x, y, z) in coordinate space and the set of momentum components (p_x, p_y, p_z) in momentum space; that is, by the vector variables $(\vec{r}, \vec{p})$ in the phase space. A volume element in phase space is $dv = dv_r dv_p = (dxdydz)(dp_x dp_y dp_z)$. In phase space, the volume assigned to a free electron (to a free electronic state) is $dv_v = dxdydz$. As a result, in the unit crystal volume, there is a number of free electronic states equal to:

$$g = \frac{1}{dv_v} = \frac{1}{dxdydz} \tag{1.96}$$

However, according to the relations of uncertainty, one can write the following for each direction:

$$\begin{aligned} dxdp_x &\geqslant 2\pi\hbar = h \\ dydp_y &\geqslant 2\pi\hbar = h \\ dzdp_z &\geqslant 2\pi\hbar = h \end{aligned} \tag{1.97}$$

So:

$$dv_v = dxdydz = \frac{h^3}{dp_x dp_y dp_z} \tag{1.98}$$

Taking into account Pauli's principle, the number of free electronic states in the unit volume and per unit of energy is:

$$g(w) = \frac{2}{h^3} dp_x dp_y dp_z = \frac{2m^3}{h^3} dv_x dv_y dv_z \tag{1.99}$$

If one takes into account the relations $w_e = \frac{p^2}{2m}$ and $p^2 = p_x^2 + p_y^2 + p_z^2$, and uses equation (1.99), one can write the relation that gives the energy dependence of the density of $g(w)$ states in the form of:

$$g(w) = \frac{(2m)^{3/2}}{2\pi^2 h^3} w^{1/2} = 4\pi \frac{(2m)^{3/2}}{h^3} w^{1/2} \qquad (1.100)$$

1.5.2.2 Fermi–Dirac probability function

The function $f(w)$, called the Fermi–Dirac probability function, is used in quantum statistics. It represents the probability that an electron is in energy state w at the thermal equilibrium of a metal (in other words, the probability of occupying the state with energy w). In quantum statistics, the expression for $f(w)$ is defined as:

$$f(w) = \frac{1}{1 + e^{\frac{w - w_F}{kT}}} \qquad (1.101)$$

This distribution function is valid for particles with a semi-integer spin. In equation (1.101), k is the Boltzmann constant, T is the absolute temperature, and w_F represents the Fermi energy or Fermi level. To illustrate the physical meaning of this parameter, two cases will be considered:

1. In the case where $T = 0\ K$, there are two situations, namely:
   (a) For $w < w_F$, the exponent $\frac{w - w_F}{kT} \to -\infty$, so $e^{\frac{w - w_F}{kT}} \to 0$ and $f(w) = 1$, i.e. all energy states with energy $w < w_F$ are occupied by electrons.
   (b) For $w > w_F$, the exponent $\frac{w - w_F}{kT} \to \infty$ and $f(w) \to 0$, so quantum states with an energy greater than w_F are unoccupied.

   Therefore, it can be noted that the physical quantity w_F (the Fermi level) can be defined as follows: w_F is the maximum energy of the level that can still be populated with electrons at 0 K. So, the Fermi level (w_F) is the energy level that demarcates the states completely occupied by electrons from those completely free at 0 K, see figure 1.23(a). It should also be noted that at $w = w_F$, $f(w) = 1/2$, regardless of the value of T. As a result, the Fermi level can also be defined as the energy of the occupied states with an occupancy probability of 50%, regardless of the temperature. Taking into account relation (1.100) for the effective density of states, figure 1.23(b), and the probability function $f(w)$ given by equation (1.101) in the case of $T = 0$ K, the number of electrons in the unit volume can be calculated:

   - For $w < w_F$:

$$n_0 = \int dN = \int_0^{w_F} g(w) f(w) dw = \int_0^{w_F} 4\pi \frac{(2m)^{3/2}}{h^3} w^{1/2} dw = \frac{8\pi}{3h^3}(2m)^{3/2} w_F^{3/2} \qquad (1.102)$$

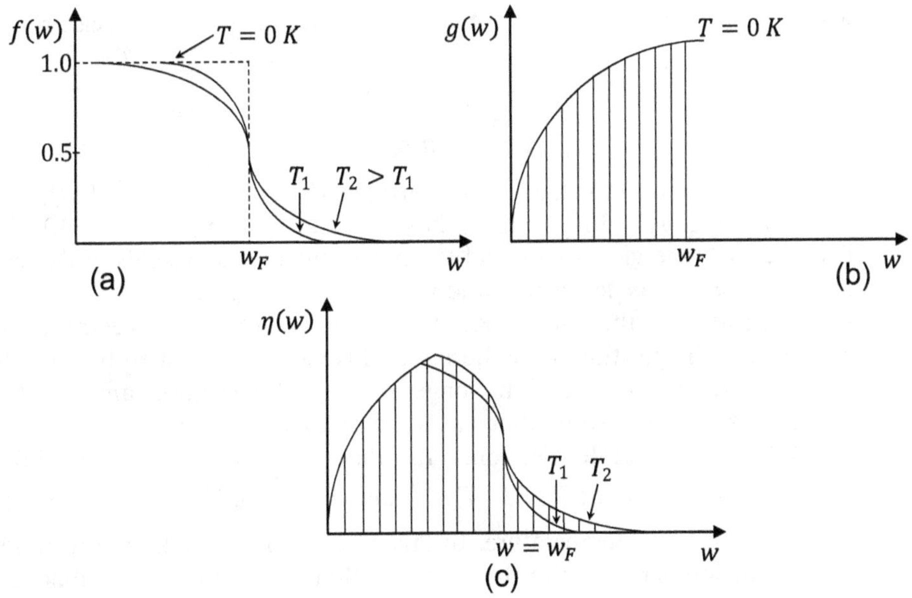

Figure 1.23. (a) The Fermi–Dirac distribution function. (b) The effective density of energetic states. (c) The distribution function for the energy of the electrons.

- For $w > w_F$:

$$n_0 = 0 \tag{1.103}$$

Using relation (1.102), one can calculate the Fermi level when the concentration of electrons in metals is known:

$$w_F = \left(\frac{3h^3}{8\pi}h_0\right)^{3/2}\frac{1}{2m} = \frac{h^2}{2m}\left(\frac{3n_0}{8\pi}\right)^{2/3} \tag{1.104}$$

If $w_F = \frac{3kT_F}{2} \Rightarrow T_F = \frac{2}{3}\frac{w_F}{k} \rightarrow \sim 10^4$ K. The Fermi level determined at $T = 0$ K is called the electrochemical potential.

The total energy of all electrons below w_F at $T = 0$ K can be calculated as follows:

$$\begin{aligned} w_0 &= \int_0^{w_F} g(w)f(w)w\,dw = 4\pi\frac{(2m)^{3/2}}{h^3}\int_0^{w_F} w^{3/2}dw \\ &= \frac{8\pi}{5h^3}(2m)^{3/2}w_F^{5/2} = \frac{3}{5}n_0 w_F \end{aligned} \tag{1.105}$$

which is the zero energy of a Fermi gas. The average energy of an electron at 0 K is:

$$w_m = \frac{3}{5}w_F = \frac{3}{5}\frac{h^2}{2m}\frac{3n_0}{8\pi}^{2/3} \tag{1.106}$$

which, for the current values of the constants ($h = 6.625 \times 10^{-34}$ J s^{-1}, $m = 9.1 \times 10^{-31}$ Kg, $n_0 = 10^{28}$ mm^{-3}), has the value $w_m = 5.4\, eV \cong 9 \cdot 10^{-19}$ J. Because such energies correspond to temperatures of thousands of degrees, the electronic gas is degenerate and is called a ***Fermi gas.***

2. At temperatures other than 0 K, the calculation of the concentration of carriers and the position of the Fermi level is more complicated because the Fermi–Dirac function itself is no longer reduced to the values 1 and 0. In this situation, as a rule, two limiting cases are analyzed, namely:
   (a) A nondegenerate electronic gas that satisfies either the condition $w - w_F \gg kT$ or $T > T_c$ ($T = \frac{w}{k}$ and $T_c = \frac{w_F}{k}$). Here, T_c is called the degeneration temperature. In this case, the Fermi–Dirac distribution function is reduced to the Maxwell–Boltzmann distribution function:

$$f(w) = \frac{1}{1 + e^{\frac{w-w_F}{kT}}} \cong e^{-\frac{w-w_F}{kT}} \tag{1.107}$$

and the number of electrons per unit volume is:

$$\begin{aligned} n_0 &= \int_0^\infty g(w)f(w)dw \\ &= \int_0^\infty \frac{4\pi(2m)^{3/2}}{h^3} w^{1/2} e^{-\frac{w-w_F}{kT}} dw \\ &= \frac{4\pi(2m)^{3/2}}{h^3} e^{\frac{w_F}{kT}} \int_0^\infty w^{1/2} e^{-\frac{w}{kT}} dw \end{aligned} \tag{1.108}$$

Noting that $\frac{w}{kT} = x \Rightarrow w^{1/2} = (kT)^{1/2}(x)^{1/2}$ and $dw = kTdx$, equation (1.108) becomes:

$$\begin{aligned} n_0 &= \frac{4\pi}{h^3}(2m)^{3/2} e^{\frac{w_F}{kT}} (kT)^{3/2} \int_0^\infty x^{1/2} e^{-x} dx \\ &= \frac{4\pi}{h^3}(2mkT)^{3/2} \frac{\sqrt{\pi}}{2} e^{\frac{w_F}{kT}} \end{aligned}$$

So:

$$n_0 = 2\left(\frac{2\pi mkT}{h^2}\right)^{3/2} e^{\frac{w_F}{kT}} = N_{\text{eff}}\, e^{\frac{w_F}{kT}} \tag{1.109}$$

where:

$$N_{\text{eff}} = 2\left(\frac{2\pi mkT}{h^2}\right)^{3/2} \tag{1.110}$$

represents the effective density of states, depending on the temperature and the actual mass of the charge carriers in the network. The dependence of the Fermi level on concentration and temperature can be removed from equation (1.107):

$$\begin{aligned} n_0 &= 2\left(\frac{2\pi mkT}{h^2}\right)^{3/2} e^{\frac{w_F}{kT}} \Rightarrow w_F \\ &= kT \ln\left[\frac{n_0 h^3}{2}(2\pi mkT)^{-3/2}\right] \end{aligned} \tag{1.111}$$

(b) A degenerate electronic gas when $w - w_F \ll kT$ or $T \ll T_c$. The calculation of n_0 and w_F presents particular difficulties. The approximate Fermi value is:

$$w_F = w_{F_0}\left[1 - \frac{\pi^2}{12}\left(\frac{T}{T_c}\right)^2\right] \tag{1.112}$$

Considering, for simplicity, that $w_F \cong w_{F_0}$, the energy distribution of electrons defined by $\eta = \dfrac{dN(w)}{dw} = g(w)f(w)$ is obtained by taking the product of $g(w)f(w)$; see figure 1.23(c).

References and further reading

[1] Antohe Ş and Antohe V-A 2023 *Electrostatics. Formalism of the Electrostatic Field in Vacuum and Matter* 1st edn (Bristol: IOP Publishing) 266 pages

[2] van Delft D and Kes P 2010 The discovery of superconductivity *Phys. Today* **76** 38–43

[3] Onnes H K 1991 Further experiments with liquid helium. C. On the change of electric resistance of pure metals at very low temperatures etc IV. The resistance of pure mercury at helium temperatures *Through Measurement to Knowledge: The Selected Papers of Heike Kamerlingh Onnes 1853–1926. Boston Studies in the Philosophy of Science* vol 124 ed K Gavroglu and Y Goudaroulis (Dordrecht: Springer) pp 261–3

[4] Pandey S C 2010 *Course in Physics (Vol 4): Electrostatics and Current Electricity* (London: Pearson Education) 288 pages

[5] Feynman R P, Leighton R B and Sands M 2011 *Feynman Lectures on Physics, Vol. II: The New Millennium Edition: Mainly Electromagnetism and Matter* (New York: Basic Books) (Feynman Lectures on Physics) 589 pages

[6] Barr J R 2023 *Principles of Direct-Current Electrical Engineering* (London: Palala Press) 592 pages

[7] Jackson J D 1998 *Classical Electrodynamics* 3rd edn (New York: Wiley) 832 pages

[8] Abrikosov A A 2017 *Fundamentals of the Theory of Metals* reprint edn (New York: Dover Publications) 640 pages

[9] Di Bartolo B 2004 *Classical Theory of Electromagnetism: With Companion Solution Manual* 2nd edn (Singapore: World Scientific) 780 pages

[10] Purcell E M and Morin D J 2013 *Electricity and Magnetism* 3rd edn (Cambridge: Cambridge University Press) 853 pages

IOP Publishing

Electromagnetism and Special Methods for Electric Circuits Analysis

Ştefan Antohe and Vlad-Andrei Antohe

Chapter 2

Electric charge transport in media other than metals

In the previous chapter, a detailed analysis of the mechanism of electric charge transport in metals was made. We now turn to the study of several electrical conduction mechanisms involved in the transport of electric charge carriers through media other than metal conductors. In this regard, we first discuss the electrical conduction of free electrons in vacuum. Afterwards, we focus on electrical conduction in ionized gases and liquid electrolytes.

2.1 Electric conduction of free electrons in vacuum

The conduction process of free electrons in vacuum, which leads to nonlinear current–voltage characteristics, takes place within electronic tubes such as diodes, triodes, pentodes, etc. In these devices, an electric conduction process can only occur in the presence of charge carriers that can move under the action of an electric field. These charge carriers often take the form of free electrons obtained using *thermionic emission* or *field emission* phenomena.[1]

2.1.1 Thermionic emission and the Richardson–Dushman relationship

In the case of electronic tubes, the charge carriers are typically free electrons obtained through thermionic emission. *Thermionic emission is the phenomenon of electron emission by a heated metal.* The flux of electrons, characterizing the density of the saturation current emitted by a solid body heated to a certain temperature, depends on the temperature and the nature of the emissive electrode. This behavior

[1] **Thermionic emission** (TE) and **field emission** (FE) effects are frequently exploited in the construction of electron guns for scanning electron microscopes (SEMs). The latter effect is used in high-resolution FE scanning electron microscopy.

doi:10.1088/978-0-7503-5854-5ch2

is described by the **Richardson–Dushman** relationship, which gives an expression for the density of the saturation current (the emission current) in thermoelectronic emission. In order to establish the Richardson–Dushman relationship, we use the distribution function of electron energies in the solid body. We also take the nondegenerate case ($w_0 - w_F \gg kT$), where, as shown in the previous chapter, the Fermi–Dirac function can be replaced by the classical distribution function known as the Maxwell–Boltzmann distribution. It is assumed that in the metal, the electrons are arranged two by two on energetic levels up to the Fermi level, according to the Pauli principle (see figure 2.1). The last two electrons placed on

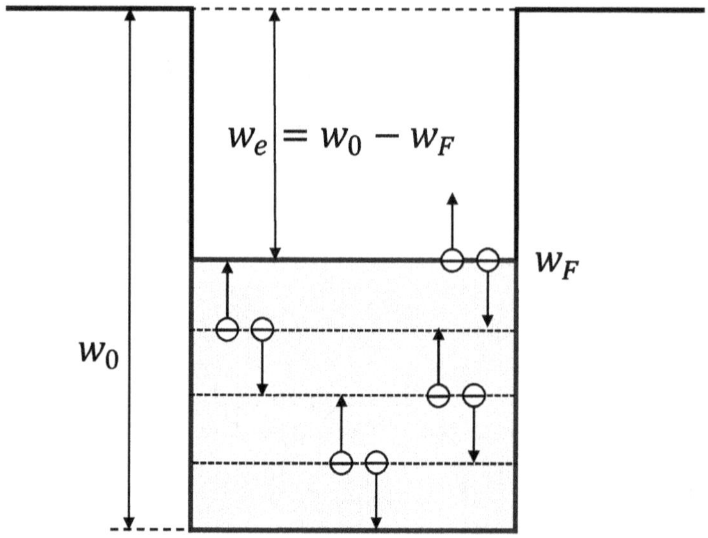

Figure 2.1. The distribution of electron energies in a metal.

the Fermi level can most easily leave the metal because they only have to overcome the barrier $w_e = w_0 - w_F$ and not the full value of w_0. Consider a volume element of a metallic conductor, of rectangular shape, with sides dp_x, dp_y, and dp_z, as shown in figure 2.2. Assuming that electrons leave the surface of the metal with a momentum

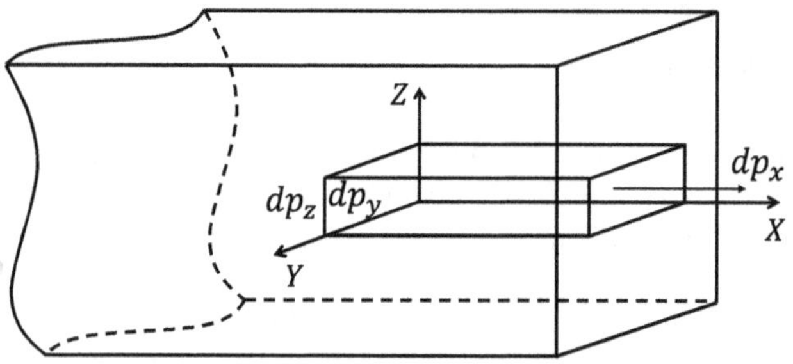

Figure 2.2. Volume element in momentum phase space.

p_x in the normal direction to the metal surface (i.e. on the OX axis), we note that only electrons with a kinetic energy larger than w_0, $\left(\frac{p_x^2(0)}{2m} \geqslant w_0 = w_e + w_F\right)$, or with a momentum:

$$p_x(0) \geqslant [2m(w_e + w_F)]^{\frac{1}{2}} \tag{2.1}$$

are able to leave the emissive cathode. The number of electrons in the metal that have a momentum between $\vec{p}$ and $\vec{p} + \overrightarrow{dp}$ (i.e. their kinetic energy falls between w and $w + dw$) is given by the product of the effective density of states and the Fermi–Dirac probability function $f(w)$; hence:

$$dn = \frac{2}{h^3} dp_x\, dp_y\, dp_z \frac{1}{1 + e^{\frac{w - w_F}{kT}}} \tag{2.2}$$

The electrons in the metal move in all three directions, but our focus here is on the number of electrons that touch the face of the metallic cathode normal to the direction OX (regardless of their momentum in the OY and OZ directions), a number denoted by $dn(p_x)$.

Thus, the number of electrons that have a momentum between p_x and $p_x + dp_x$, reaching the metal surface in unit time and being emitted from the metal with a speed of $v_x = \frac{p_x}{m}$, is:

$$n[p_x(0)] = \int_{p_x(0)}^{\infty} \frac{p_x}{m} dn(p_x) \tag{2.3}$$

where $dn(p_x)$ is the number of electrons that have their momentum oriented in the direction x, with a value between p_x and $p_x + dp_x$, regardless of the values of their components p_y and p_z. This is given by:

$$dn(p_x) = \frac{2}{h^3} dp_x \int_{-\infty}^{\infty} \int_{-\infty}^{\infty} \frac{dp_y\, dp_z}{1 + e^{\frac{w - w_F}{kT}}} \tag{2.4}$$

Under the condition $w - w_F \geqslant kT$, the Fermi–Dirac function reduces to the Maxwell–Boltzmann distribution for a nondegenerate electronic gas, and the above equation then becomes:

$$dn(p_x) = \frac{2}{h^3} dp_x \int_{-\infty}^{\infty} \int_{-\infty}^{\infty} dp_y\, dp_z\, e^{-\frac{w - w_F}{kT}} \tag{2.5}$$

Since the kinetic energy of the electron is:

$$w = \frac{p^2}{2m} = \frac{p_x^2 + p_y^2 + p_z^2}{2m} \tag{2.6}$$

we obtain:

$$dn(p_x) = \frac{2}{h^3} e^{\frac{w_F}{kT}} e^{-\frac{p_x^2}{2mkT}} dp_x \int_{-\infty}^{\infty} e^{-\frac{p_y^2}{2mkT}} dp_y \int_{-\infty}^{\infty} e^{-\frac{p_z^2}{2mkT}} dp_z \tag{2.7}$$

We can now write: $\frac{p_y^2}{2mkT} = \alpha y^2 \Rightarrow y = \frac{p_y}{(2mkT\alpha)^{1/2}} \Rightarrow dy = \frac{dp_y}{(2mkT\alpha)^{1/2}}$; hence, the integral above becomes:

$$\int_{-\infty}^{\infty} e^{-\frac{p_y^2}{2mkT}} dp_y = \int_{-\infty}^{\infty} e^{-\alpha y^2} (2mkT\alpha)^{\frac{1}{2}} dy = (2mkT\alpha)^{\frac{1}{2}} \int_{-\infty}^{\infty} e^{-\alpha y^2} dy$$

$$= (2mkT\alpha)^{1/2} \left(\frac{\pi}{\alpha}\right)^{1/2} = (2m\pi kT)^{1/2}$$

and then:

$$dn(p_x) = \frac{2}{h^3}(2m\pi kT) e^{\frac{w_F}{kT}} e^{-\frac{p_x^2}{2mkT}} dp_x \tag{2.8}$$

The current density of the electrons emitted in the OX direction is:

$$j_x = en[p_x(0)] = e \cdot \int_{p_x(0)}^{\infty} \frac{p_x}{m} dn(p_x) = \frac{4\pi ekT}{h^3} e^{\frac{w_F}{kT}} \int_{p_x(0)}^{\infty} e^{-\frac{p_x^2}{2mkT}} p_x \, dp_x \tag{2.9}$$

If the new variable $u = \frac{p_x^2}{2mkT}$ is used, $\Longrightarrow \quad 2p_x \, dp_x = 2mkTdu$, and it follows that:

$$\int_{p_x(0)}^{\infty} e^{-\frac{p_x^2}{2mkT}} p_x \, dp_x = \int_{\frac{p_x^2(0)}{2mkT}}^{\infty} e^{-u} mkTdu = -mkTe^{-u} \Big|_{\frac{p_x^2(0)}{2mkT}}^{\infty} =$$

$$= mkTe^{-\frac{p_x^2(0)}{2mkT}}$$

Furthermore:

$$j_x = \frac{4\pi ekT}{h^3}(mkT) e^{\frac{W_F}{kT}} e^{-\frac{p_x^2(0)}{2mkT}} = \frac{4\pi mek^2}{h^3} T^2 e^{\frac{w_F}{kT}} e^{-\frac{w_e + w_F}{kT}} =$$

$$\frac{4\pi mek^2}{h^3} T^2 e^{-\frac{w_e}{kT}} = RT^2 e^{-\frac{w_e}{kT}} \Longrightarrow j = RT^2 e^{-\frac{qV_b}{kT}} \tag{2.10}$$

Equation (2.10) is called the **Richardson–Dushman** relationship. In this relationship, the quantity $R = \frac{4\pi mek^2}{h^3}$ appears. This is called the ***Richardson constant*** and has a value of 120 A cm^{-2} K^{-2} in the case of emission from a metallic electrode in vacuum. Furthermore, w_e is the energy of extraction of the electron from the metal (and is basically the energy needed for the emission of one electron from the Fermi level to vacuum), being the constant of a material called the ***'work function'***. Relation (2.10) shows that the density of the thermionic emission current is proportional to the square of the absolute temperature and depends on a single material constant w_e. In the saturation region of a vacuum diode, the current has this dependence on temperature and from a graph of $\ln \frac{j}{T^2} = f\left(\frac{1}{T}\right)$, we can obtain the linear regression:

$$\ln\left(\frac{j}{T^2}\right) = \ln R - \frac{w_e}{kT} \tag{2.11}$$

The slope of this linear regression gives us the thermodynamic work function of the emissive cathode, while its intercept with the vertical axis also allows us to determine the Richardson constant. Experimental measurements have found values of R in the range of 15–35 A cm^{-2} K^{-2}, which differs markedly from the theoretical value of 120 A cm^{-2} K^{-2}. This difference arises because a few factors were neglected when establishing the Richardson–Dushman relationship, such as: (i) reflections at the emissive surface, (ii) the effect of the space charge, and (iii) the presence of an external electric field.

(a) **Reflections at the emissive surface**

When we consider the reaction of free electrons in metals to a potential barrier, quantum mechanics states there is a nonzero probability that the electrons will return into the metal when they touch the wall of the barrier; as a result, only some of them will leave the metal. On the barrier's surface, reflection occurs with an average reflection coefficient of $\langle R \rangle$. For example, in the case of a high barrier of $w_e = 10$ eVand an electron energy of 0.1 eV, the reflection coefficient is $\langle R \rangle = 67\%$, so only 33% of the electrons incident on the surface can leave the metal.

Taking account of this effect, we can correct the Richardson–Dushman relationship by introducing the mean transparency coefficient $\langle D \rangle = 1 - \langle R \rangle$, so that relation (2.10) becomes:

$$j = \langle D \rangle R T^2 e^{\frac{-w_e}{kT}} \tag{2.12}$$

(b) **The effect of the presence of an electric space charge**

In deducing the Richardson–Dushman formula, it was considered that all electrons leaving the metal immediately depart from its surface, i.e. without influencing the output of other electrons from the metal. This process occurs only in the presence of an external electric field. However, if no electric field acts on the emitted electrons, they remain located in the immediate vicinity of the metal's emissive surface, forming a negative space charge that opposes the exit of other electrons from the metal, thus diminishing the emission current.

(c) **The presence of an electric field**

When an electric field is applied to the emission surface of the metal, both the intensity and the nature of the emission current change. On the one hand, the action of the external electric field can reduce the value of the work function (*the Schottky effect*). On the other hand, the electric field increases the probability of the potential barrier being crossed through *the tunneling effect.* This effect occurs when some free electrons in a metal, despite having insufficient energy to overcome the potential barrier, are able to cross it. The latter process is a phenomenon known as *'field emission'.*

2.1.2 The Schottky effect

In establishing the Richardson–Dushman relationship, we did not consider that the emission of electrons from the metal charges it positively, exerting a force of attraction on the emitted electrons. The force of interaction between the electron emitted in vacuum at a distance x from the emissive cathode and its *'image electric charge'* left in the metal is:

$$\vec{F} = -\frac{e^2}{16\pi\varepsilon_0 x^2}\frac{\vec{x}}{x} \tag{2.13}$$

Furthermore, the energy of interaction between the electron and its *'image'* in the metal is:

$$w_p = -\int_{-\infty}^{x} \frac{-e^2}{16\pi\varepsilon_0 x^2}dx = -\frac{e^2}{16\pi\varepsilon_0 x} \tag{2.14}$$

The potential energy of the electron in the field of the 'image' force depends on x, as can be seen in figure 2.3. If an electric field applied between the electrodes is oriented from the anode to the cathode, then this field pulls out the electron; in other words, it tries to remove it from the cathode. The energy of the electron in this field is:

$$w_e = -eEx \tag{2.15}$$

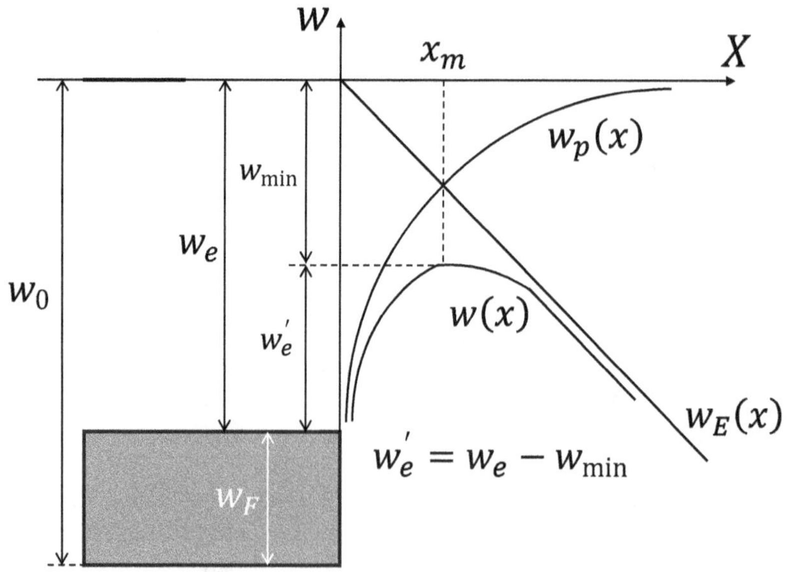

Figure 2.3. Potential energy of an electron in the field of the *'image force'* and in the applied electric field.

which varies linearly with distance (see figure 2.3). The total energy of the electron in the electric field of the *'image force'* and in the applied electric field is:

$$w(x) = -\frac{e^2}{16\pi\varepsilon_0 x} - eEx \tag{2.16}$$

The *'image force'* tends to draw the electron into the metal. The electric force required to remove it from the metal to a distance x_m, at which these forces reach equilibrium and the total energy is minimal, is given by:

$$\frac{dw(x)}{dx} = 0 \Rightarrow \frac{e^2}{16\pi\varepsilon_0 x^2} - eE = 0 \Rightarrow x_m = \sqrt{\frac{e}{16\pi\varepsilon_0 E}} \tag{2.17}$$

thus:

$$w_{\min} = -\frac{e^2}{16\pi\varepsilon_0\sqrt{\frac{e}{16\pi\varepsilon_0 E}}} - eE\sqrt{\frac{e}{16\pi\varepsilon_0 E}} = -\sqrt{\frac{e^3}{4\pi\varepsilon_0}}\sqrt{E} \tag{2.18}$$

If, in the absence of the electric field, in order to reach the anode, the electrons had to overcome an energy barrier of height:

$$w_e = w_0 - w_F \tag{2.19}$$

then, in the presence of the electric field, they must overcome a barrier of $w_e' = w_0 - w_F - w_{\min} = w_e - w_{\min}$, which is reduced by $w_{\min}$. This is therefore equivalent to saying that in the presence of the electric field, the *'work function'* of the emissive electrode is reduced from w_e to w_e' by $w_{\min}$. As a result, the density of the saturation current given by the Richardson–Dushman relationship becomes:

$$j_x = \langle D\rangle RT^2 e^{-\frac{w_e - w_{\min}}{kT}} = \langle D\rangle RT^2 e^{-\frac{w_e}{kT}} e^{\frac{w_{\min}}{kT}} = j_{x_0} e^{\frac{\sqrt{\frac{e^3}{4\pi\varepsilon_0}}\sqrt{E}}{kT}}$$

$$= j_{x_0} e^{\frac{\beta_S\sqrt{E}}{kT}} = j_{x_0} e^{\frac{\beta_S\sqrt{U}}{kT\sqrt{d}}} \tag{2.20}$$

Relation (2.20) shows that in the presence of an intense electric field, the saturation (emission) current increases with the intensity of the electric field. This phenomenon explains the behavior of the saturation current in a vacuum diode (see section 2.1.5), where at voltages greater than the saturation voltage, the emission current increases exponentially with $\sqrt{U}$. The increase in the applied field causes a decrease in the barrier height but also its narrowing (see figure 2.4), also enabling the tunneling phenomenon through the barrier at the metal–vacuum interface.

2.1.3 Field emission

The emission of electrons from metals that occurs under the action of an intense electric field alone, i.e. without the metal being heated, is called ***'field emission'*** or

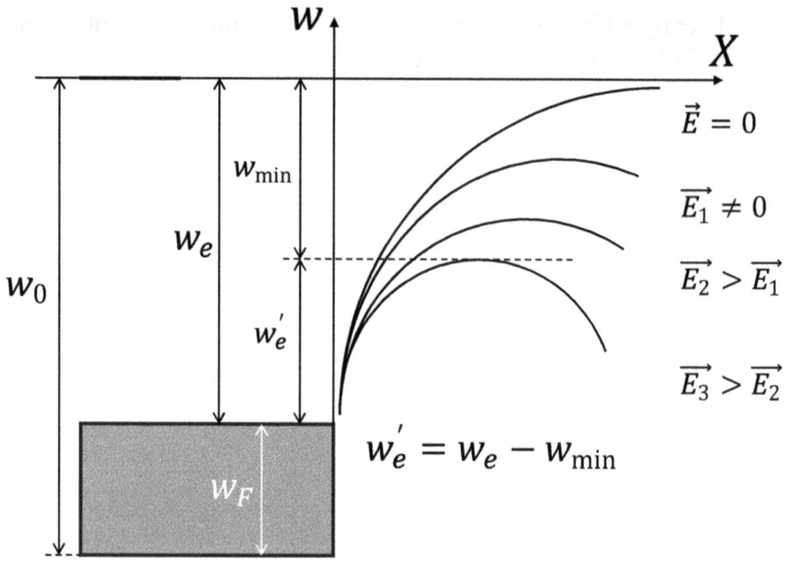

Figure 2.4. Changes of the barrier at the metal–vacuum interface with applied electric field.

'cold emission'. When the cold emission current produced under intense fields was measured and compared to the value calculated on the basis of the Schottky relationship, the experimental values were found to be two orders of magnitude larger than the theoretical ones. This observation led to the conclusion that under intense applied electric fields, the energy barrier narrows greatly, so that it can also be 'tunneled' by electrons from the metal that do not have enough kinetic energy to pass over the potential barrier by thermionic emission (see figure 2.5).

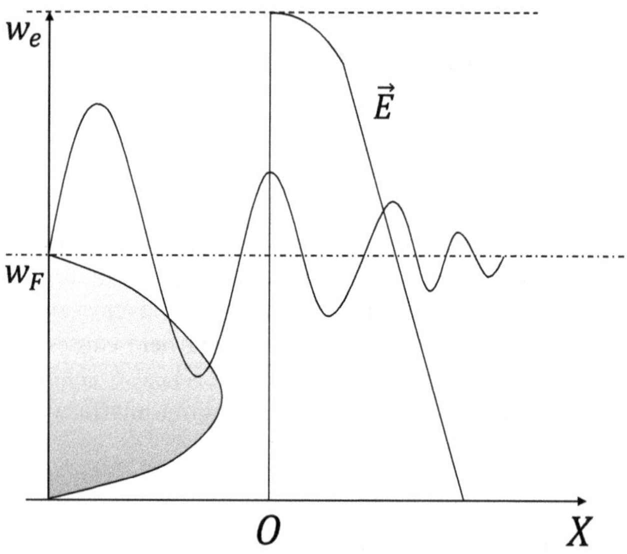

Figure 2.5. Tunnelling through the barrier at the metal–vacuum interface under an intense electric field.

The tunneling emission is explained by the Fowler–Nordheim model, using a rectangular energy barrier. If $D_t(E)$ represents the probability that the electrons will cross the potential barrier by tunneling (barrier transparency), then, in a treatment similar to that used for thermionic emission, the density of the electronic current will be:

$$j_t = \frac{e}{m}\int_0^{\infty} D_t(E) p_x \, dv(p_x) \tag{2.21}$$

Finally, Fowler and Nordheim obtained the relation:

$$j_t = A_1 E^2 e^{-\frac{C}{E}} \tag{2.22}$$

where:

$$A_1 = \frac{e w_F^{1/2}}{2\pi h w_0 w_e^{3/2}} \text{ and } C = \frac{8\pi}{3he}(2m)^{1/2} w_e^{3/2} \tag{2.23}$$

The Fowler–Nordheim relationship given in equation (2.22) has a resemblance to the Richardson–Dushman relationship if the temperature T is replaced by E.

2.1.4 Secondary electronic emission

The process of electron emission from a solid body due to irradiation by a beam of neutral or charged particles (electrons, protons, ions, neutrons, atoms, etc) that have a kinetic energy at least equal to the extraction work function of the metal is called ***'secondary emission'***. After the irradiation of a solid body with a beam of primary electrons produced by a heated filament, secondary electrons are emitted from the target metal. The secondary emission coefficient (σ_S), which represents the ratio between the intensity of the secondary current (I_S) and the intensity of the primary current (I_P), characterizes the emission process; it is not a constant specific to each metal but has a monotonous variation with the incident electric energy (see figure 2.6). Some values for σ_S and w_e are given in table 2.1. It can be observed that there is a proportionality between $\sigma_{S\,\max}$ and w_e.

2.1.5 Electronic tube diode

We previously discussed how free electrons can be produced. We now present a further analysis of the electric conduction mechanism associated with their movement in vacuum. The electric conduction mechanism of free electrons in vacuum is essentially different from that of electrons in metals, which was previously discussed; as presented later, it leads to a nonlinear current–voltage (I–U) characteristic. This conduction mechanism can be analyzed by studying the current–voltage characteristics of electronic tubes. These tubes, particularly electronic diodes, operate using the processes of displacement of free electrons emitted through thermionic emission in vacuum. The vacuum diode consists of two electrodes, the cathode and the anode, placed in a glass tube that has an internal pressure of 10^{-7} torr. The electrodes usually have a cylindrical shape. The cathode is indirectly heated, the filament being

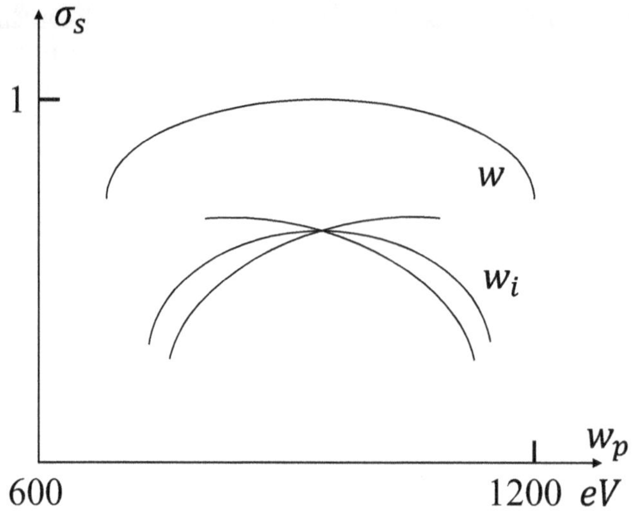

Figure 2.6. Dependence of the secondary emission current on the energy of the incident electrons.

Table 2.1. σ_S and w_e for some common metals.

Metal	$\sigma_{S\ max}$	w_e (eV)
Cu	1.3	4.3
Ag	1.5	4.55
Al	1	4.2
Ni	1.31	4.61

made of tungsten (W) or molybdenum (Mo), while the outer cylinder is made of nickel (Ni) or tantalum (Ta) activated with oxides of alkaline earth metals (see figure 2.7).

The operation of the diode can be plotted using curves that represent the variation of the potential and the electric field between the electrodes (see figure 2.8). If the cathode is unheated, that is, no electrons are emitted, the potential varies linearly with the distance between the cathode and the anode, the field remaining constant (curve 1 in figure 2.8). As the temperature rises, the emitted electrons form a space charge (an electron cloud) that deforms the potential and field variation curves. The emitted electrons form a spatial charge of maximum density near the cathode, so that the potential becomes negative in this region. A potential minimum (V_{min}) is formed at a distance of r_{min}, where the total field is zero, and the electrons reach the anode if they have enough energy to overcome this potential barrier. Electrons that are able to overcome the barrier enter an acceleration field that moves them toward the anode, forming the anodic current.

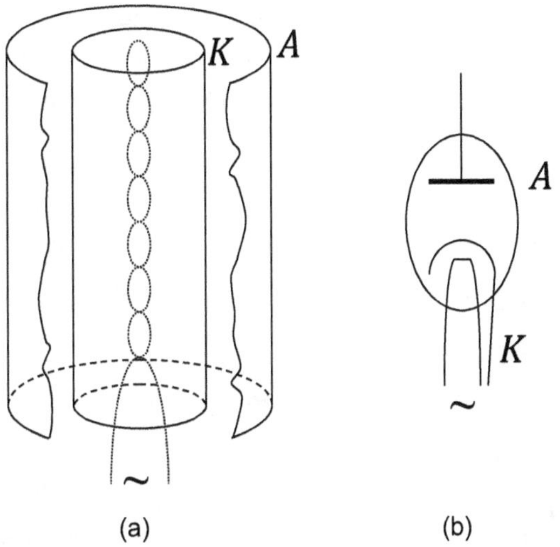

Figure 2.7. Section through an electronic diode (a) and its schematic representation (b).

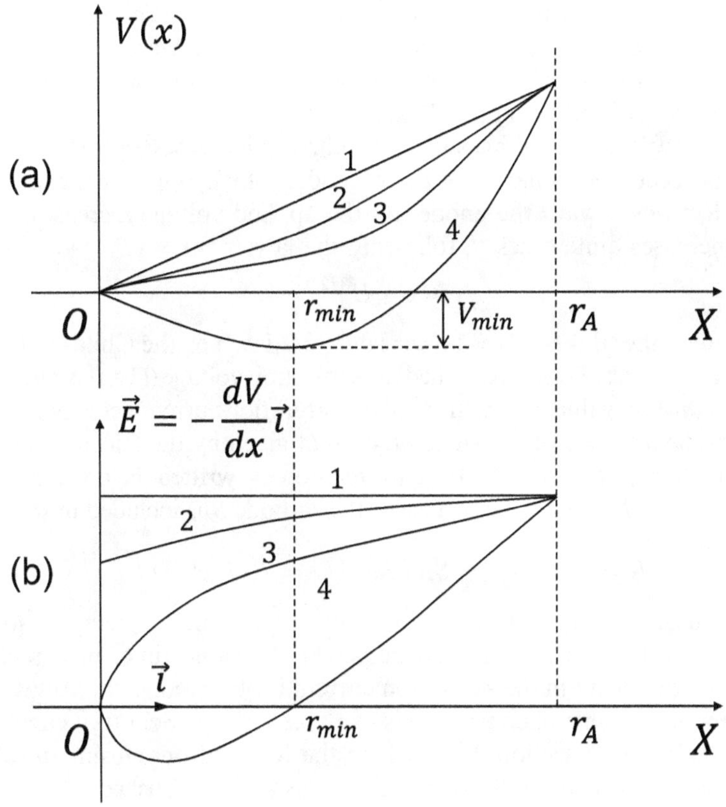

Figure 2.8. (a) The dependence of the potential $V(x)$ and (b) of the electric field $\vec{E} = -\frac{dV}{dx}\vec{i}$ on distance in the cathode–anode space of a vacuum electronic diode.

Of course, in practice, the current–voltage (*I–U*) characteristics are of interest, which are usually measured with the help of the experimental setup shown in figure 2.9(a).

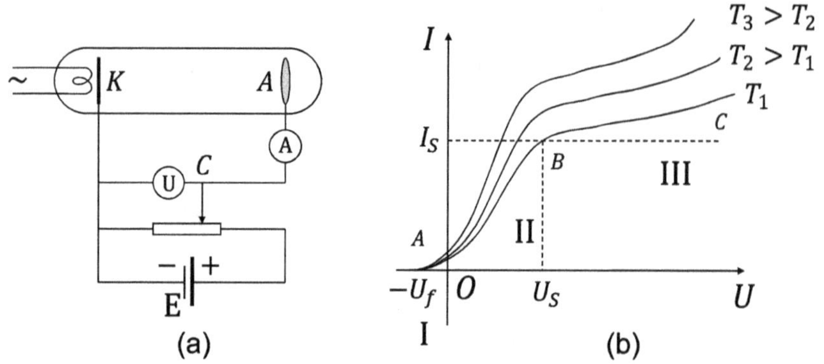

Figure 2.9. (a) Potentiometric setup for measuring the *I–U* characteristics; (b) *I–U* characteristics at different filament temperatures of an electronic diode.

The *I–U* characteristics for different filament temperatures are shown in figure 2.9(b). Along the *I–U* characteristics, several regions can be observed at a given temperature:

(I) At a given temperature, there may be a small nonzero anodic current. This is produced by electrons emitted with enough kinetic energy to reach the anode without being accelerated.

(II) At low applied voltages (U_A), the current increases very slowly due to the space charge cloud near the cathode, which opposes the movement of electrons toward the anode. As the applied voltage increases, the current increases quite quickly, following the law:

$$I = \alpha \cdot U_A^{3/2} \tag{2.24}$$

called the 'three-halves power law' ('3/2 law') or the Child–Langmuir law.

(III) Above a certain voltage called the saturation voltage (U_S), the current acquires a constant value, the intensity of the saturation current being practically equal to the intensity of the emission current given by the Richardson–Dushman relation, equation (2.10). This relation is written below for the current intensity I_E (where the surface of the cathode S is included in R_0):

$$I_S = I_E = Sj_E = SRT^2 e^{\left(-\frac{W_e}{kT}\right)} = R_0 T^2 e^{\left(-\frac{W_e}{kT}\right)} \tag{2.25}$$

The saturation current is strongly dependent on temperature, as the Richardson relationship suggests. At the same time, in a real diode, a slight increase in the saturation current is observed as the voltage increases above U_S. This phenomenon is explained by the Schottky effect described previously in section 2.1.2, where the Richardson–Dushman relationship correlated with the Schottky effect was also established:

$$I_S = S\langle D\rangle RT^2 e^{-\frac{W_e - W_{\min}}{kT}} = S\langle D\rangle RT^2 e^{-\frac{W_e}{kT}} e^{\frac{\beta_S\sqrt{U}}{kT\sqrt{d}}} \tag{2.26}$$

For voltages smaller than the saturation voltage (U_S), the I–U characteristic is given by the *three-halves power law* (or *3/2 law*), which will also be derived here. First, we consider a diode with parallel plane electrodes, neglecting the influence of the minimum potential ($V_{\min}$) introduced above and considering that electrons are emitted without an initial velocity ($v_0 = 0$). The electrons emitted by the cathode form a space charge cloud in the cathode–anode space (K–A), as shown in figure 2.10. The determination of the I–U characteristics starts from the definition

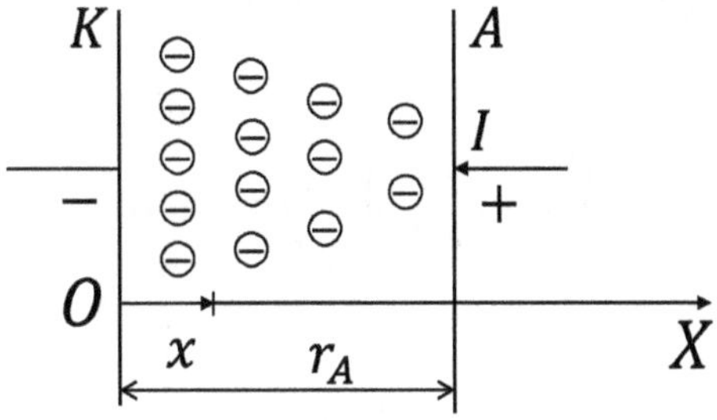

Figure 2.10. The electronic space charge cloud in the K–A space of an electronic tube diode.

of the current intensity:

$$I = \frac{dq}{dt} = Sne\frac{dx}{dt} = Snev \tag{2.27}$$

In relation (2.27), both quantities, i.e. n and v, depend on the applied voltage U_A.

At any point in the K–A space, the kinetic energy of electrons is related to the accelerating voltage $U(x)$ by the relation:

$$\frac{mv^2}{2} = eU(x) \quad \Rightarrow \quad v = \left(\frac{2e}{m}\right)^{1/2} U^{1/2}(x) \tag{2.28}$$

On the other hand, the dependence of concentration $n(x)$ on voltage $U(x)$ is given by the Poisson equation valid in K–A space:

$$\frac{d^2U}{dx^2} = \frac{en}{\varepsilon_0} \tag{2.29}$$

To find the dependence $n = f(U(x))$, equation (2.29) will be solved, taking into account the boundary and initial conditions below:

(a) $x = 0$: $v_0 = 0$, $E_x = 0$; $U(0) = 0$, $E_x = -\frac{dU}{dx}$;
(b) $x = r_A$: $U(r_A) = U_A$;
(c) $I = Snev$ is less than the saturation current and is constant in any plane parallel to the cathode;
(d) Collisions between electrons are neglected.

Using the current density $j = \frac{I}{S} = nev$ and multiplying both sides of equation (2.29) by $\frac{dU}{dx}dx$, we obtain:

$$\frac{d^2U}{dx^2}\frac{dU}{dx}dx = \frac{j}{v\,\varepsilon_0} = \frac{j}{\varepsilon_0}\left(\frac{m}{2e}\right)^{1/2} U(x)^{-\frac{1}{2}}\frac{dU}{dx}dx \tag{2.30}$$

By integrating equation (2.30), we obtain:

$$\int \frac{1}{2}\frac{d}{dx}\left[\left(\frac{dU}{dx}\right)^2\right]dx = \int \frac{j}{\varepsilon_0}\left(\frac{m}{2e}\right)^{1/2} U^{-1/2}dU \tag{2.31}$$

or:

$$\left(\frac{dU}{dx}\right)^2 = \frac{2j}{\varepsilon_0}\left(\frac{m}{2e}\right)^{1/2} 2U^{1/2} + K \tag{2.32}$$

From equation (2.32), it follows that:

$$\frac{dU}{dx} = -E(x) = 2\left(\frac{j}{\varepsilon_0}\right)^{1/2}\left(\frac{m}{2e}\right)^{1/4} U^{1/4} + K \tag{2.33}$$

Taking into account the initial and boundary conditions, the constant K is equal to zero. Subsequent integration of equation (2.33) leads to:

$$\int U^{-1/4}du = \int 2\left(\frac{j}{\varepsilon_0}\right)^{1/2}\left(\frac{m}{2e}\right)^{1/4} dx$$

and, respectively:

$$\frac{4}{3}U^{\frac{3}{4}} = 2\left(\frac{j}{\varepsilon_0}\right)^{1/2}\left(\frac{m}{2e}\right)^{1/4} x + C$$

At $x = 0$, $U = 0$, and $C = 0$, the above equation can be rewritten as:

$$\frac{16}{9}U^{3/2} = 4\frac{j}{\varepsilon_0}\left(\frac{m}{2e}\right)^{1/2} x^2 \tag{2.34}$$

while, at $x = r_A$, $U = U_A$ results in:

$$\frac{4}{9}U_A^{3/2} = \frac{j}{\varepsilon_0}\left(\frac{m}{2e}\right)^{1/2} r_A^2 \quad \Rightarrow j = \frac{4}{9}\varepsilon_0\left(\frac{2e}{m}\right)^{1/2} r_A^{-2}U_A^{3/2} \tag{2.35}$$

Basically, the current density is constant in any plane of the diode that has the value $j = \alpha' U_A^{3/2}$, where $\alpha' = \frac{4}{9}\frac{\varepsilon_0}{r_A^2}\left(\frac{2e}{m}\right)^{1/2}$. If the area of the electrodes is S, the current intensity is:

$$I = S\frac{4}{9}\frac{\varepsilon_0}{r_A^2}\left(\frac{2e}{m}\right)^{1/2} U_A^{3/2} = \alpha U_A^{3/2} \tag{2.36}$$

where: α (called ***'tube perveance'***) is the proportionality coefficient between the intensity of the anode current and $U_A^{3/2}$; this includes both universal constants and parameters that characterize the geometry of the tube. Relation (2.34) allows us to determine the relationship $U(x)$ for $j =$ constant:

$$U^{3/2} = \frac{9}{4}\frac{j}{\varepsilon_0}\left(\frac{m}{2e}\right)^{1/2} x^2 \quad \Rightarrow U(x) = \left(\frac{9}{4}\frac{j}{\varepsilon_0}\right)^{2/3}\left(\frac{m}{2e}\right)^{1/3} x^{4/3} \tag{2.37}$$

Introducing j from equation (2.35) into equation (2.36) gives:

$$U(x) = U_A\left(\frac{x}{r_A}\right)^{4/3} \tag{2.38}$$

Using this result, we can obtain the relationship between the electric field $E(x)$ and the distance x in the K–A:

$$E(x) = -\frac{dU}{dx} = -U_A \cdot \frac{4}{3}x^{1/3}r_A^{-4/3} = -\frac{U_A}{r_A}\left(\frac{x}{r_A}\right)^{1/3} \tag{2.39}$$

The dependence of the electron concentration $n(x)$ on the distance x in the K–A space is:

$$n(x) = \frac{j}{ev} = \frac{j}{e\left(\frac{2e}{m}\right)^{1/2}U(x)^{1/2}} \Rightarrow n(x) = \frac{\frac{4}{9}\frac{\varepsilon_0}{r_A^2}\left(\frac{2e}{m}\right)^{1/2}U_A^{3/2}}{e\left(\frac{2e}{m}\right)^{1/2}U_A^{1/2}\left(\frac{x}{r_A}\right)^{2/3}} =$$

$$= \frac{4}{9}\varepsilon_0\frac{U_A}{r_A^2}\left(\frac{x}{r_A}\right)^{-2/3} \tag{2.40}$$

Graphical representations of $U(x)$, $E(x)$, and $n(x)$ are shown in figure 2.11, which illustrates their dependence on the dimensionless variable x/r_A. Coming back to

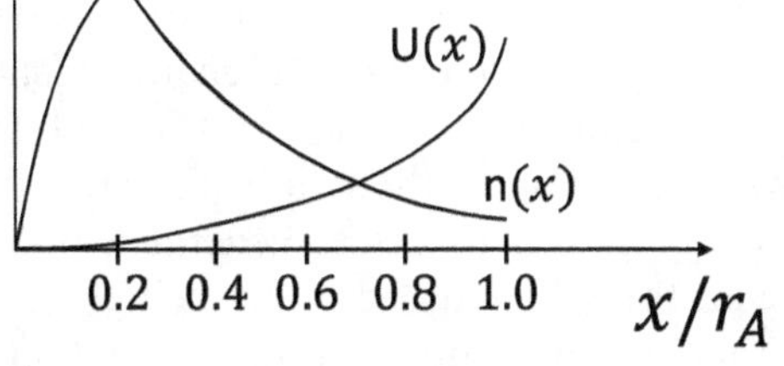

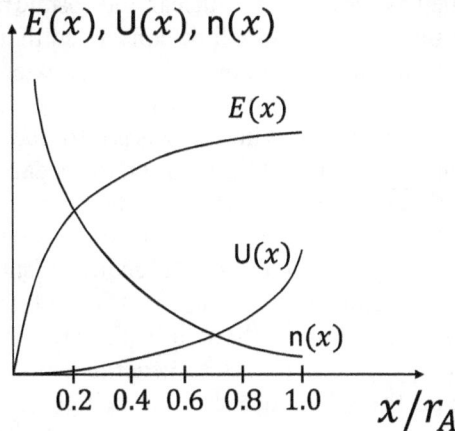

Figure 2.11. Positional dependence of the electric field $E(x)$, the potential $U(x)$, and the electron concentration $n(x)$ in the K–A space.

equation (2.36), we can see that the parameter α can be rewritten as a function of the electrodes' surfaces and the distance between them, as follows:

$$\alpha = 2.33 \cdot 10^{-6} \frac{S}{r_A^2} \tag{2.41}$$

which is valid in the case of planar electrodes. In the case of cylindrical electrodes, the coefficient $\beta(r_k, r_a)$ must be added, which depends on the radius of the cathode and the anode.

The semi-cubic parabola described by the $I = \alpha U_A^{3/2}$ law, which is valid for currents smaller than the saturation current ($I_A < I_{sat}$), was derived without considering the minimum potential V_{min} located at a distance of r_{min} from the cathode. If this is taken into account, the electrons are blocked at a distance of $0 - r_{min}$ and they do not reach the anode, except for those with a velocity $v \geqslant v_1 = \left(\frac{2eV_{min}}{m}\right)^{1/2}$, leading to the relation:

$$I_A = 2.33 \cdot 10^{-6} \frac{S}{(r_A - r_{min})^2} (U_A + V_{min})^{3/2} \tag{2.42}$$

In other words, the electric field is zero in the region at a distance r_{min} from the cathode. It can thus be considered that at this point, there is a *'virtual cathode'*.

2.2 Electrical conduction in ionized gases

A gas contains both electrically neutral particles (i.e. molecules and atoms) as well as positive and negative ions that generally occur as a result of the emission of electrons from the molecules or atoms of the gas under the action of various external factors. Depending on the nature of the external factors causing them, the ionization processes can be classified into: (i) thermo-ionization (ionization produced by heating the gas); (ii) photo-ionization (ionization determined by the interaction of light with the gas); or (iii) ionization determined by the action of nuclear and cosmic radiation on molecules or atoms of the gas. In the following, it will be considered that in the act of primary ionization, only monovalent ions are obtained. When an external electric field is not applied, the ions generated in the primary generation processes move randomly through the gas until they recombine. When the gas is introduced into an electric field, the positive ions move in the direction of the field and the negative ones in the opposite direction, giving rise to an electric current. With the help of the scheme shown in figure 2.12(a), the current–voltage (*I–U*) characteristic for a given gas can be obtained (see figure 2.12(b)). The interpretation of the *I–U* characteristic requires clarification of the processes that occur in the gas when an electric field is applied. The primary ionization processes (which occur under the action of external factors) are followed on one hand by secondary ionization processes generated by collisions between the primary ions (which gain the required energy by being accelerated in the applied electric field) and other molecules or atoms in the gas, and on the other hand by the recombination of positive and negative ions in the ionized gas. When we consider the

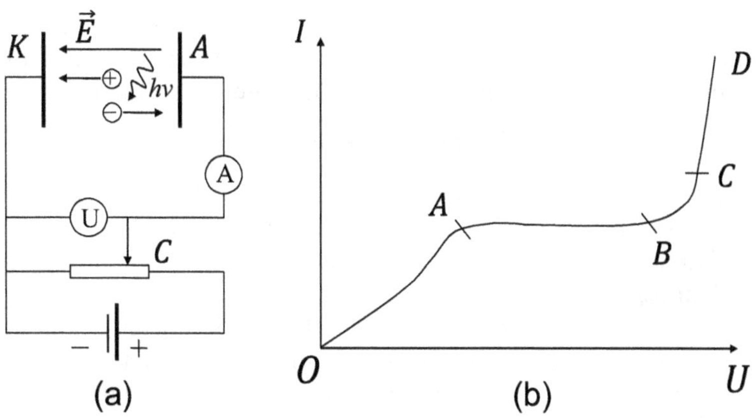

Figure 2.12. (a) Potentiometric scheme used to measure the current–voltage characteristic of an ionized gas; (b) corresponding current–voltage characteristic.

processes of secondary ionization and recombination, we realize that the number of ions of a given type that reach one of the electrodes cannot always equal the number of ions of the same type generated in the primary ionization processes. Thus, within the *OA* portion of the *I*–*U* characteristic, fewer ions arrive at an electrode than those of the same type generated per unit time by the processes of primary ionization. In other words, the recombination process is dominant over the secondary ionization process. Within the *AB* portion, the number of ions generated in primary ionization processes equals the number of ions that reach the electrodes, since the recombination rate equals the secondary ionization rate. The *OA* and *AB* portions represent an externally maintained electrical discharge. Within the *BC* portion, the rate of secondary ionization processes is higher than the rate of recombination processes, and the discharge begins to be self-sustaining. Within the *CD* portion, secondary ionization develops rapidly, the self-sustaining discharge being of the '*Townsend*' type.[2] In the latter, secondary ionization takes place in an avalanche and is clearly dominant over recombination. Within the *OA* portion, the *I*–*U* characteristic is ohmic:

$$\begin{cases} j = n_+ e v_+ + n_- e v_- = ne\left(\mu_+ + \mu_-\right)E = \sigma E \\ I = \sigma E S = \dfrac{\sigma S U}{d} = \dfrac{S}{\rho d} U = \dfrac{U}{R} \\ \sigma = ne(\mu_+ + \mu_-) = \dfrac{1}{\rho} \end{cases} \tag{2.43}$$

[2] In electromagnetism, the **Townsend discharge** or **Townsend avalanche** is an ionization process in gases where free electrons are accelerated by an electric field, collide with gas molecules, and consequently generate additional electrons that are further accelerated and generate more electrons. The final result is an avalanche multiplication that causes a significant increase in electric conduction through the gas. The phenomenon was discovered in 1897 by **Sir John S E Townsend** (1868–1957), an Irish-British mathematical physicist and a professor of physics at Oxford University.

Within the *AB* portion, the current remains constant, being proportional to the concentration of ions generated in the primary ionization processes, the rate of secondary ionization processes being equal to the rate of recombination processes:

$$j = n_s ev \Rightarrow I = Nev = n_s eSdv = \text{constant} \tag{2.44}$$

where n_s is the concentration of ions generated in the primary ionization processes. Knowledge of these processes is very useful when we want to understand the operation of radiation detectors such as the Geiger–Müller (GM) counter and the ionization chamber (IC). The GM counter operates at a high voltage (in the *CD* portion of the *I*–*U* characteristics in figure 2.12(b)). Under these conditions, the rate of secondary ionization processes is very high, giving rise to a self-sustaining discharge, accompanied by a high current pulse. This current pulse creates a high-voltage pulse across a load resistance connected in series with the counter, which is easily recorded regardless of the energy of the emergent particle when it arrives at the counter. In simple terms, the GM detector counts and records any incident particle with high accuracy. In the case of the IC, the operational voltage used is moderate (in the *AB* portion of the *I*–*U* characteristics in figure 2.12(b)), where the number of ions generated in primary ionization processes is equal to the number of ions that reach the electrodes, the recombination rate being equal to the secondary ionization rate. Under these conditions, the current pulse is proportional to the concentration of ions generated in primary ionization processes; therefore, it depends on the ionization power of the incident particle arriving at the detector. The IC not only counts the incident radioactive particles with good accuracy but also gives information about their energy (nature).

2.3 Electric conduction in electrolytic solutions

2.3.1 Electrolysis with a soluble anode

Many pure liquids (water, oil, etc) are not electrically conductive (they behave as insulators), while acids or bases in water and solutions of salts are good conductors of electricity, called electrolytes. An electrolyte contains both positive and negative ions, which are dissociated molecules in the solution. Applying a potential difference to the electrolyte causes positive ions to move in the direction of the electric field, while negative ions move in the opposite direction; both types of ions participate in conduction, thereby creating an ionic current. This ionic current is also accompanied by mass transport. The processes that take place at the molecular scale can be understood by analyzing the following example. Let us consider an aqueous solution of copper sulfate ($CuSO_4$) in a vessel in which two electrodes are immersed. Between the immersed electrodes, a potential difference is applied by a galvanic source (see figure 2.13). The electrode connected to the (+) terminal of the source is made from Cu (and is called the anode, A). The electrode connected to the (−) terminal of the source is made from graphite, C (and is called the cathode, K). The presence of ions can be demonstrated by the passage of an ionic current through the electrolytic solution. This shows that the molecules of the active substance have dissociated, so that the solution contains positive ions, Cu^{2+}, and negative ions, SO_4^{2-}. When an

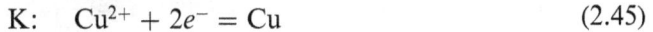

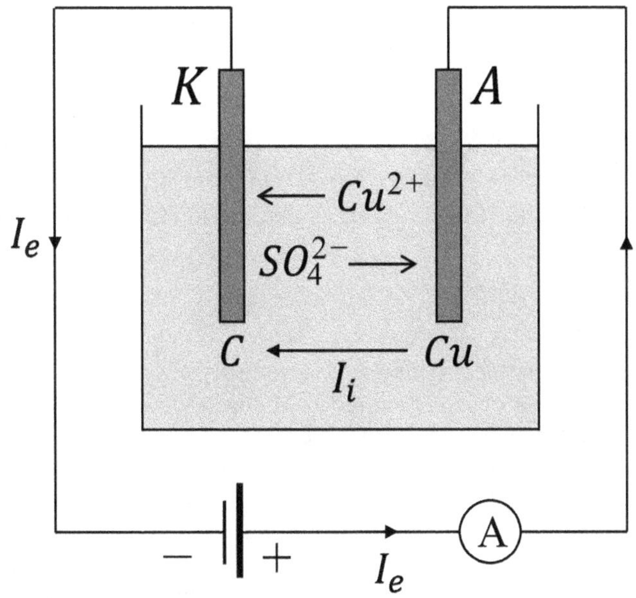

Figure 2.13. Electrolysis with a soluble anode.

electric field is applied, the positive ions move toward the cathode (hence, they are called *cations*), while the negative ions move toward the anode (these are called *anions*). The reaction taking place at the cathode is:

$$\text{K:} \quad Cu^{2+} + 2e^- = Cu \tag{2.45}$$

through which Cu^{2+} ions capture $2e^-$, becoming electrically neutral atoms of Cu which are deposited on the cathode. In electrochemical terms, such a reaction in which an ion receives electrons is called a *reduction reaction.*[3] The reaction occurring at the anode is:

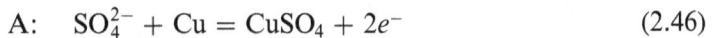

$$\text{A:} \quad SO_4^{2-} + Cu = CuSO_4 + 2e^- \tag{2.46}$$

whereby SO_4^{2-} reacts with Cu from the anode by restoring the $CuSO_4$ molecule. In electrochemical terms, such a reaction in which an ion loses electrons is called an *oxidation reaction*. These $CuSO_4$ molecules pass back into the solution, where they dissociate, and the process starts again, resulting in a couple of reduction and oxidation reactions that can sustain each other, or in short, a *redox* couple. It can be observed in this example that the mass of $CuSO_4$ remains constant, and the Cu from the anode is transferred to the cathode. This type of electrolysis, i.e. with a soluble anode, is used in galvanoplasty (or electroplating) to cover various objects with

[3] In most cases, this is the main mechanism governing the process of **electrochemical deposition** of thin films from liquid electrolytes, i.e. the reduction of cations of a diluted specific material from a liquid solution in order to crystallize them onto a conducting layer or substrate.

different metal layers. Analysis of various cases of electrolysis shows that metals and hydrogen always form positive ions.

2.3.2 Faraday's laws of electrolysis

First law: The mass M of the substance deposited on the electrode is proportional to the intensity I of the current and the duration t of the current flowing through the electrolyte:

$$M = C_1 It = C_1 Q = C_1 \int_{t_0}^{t_0+t} idt \tag{2.47}$$

where C_1 is the electrochemical equivalent, a quantity equal to the mass deposited on an electrode carried by a charge Q equal to the unit charge.
Second law: The electrochemical equivalent C_1 is proportional to the chemical equivalent (A/n) of the substance deposited on the electrode:

$$C_1 = C_2 \frac{A}{n} \tag{2.48}$$

where A is the atomic mass of the element (in atom-grams) and n is the valence of the substance. C_2 is a proportionality coefficient usually expressed as $C_2 = \frac{1}{F}$, where $F = 96\,494\ C/\text{gram-equivalent}$ is Faraday's number. Combining the two expressions above, we obtain:

$$M = \frac{1}{F}\frac{A}{n} It = \frac{1}{F}\frac{A}{n} Q \tag{2.49}$$

From relation (2.49), it can be observed that $M = \frac{A}{n}$; therefore, if an equivalent gram of a substance is deposited on an electrode, then $F = Q$. As a result, *Faraday's number (F) represents the amount of electricity (i.e. charge Q) whose passage through the electrolyte deposits an equivalent gram of substance on the electrode.* Measurements of electrochemical equivalents give Faraday's number the following value:

$$F = 96\,494 \frac{C}{\text{gram-equivalent}} \tag{2.50}$$

Faraday's laws, in addition to their direct benefit in understanding electrolysis, helped to prove the particulate nature of electricity and determine the elementary (electron) charge.

2.3.3 Microscopic interpretation of the conductivity of electrolytes

In the absence of an external electric field, the ions in electrolytes undergo movement due to thermal agitation, so the resulting current is equal to zero. In the presence of an electric field, positive ions of mass m^+ and charge q^+ move in the direction of the field with a velocity v^+, while the negative ions of mass m^- and charge q^- move in the opposite direction with velocity v^-. In treatments based on classical mechanics, ions moving into the electrolyte are not only subjected to the electric force but also to a frictional force with the electrolyte, proportional to their speed, resulting in the following equations of motion for the two types of ions:

$$\begin{cases} m_+a_+ = q_+E - k_+v_+ \\ m_-a_- = q_-E - k_-v_- \end{cases} \tag{2.51}$$

Under conditions of uniform movement of the two ions, equations (2.51) lead to:

$$\begin{cases} a_+ = 0 \Longrightarrow v_+ = \dfrac{q_+E}{k_+} = \mu_+E \\ a_- = 0 \quad \Longrightarrow \quad v_- = \dfrac{q_-E}{k_-} = \mu_-E \end{cases} \tag{2.52}$$

It can be seen that the velocities of the positive and negative ions, also called drift speeds, are proportional to the intensity of the electric field, the proportionality coefficients being the mobilities of the positive and negative ions (μ_+ and μ_-). The density of the electric current that arises due to the displacement of the positive and negative ions in the electric field is:

$$j = j_+ + j_- = q_+n_+v_+ + q_-n_-v_- \tag{2.53}$$

where n_+ and n_- are the concentrations of the positive and negative ions in the solution, respectively. If we define $\alpha = \frac{n_+}{n_0} = \frac{n_-}{n_0}$, the *dissociation coefficient* (which represents the number of dissociated molecules in the unit volume relative to the number of molecules n_0 in the unit volume of the dissolved substance), relation (2.53) becomes:

$$j = \alpha n_0(q_+v_+ + q_-v_-) = q\alpha\, n_0(v_+ + v_-) \tag{2.54}$$

where we assumed that $q_+ = q_- = q$. The dissociation coefficient takes values between zero and one ($\alpha = 1$ when all molecules are dissociated, i.e. $n_+ = n_- = n_0$, while $\alpha = 0$ when no molecule is dissociated, i.e. $n_+ = n_- = 0$). The intensity of the electronic current in the external circuit is determined by the intensity of the ionic current in the electrolytic solution. Indeed, as shown previously, the negative ions move toward the anode, where they neutralize, their charge (electrons) passing into the external circuit so that the electronic current equals the anionic one. Similarly, electrons from the external circuit move toward the cathode, which allows the neutralization of positive ions (cations) that are discharged at the cathode. Introducing the quantity $\eta = \frac{n_0}{N'}$, called the *equivalent concentration*, representing the number of gram equivalents in the unit volume, we obtain the following relationship for the electric current density:

$$j = \alpha N' q\eta\, (v_+ + v_-) \tag{2.55}$$

in which n_0 is the concentration of molecules in the solution and N' is the number of molecules in a gram equivalent. However, since N' represents the number of molecules in a gram equivalent, it follows that $N'q$ is exactly the electric charge transported through the electrolyte to deposit one gram equivalent at the electrode;

in other words, it is precisely Faraday's number. At this point, the current density can be rewritten as:

$$j = F\alpha\eta(v_+ + v_-) \tag{2.56}$$

or, taking into account relationships (2.52) for the definition of drift velocities, we obtain:

$$j = F\alpha\eta(\mu_+ + \mu_-)E \tag{2.57}$$

For a given solution, the value of $F\alpha\eta(\mu_+ + \mu_-)$ is constant; taking into account Ohm's law in its local form, i.e. $j = \sigma E$, this coefficient is effectively the electrolytic conductivity:

$$\sigma = F\alpha\eta(\mu_+ + \mu_-) \tag{2.58}$$

The notion of *equivalent conductivity* $\Lambda = \frac{\sigma}{\eta}$, defined as the ratio between the conductivity of the electrolyte and its equivalent concentration, can also be introduced. If we use this definition, relation (2.58) for the electrolyte's conductivity becomes:

$$\Lambda = \frac{\sigma}{\eta} = F\alpha\,(\mu_+ + \mu_-) \tag{2.59}$$

In the case of weak solutions (i.e. highly diluted), when practically all the molecules of the dissolved substance are dissociated ($\alpha = 1$), the result is $\Lambda_\infty = F(\mu_+ + \mu_-)$. So, for solutions where $0 < \alpha < 1$, by determining Λ, one can determine the dissociation coefficient α, using:

$$\alpha = \frac{\Lambda}{\Lambda_\infty} \tag{2.60}$$

Λ and σ both depend on the concentration of the dissolved substance and the value of the dissociation coefficient (α). Figure 2.14 shows the dependence of the electrical

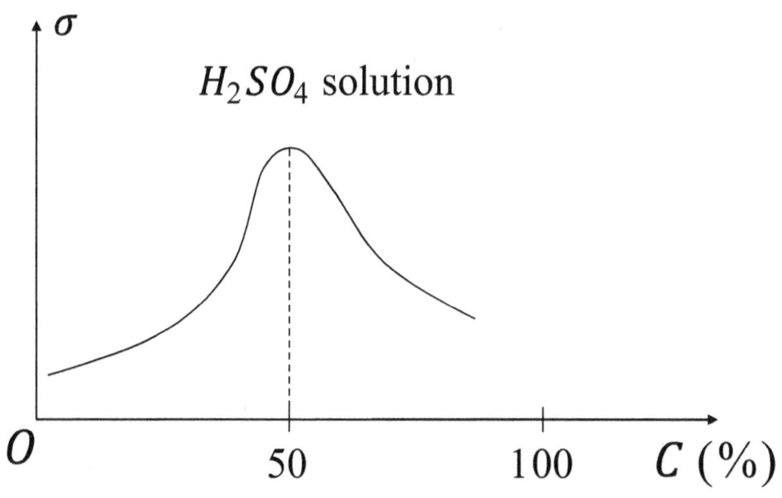

Figure 2.14. The dependence of the electric conductivity of an aqueous solution of sulfuric acid on the acid concentration.

conductivity (σ) on the concentration of molecules of the dissolved substance for an aqueous solution of sulfuric acid. At concentrations close to 50%, it reaches a maximum. At higher concentrations, it decreases, due to a decrease in the dissociation coefficient. The study of electrolytic solutions has great practical importance.

2.4 Thermoelectric effects

Thermoelectric effects refer typically to those phenomena related to the direct conversion of a temperature difference into an electric voltage and vice versa.[4]

2.4.1 Potential difference at the contact between two different metals

By the end of the 18th century, Volta[5] had noticed that a potential difference occurs at the contact between two different metals; in other words, the electrical potential changes its value when passing from one metal to the other. This difference in potential is called a *contact potential difference*. The emergence of this difference in contact potential can be explained using the classical theory of the electronic gas in metals, a theory that was also used in the analysis of electronic conduction in metals. In this context, we consider two different metals in close contact; their concentrations of free electrons are n_1 and n_2, respectively, with $n_1 > n_2$ (see figure 2.15). Whatever the initial temperature of each metal, after they make

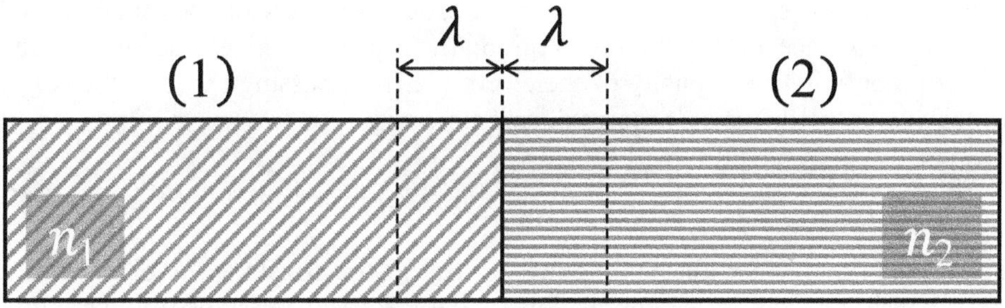

Figure 2.15. Contact between two different metals.

contact, their temperatures become equal; hence, it can be considered that the average thermal velocity of electrons is the same on both sides of the contact. For simplicity, we assume that the average free path of electrons is the same for both metals in contact. (However, this hypothesis is not perfectly correct, due to the kinetic–molecular theory of gases. According to this theory, $\lambda = \dfrac{1}{\sqrt{2}n\sigma}$ depends on the particle density and thus differs if the concentrations of electrons in the two

---

[4] A thermoelectric device generates a voltage when there is a temperature gradient between its terminals.

[5] **Alessandro Volta** (1745–1827) was an Italian physicist and chemist who invented the voltaic pile in 1799, proving that electricity could be generated chemically.

metals are different, i.e. $n_1 \neq n_2$.) Due to thermal motion, electrons diffuse through the contact surface from one metal to the other. During the average time (τ) between two consecutive collisions, only those electrons that are located at a distance less than or equal to λ relative to the contact surface can pass from one metal to the other. If the contact surface area is A, the number of electrons that can diffuse from metal (1) to metal (2) is $N_1 = n_1 \lambda A$, while the number of electrons that can diffuse from metal (2) to metal (1) is $N_2 = n_2 \lambda A$. Since electrons can move identically in the three directions of space, only a third of them move along the metals, and only one-sixth, i.e. $N_x = \frac{n\lambda A}{6}$, move toward the contact surface. So, $N_{1x} = \frac{n_1 \lambda A}{6}$ electrons diffuse from metal (1) to metal (2), while $N_{2x} = \frac{n_2 \lambda A}{6}$ electrons diffuse from metal (2) to metal (1). As $n_1 > n_2$, more electrons diffuse from (1) to (2), so that metal (2) becomes negatively charged, and metal (1) becomes positively charged. This causes the generation of an electric contact field between the two metals, oriented from metal (1) to metal (2). This field acts on the electrons with a force oriented in the opposite direction to that of the electric contact field, and as a result, it opposes further diffusion of electrons from metal (1) to metal (2). This leads to a dynamic equilibrium, a state in which the net number of electrons passing through the contact surface is equal to zero. From this moment, the value of the contact field no longer increases (it remains constant), and a difference in contact potential is established between the two metals. The order of magnitude of the difference in contact potential can be established by calculating the intensity of the current that crosses the contact due to the process of diffusion over time τ and the electrical resistance of the contact region; their product gives the difference in contact potential. The intensity of the electric current passing through the contact surface due to the diffusion process during time τ is:

$$I = \frac{\Delta Q}{\Delta t} = \frac{e(N_{1x} - N_{2x})}{\tau} = \frac{e\lambda A}{6\tau}(n_1 - n_2) = \frac{e\overline{v_T} A}{6}(n_1 - n_2) \tag{2.61}$$

where $\overline{v_T}$ is the average thermal velocity. The electric resistance R of the contact region consists of the series connection of the electrical resistance of the contact region $R_1 = \rho_1 \frac{\overline{\lambda}}{A}$ in metal (1) and the electrical resistance of the contact region $R_2 = \rho_2 \frac{\overline{\lambda}}{A}$ of metal (2):

$$R = R_1 + R_2 = \rho_1 \frac{\overline{\lambda}}{A} + \rho_2 \frac{\overline{\lambda}}{A} = \frac{\overline{\lambda}}{A}(\rho_1 + \rho_2) \tag{2.62}$$

However, according to the classical theory of electronic conduction in metals, i.e. $\rho = \frac{2m\overline{v_T}}{ne^2\overline{\lambda}}$, equation (2.62) becomes:

$$R = \frac{\overline{\lambda}}{A} \cdot \frac{2m\,\overline{v_T}}{e^2\,\overline{\lambda}}\left(\frac{1}{n_1} + \frac{1}{n_2}\right) = \frac{2m\overline{v_T}}{Ae^2}\left(\frac{1}{n_1} + \frac{1}{n_2}\right) \tag{2.63}$$

Given these results, we can calculate the difference in contact potential, which is essentially caused by the different concentrations of charge carriers in the two metals, as follows:

$$\Delta V_{i_{1,2}} = I \cdot R = \frac{e\overline{v_T}A}{6}(n_1 - n_2) \cdot \frac{2m\overline{v_T}}{Ae^2}\left(\frac{1}{n_1} + \frac{1}{n_2}\right)$$
$$= \frac{m\overline{v_T}^2}{3e}\left(1 - \frac{n_2}{n_1} + \frac{n_1}{n_2} - 1\right) = \frac{m\overline{v_T}^2}{3e}\left(\frac{n_1}{n_2} - \frac{n_2}{n_1}\right) \tag{2.64}$$

However, the thermal velocity is $\overline{v_T} = \sqrt{\frac{3kT}{m}}$, which leads to:

$$\Delta V_{i_{1,2}} = \frac{kT}{e}\left(\frac{n_1}{n_2} - \frac{n_2}{n_1}\right) \tag{2.65}$$

This difference in contact potential, described as **'intrinsic'**, is directly proportional to the contact temperature and depends on the nature of the metals brought into contact in terms of their electron concentrations n_1 and n_2. In a quantum treatment of electronic gas, it takes another form; however, this obviously depends on the absolute temperature (T) and the concentrations n_1 and n_2. We recall that the expression for the saturation current in the case of thermionic emission was established using quantum statistics, according to which, electrons were placed on energy levels in agreement with the Pauli principle, up to the Fermi level, at 0 K. In this scenario, the work function was not actually the total depth w of the potential barrier but rather the difference, i.e. $w_e = w_0 - w_F$. As a result, the density of the emission current was:

$$j_s = \mathrm{RT}^2 e^{-\frac{w_e}{kT}} = \frac{4\pi \mathrm{me}k^2}{h^3} T^2 e^{-\frac{w_e}{kT}} \tag{2.66}$$

If one takes into account the average velocity $\overline{v} = \left(\frac{8kT}{\pi m}\right)^{1/2}$ and the effective density of states $N_{ef} = 2\left(\frac{2\pi mkT}{h^2}\right)^{3/2}$, also deduced within the framework of quantum statistics, relation (2.66) can be written as $j_s = \frac{ne\overline{v}}{4}$, noting that in fact the concentration of carriers leaving a metal can be correlated with the work function w_e, and, as a result, with the Fermi level w_F of the metal. Thus, we can write:

$$\begin{cases} n_1 = N_{ef}\, e^{\frac{w_{F_1}}{kT}} \\ n_2 = N_{ef}\, e^{\frac{w_{F_2}}{kT}} \end{cases} \tag{2.67}$$

where:

$$n = \int_0^\infty g(w)f(w)dw = \frac{4\pi(2m)^{3/2}}{h^3}\int_0^\infty w^{1/2}e^{-\frac{w-w_F}{kT}}dw$$
$$= 2\left(\frac{2\pi mkT}{h^2}\right)^{3/2} e^{\frac{w_F}{kT}} = N_{ef}\, e^{\frac{w_F}{kT}} \tag{2.68}$$

At the same temperature, $N_{ef_1} = N_{ef_2}$, and therefore the ratio of concentrations in the two metals can be correlated with the difference in their Fermi energies:

$$\ln \frac{n_1}{n_2} = \frac{-w_{F_2} + w_{F_1}}{kT} = \frac{-e(V_2 - V_1)}{kT} = \frac{e\Delta V_i}{kT} \tag{2.69}$$

resulting in:

$$\Delta V_{i_{12}} = \frac{kT}{e} \ln \frac{n_1}{n_2} \tag{2.70}$$

This relationship expresses *the intrinsic contact potential difference* associated with the difference in Fermi energies, which is obviously correlated with the difference in concentrations. However, the intrinsic contact potential difference is not the only potential difference that occurs at the contact between two different metals. In the case where the two metals have the same electron concentration ($n_{01} = n_{02}$) but different work functions, a difference in contact potential occurs between them called *the extrinsic contact potential difference*, which is given by the difference in the work functions.

Let us now consider two metals with work functions $w_{e_1} < w_{e_2}$, the number of electrons in the unit volume being the same ($n_1 = n_2$) as in figure 2.16. If the two

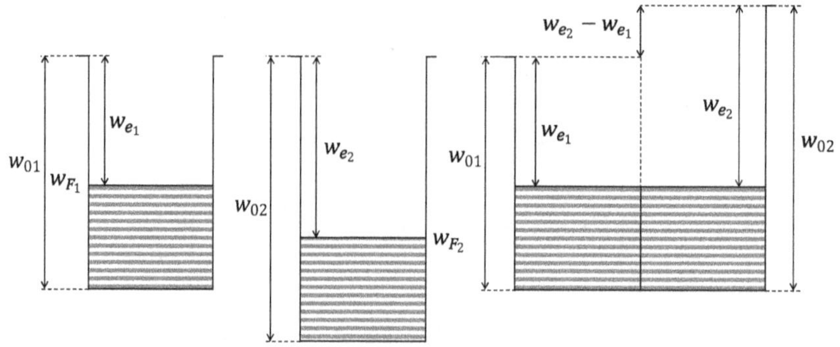

Figure 2.16. Energy diagrams illustrating the presence of an extrinsic contact potential difference arising at the interface between two metals with different work functions.

conductors are brought into contact, an exchange of electrons occurs between them; namely, several electrons from metal (1) (with the smaller work function) pass into metal (2), charging it negatively, while metal (1) remains positively charged. This process continues until the Fermi levels are equalized. A potential difference of ΔV_e appears between the two metals, i.e. an electric field oriented from metal (1) to metal (2) that prevents the subsequent passage of electrons from metal (1) to metal (2). The electric field $\overrightarrow{E_i}$ that prevents the passage of electrons from metal (1) to metal (2) is called *the induced contact electric field*. In this case, we assume that the two metals are at the same temperature and are not subject to the action of any external agent capable of transmitting energy to them. The induced contact field that occurs upon contact between two energy-insulated metals at the same temperature is called the *voltaic induced field.* The difference in contact potential generated by the difference between work functions is called the *extrinsic contact potential difference* or the *Volta contact voltage.* It is independent of temperature and is given by the relation:

$$\Delta V_{e_{1,2}} = \frac{w_{e_2} - w_{e_1}}{e} = V_1 - V_2 \tag{2.71}$$

In the case of contact between two metals that differ both by thermodynamically extracted work ($w_{e_1} \neq w_{e_2}$) and by concentration ($n_1 \neq n_2$), the difference in contact potential is given by the sum of the differences between the extrinsic contact potentials ΔV_e and the intrinsic contact potential differences ΔV_i:

$$\Delta V_{12} = \Delta V_{e_{1,2}} + \Delta V_{i_{1,2}} = \frac{w_{e_2} - w_{e_1}}{e} + \frac{kT}{e} \ln \frac{n_1}{n_2} \tag{2.72}$$

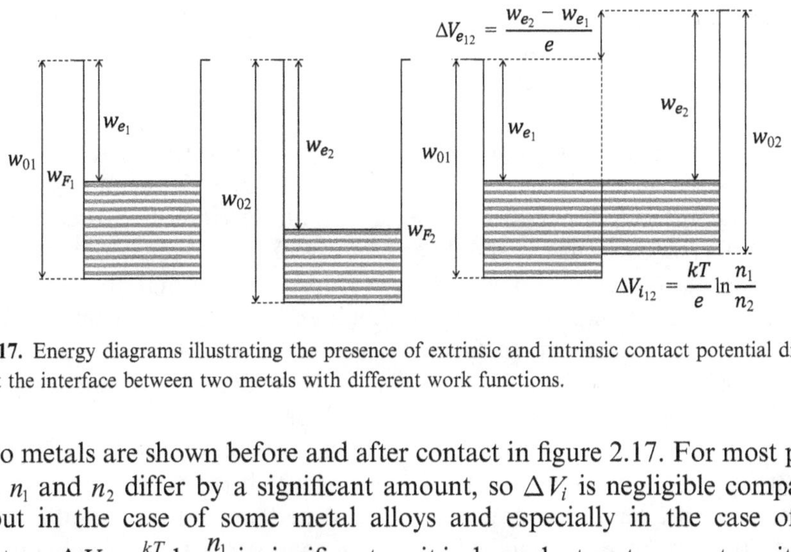

Figure 2.17. Energy diagrams illustrating the presence of extrinsic and intrinsic contact potential differences arising at the interface between two metals with different work functions.

The two metals are shown before and after contact in figure 2.17. For most pairs of metals, n_1 and n_2 differ by a significant amount, so ΔV_i is negligible compared to ΔV_e, but in the case of some metal alloys and especially in the case of semiconductors, $\Delta V_i = \frac{kT}{e} \ln \frac{n_1}{n_2}$ is significant; as it is dependent on temperature, it has an important role in explaining thermoelectric effects.

2.4.2 Potential difference at the ends of a series of metals

Let us consider a closed circuit consisting of three metals M_1, M_2, and M_3 that have common contact surfaces (see figure 2.18). The difference in contact potential at the ends of such an interrupted circuit between M_1 and M_3 is the sum of the contact voltages produced by each pair of metals:

$$\Delta V_{13} = \Delta V_{12} + \Delta V_{23} = V_1 - V_2 + V_2 - V_3 = V_1 - V_3 \tag{2.73}$$

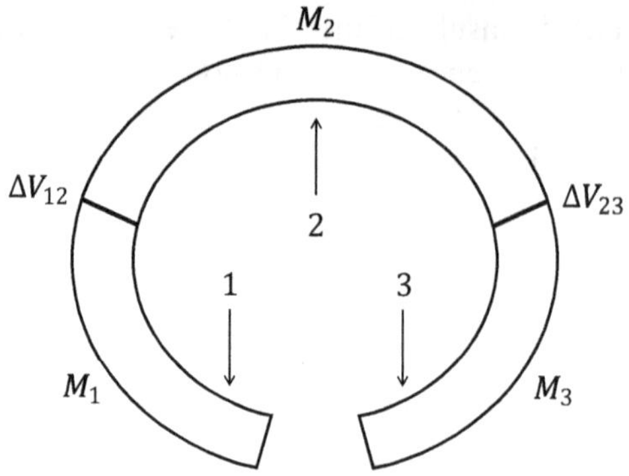

Figure 2.18. Contact potential difference in a ring of different metals.

We therefore note that it depends only on the potentials of the extreme metals M_1 and M_3. *Thus, we can generalize this finding by saying that the difference in potential at the ends of a string of conductors depends only on the nature of the metals at the extremities.* In the case where a circuit of three conductors is closed, a difference no longer appears in the resultant potential:

$$\Delta V_{12} + \Delta V_{23} + \Delta V_{31} = V_1 - V_2 + V_2 - V_3 + V_3 - V_1 = 0 \tag{2.74}$$

This can also be derived from general thermodynamic considerations. As already known, the flow of an electric current through a metal is not accompanied by chemical phenomena that may produce energy. If an electromotive force (EMF) and a corresponding electric current should appear in a closed-circuit conducting system, the circuit develops a certain power. However, it is known that an electric current can arise only at the expense of energy obtained from an agent external to the system; for example, heat taken from the surrounding bodies, or from light, etc. However, such energy transmission does not take place at constant temperature. Due to the fact that the temperature of the entire series of conductors is constant, the production of electrical energy without the consumption of another energy would represent a *perpetuum mobile*, the realization of which is impossible. To create an EMF in a closed circuit of conductors at the same temperature, type I conductors (i.e. metals) must be in contact with type II conductors (i.e. semiconductors or electrolytes). This inhomogeneity within the closed circuit, due to the presence of conductors of distinct natures, leads to chemical processes that transform chemical energy into electrical energy, as in the case of electric piles (as discussed in the previous paragraph). Otherwise, if the system consists of conductors of the same nature, an EMF can only occur if there is a temperature gradient across the system, as in the case of thermoelectric devices.

2.4.3 The Seebeck effect

The Seebeck effect *consists of the appearance of a voltage, called a thermoelectromotive voltage (t.e.m.v.), in a closed circuit formed of conductors of different natures* *when the contacts between the conductors are maintained at different temperatures.* Let us consider a closed circuit consisting of two conductors (1) and (2), made from Cu and Fe, respectively, as shown in figure 2.19. When the two contacts are held at

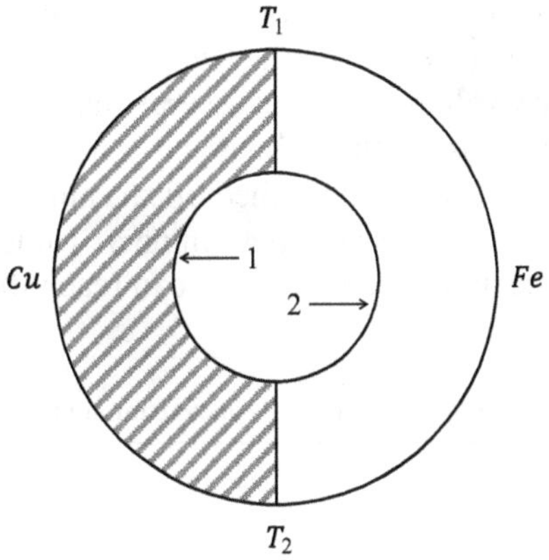

Figure 2.19. System of two metallic conductors in contact; the contacts are kept at different temperatures.

the same temperature, the electromotive voltage (equal to the sum of the voltage drops across the entire circuit) is zero. Indeed, according to (2.65) and (2.71):

$$\Delta V = \Delta V_{12} + \Delta V_{21} = \frac{w_2 - w_1}{e} + \frac{kT}{e}\left(\frac{n_1}{n_2} - \frac{n_2}{n_1}\right) + \frac{w_1 - w_2}{e} + \frac{kT}{e}\left(\frac{n_2}{n_1} - \frac{n_1}{n_2}\right) = 0. \tag{2.75}$$

If the two conductors are maintained at different temperatures T_1 and T_2, then:

$$\Delta V = \frac{w_2 - w_1}{e} + \frac{kT_1}{e}\left(\frac{n_1}{n_2} - \frac{n_2}{n_1}\right) + \frac{w_1 - w_2}{e} + \frac{kT}{e}\left(\frac{n_2}{n_1} - \frac{n_1}{n_2}\right)$$
$$= \frac{k}{e}\left(\frac{n_1}{n_2} - \frac{n_2}{n_1}\right)(T_1 - T_2) = \alpha(T_1 - T_2) \Longrightarrow$$

$$\Delta V = \alpha(T_1 - T_2) = \frac{k}{e}\ln\frac{n_1}{n_2}(T_1 - T_2) \tag{2.76}$$

This voltage is called the *thermoelectromotive voltage* (t.e.m.v.). The coefficient:

$$\alpha = \frac{k}{e}\left(\frac{n_1}{n_2} - \frac{n_2}{n_1}\right) \text{ or } \alpha = \frac{k}{e}\ln\frac{n_1}{n_2} \tag{2.77}$$

which depends on the nature of the two conductors as expressed by n_1 and n_2, as well as on the temperature, is called the *specific electromotive voltage* or *Seebeck coefficient.* As the circuit is closed, a current called the *'thermoelectric current'* appears, due to the nonzero thermoelectromotive voltage discovered in 1821 by Seebeck.[6] The explanation for the Seebeck phenomenon lies precisely in the fact that the difference in contact potential depends on the temperature included in the term $\Delta V_1 = \frac{kT}{e} \ln \frac{n_1}{n_2}$. Regarding the coefficient α, it should be noted that its value differs from conductor to conductor and is given for each metal in contact with a reference metal, namely platinum (Pt). Table 2.2 gives some values of α for several materials (conductors and semiconductors) in relation to Pt. It is worth noting that α can also be calculated theoretically. There is a substantial difference between the theoretical value and the experimental value, which is due to the classical theory used to determine it. Table 2.2 shows that in metals, $\alpha \in (1 - 100)$ μV K^{-1}, while in semiconductors it can reach thousands of μV K^{-1}.

Table 2.2. Seebeck coefficients (α) of some common materials.

Material	α (μV K^{-1})
Al	+4
Alumel (95% Ni, 5% Al)	−13.8
Kopel (56% Cu, 44% Ni)	−35
Chromel (90% Ni, 10% Cr)	+27
Cu_2O	+1000
FeO	−500
ZnO	−714
Sb_2Te_3	+100

2.4.3.1 Applications of the Seebeck effect

1. The Seebeck effect is used for temperature measurement within devices called thermocouples. Thermocouples consist of two different metals that are in contact with each other. As shown in figure 2.20, one of the contacts is

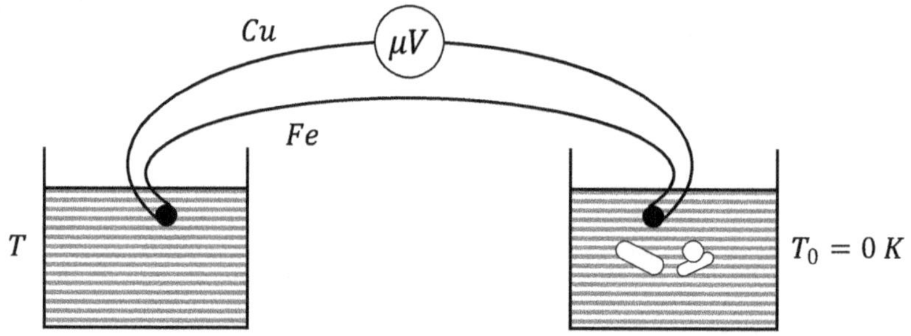

Figure 2.20. Cu/Fe thermocouple.

[6] **Thomas Johann Seebeck** (1770–1831) was a Baltic German physicist who made important discoveries in the field of thermoelectricity, as well as discoveries related to the relationship between heat and magnetism.

inserted into the substance whose temperature is to be determined, and the other contact is placed in a mixture of water and ice. If the calibration curve of the thermocouple is known, then for each value of the thermoelectric EMF indicated by the microvoltmeter, the exact temperature value can be determined.

2. Another application of the Seebeck effect is its use in converting heat into electricity. For this purpose, several thermocouples made of semiconductor materials are connected in series. In this way, thermoelectric sources reaching voltages on the order of tens of volts can be obtained. We note that from a thermodynamic point of view, the thermocouple works similarly to a thermal motor. The working substance in this case is the electronic gas present in metals, which takes thermal energy from the hot contact and transfers it to the environment at the cold contact. Unlike thermal engines that convert thermal energy into mechanical energy, thermocouples transform heat directly into electrical energy. Currently, intensive research is being carried out with the aim of increasing the efficiency of thermoelectric sources [1]. It is also known from thermodynamics that if we supply a thermal engine with external mechanical work (energy), it can function as a heat pump by taking heat from a cold source and transferring it to a sink at a higher temperature, thus functioning as a refrigerator.

2.4.4 The Peltier effect

In contrast to the Seebeck effect (see section 2.4.3), it is expected that when current passes from an external source through a circuit in which there are two contacts, initially at the same temperature, a difference in temperature will appear between the two contacts as a result of the current passing through the system. Such an effect was highlighted in 1834 by J Peltier.[7] Hence, *the Peltier effect consists of the release or absorption of heat at the contact between two metals or at a metal/semiconductor contact when a current is established through the contact.* The heat released or absorbed, called *Peltier heat,* is given by:

$$Q_p = \Pi I t \tag{2.78}$$

where Π is the Peltier coefficient, defined as the *amount of heat released or absorbed in unit time when a current of* 1 *A flows.* In equation (2.78), I is the intensity of the electric current and t is the duration of current flow. To explain this phenomenon, we consider the contact between two metals, Fe and Cu, as shown in figure 2.21. Given a sense for the intensity of the electric current and the polarity corresponding to the two contacts, an equal potential difference appears at the two contacts but in the opposite direction, as depicted in the lower part of figure 2.21. Thus, when the electric current passes from right to left, the electrons move in the opposite direction.

[7] **Jean C A Peltier** (1785–1845) was a French physicist. He was originally a watch dealer, but at the age of 30 began experiments and observations in physics.

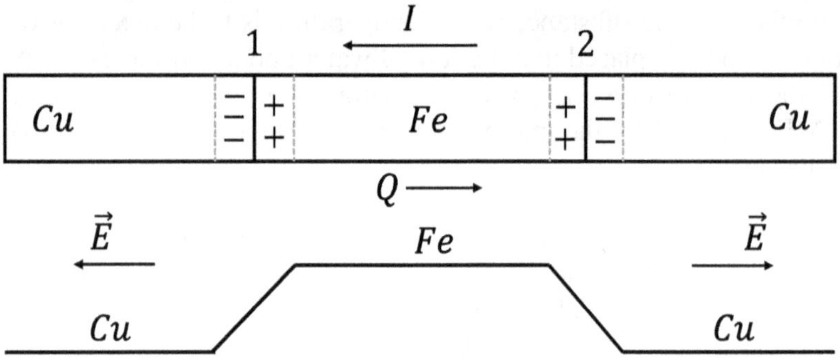

Figure 2.21. Illustration of the Peltier effect.

At contact 1, they move in the opposite direction to that of the electric field, so they are accelerated and their kinetic energy increases. The surplus of kinetic energy acquired during acceleration is transferred by repeated collisions with the nodes of the crystal lattice around the contact, as a result of which contact 1 heats up. At contact 2, the electrons move in the direction of the electric field, so they are slowed by the electric field, and their kinetic energy decreases. To reach thermodynamic equilibrium with the crystalline network, they absorb energy from the network, causing contact 2 to cool down. Changing the direction of the current cools contact 1 and warms contact 2. As already discussed, when electric current passes through a conductor, the conductor is always heated due to the Joule effect. The heat released into the conductor by the Joule effect is $Q = RI^2t$, where R is the resistance of the conductor, I is the current intensity, and t is the time. The Joule effect should not be confused with the Peltier effect. Due to the Joule effect, both contacts warm up to the same extent. Due to the Peltier effect, at one of the contacts, additional heat $Q_p = \Pi It$ is released, while at the other contact, an equal amount of heat is absorbed. So, if we insert the two contacts from figure 2.21 into two micro-calorimeters, as shown in figure 2.22, then at contact 1, the total heat $Q_j + Q_p$ is released, and at contact 2, the heat $Q_j - Q_p$ is released. By measuring these amounts

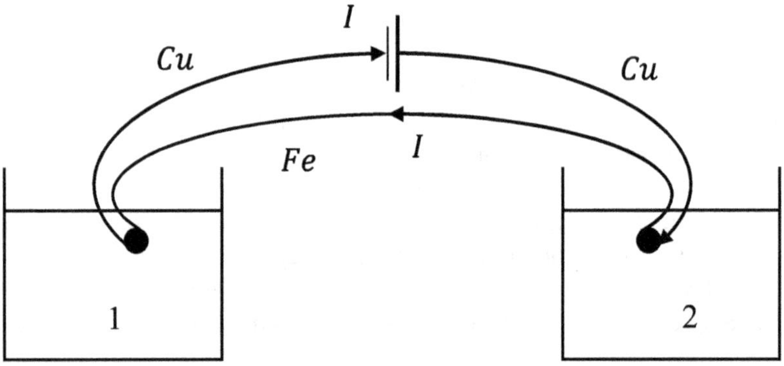

Figure 2.22. Due to the Peltier effect, contact 1 heats up and contact 2 cools down.

of heat and taking their difference, we get $Q_j + Q_p - Q_j + Q_p = 2Q_p$, so the Peltier heat can be determined in this way. It is worth noting that the Peltier coefficient (Π) is a function of the nature of the materials that make up the contacts and is also temperature dependent. The Peltier heat is independent of the electrical resistance of the contact and is a first-degree function in relation to the current intensity, so the Peltier effect is a polar effect. The heat produced by the Joule effect depends on the resistance of the contact and is a second-degree function with respect to the current, so the Joule (or electrocaloric) effect is a nonpolar effect. The Peltier effect is used, among other things, in the construction of refrigerators (see figure 2.23). If a battery of thermocouples connected in series is formed so that all contacts 1–2 are inside the refrigerator and all contacts 2–1 are outside the refrigerator (see figure 2.23), then when the electric current passes in the right direction, contacts 1–2 will cool down and contacts 2–1 will warm up. The temperature in the enclosure will drop and become equal to that of the cold contacts.[8]

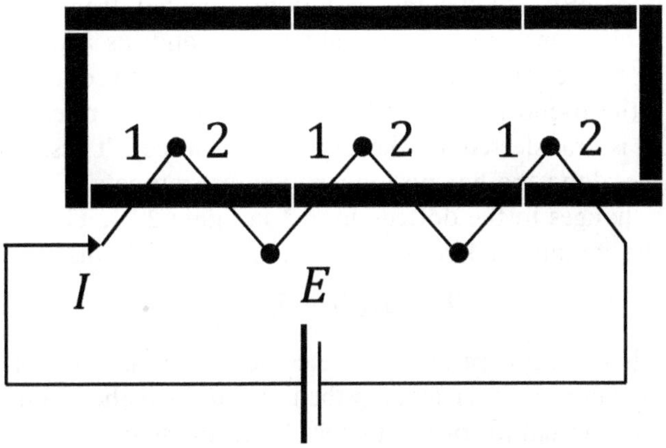

Figure 2.23. Small refrigerator based on the Peltier effect.

2.4.5 The Thomson effect

If a temperature difference is produced across a portion of a conductor, a potential difference (the Thomson[9] effect) appears in that portion of the conductor. By heating a Cu bar in the middle, we can obtain a temperature distribution that covers the full curve 1 in figure 2.24. At the same time, due to the diffusion of electrons from the warm zone to the cold ends, a potential difference occurs between the middle of the bar and either of its ends. Thus, we have two induced electric fields acting in

[8] To increase the cooling efficiency of the refrigerator, the warm contacts can be cooled down using radiators with fans or with additional cold water flow circuits.

[9] **William Thomson** (1824–1907) was a British mathematician, physicist, and engineer born in Belfast. He was the professor of natural philosophy at the University of Glasgow for 53 years, where he undertook significant research, a mathematical analysis of electricity, the formulation of the first and second laws of thermodynamics, and contributed significantly to unifying physics.

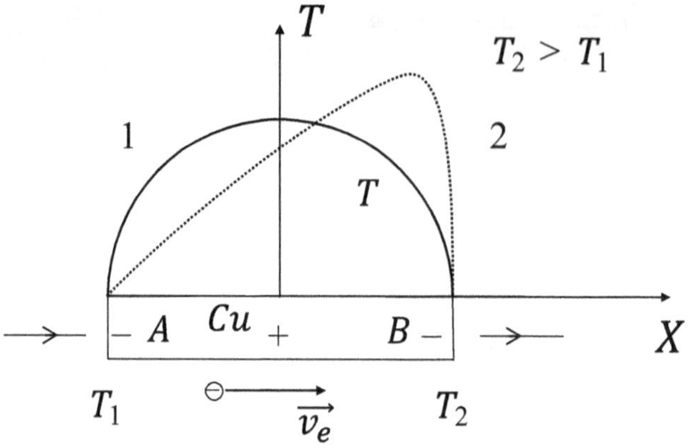

Figure 2.24. Illustration of the Thomson effect.

opposite directions. By establishing a current oriented from left to right (see figure 2.24), the electrons move from right to left, and, as a result, the A-end is cooled according to the Peltier effect (the electrons are slowed, taking energy from the lattice), and the B-end heats up (the electrons are accelerated, and the surplus energy acquired is transferred to the crystalline lattice). Thus, the temperature distribution curve along the bar no longer remains symmetrical (as per curve 1 in figure 2.24) but changes to the dotted curve 2 in figure 2.24. The heat absorbed or released under the conditions described above, known as Thomson heat, is given by:

$$Q = \tau It(T_2 - T_1) \tag{2.79}$$

where $(T_2 - T_1)$ is the temperature difference across the portion of the conductor considered, I is the current intensity, t is the time, and τ is the Thomson coefficient, which depends on the nature of the material and the temperature. The Thomson effect is important from a theoretical point of view, especially in the study of condensed matter.

References and further reading

[1] Piraux L, Marchal N, Van Velthem P, da Câmara Santa Clara Gomes T, Ferain E, Issi J-P and Antohe V-A 2023 Polycrystalline bismuth nanowire networks for flexible longitudinal and transverse thermoelectrics *Nanoscale* **15** 13708–17

[2] Antohe Ş and Antohe V-A 2023 *Electrostatics. Formalism of the Electrostatic Field in Vacuum and Matter* 1st edn (Bristol, UK: IOP Publishing) 266 pages

[3] Van Der Bijl H J 2022 *The Thermionic Vacuum Tube And Its Applications* (London: Legare Street Press) 420 pages

[4] Reich H J 2013 *Theory and Applications of Electron Tubes* (Whitefish, MT: Literary Licensing, LLC) 736 pages

[5] Smirnov B M 2011 *Fundamentals of Ionized Gases: Basic Topics in Plasma Physics* (Weinheim: Wiley-VCH) 488 pages

[6] Cavaliere P 2023 *Water Electrolysis for Hydrogen Production* (New York: Springer) 866 pages
[7] Pandey S C 2010 *Course in Physics (Vol 4): Electrostatics and Current Electricity* (London: Pearson Education) 288 pages
[8] Feynman R P, Leighton R B and Sands M 2011 *Feynman Lectures on Physics, Vol. II: The New Millennium Edition: Mainly Electromagnetism and Matter* (New York: Basic Books) (Feynman Lectures on Physics) 589 pages
[9] Purcell E M and Morin D J 2013 *Electricity And Magnetism* 3rd edn (Cambridge: Cambridge University Press) 853 pages
[10] Spitzer L 2006 *Physics of Fully Ionized Gases* 2nd edn (New York: Dover) 192 pages

Chapter 3

Direct current circuits

Direct current (DC) circuits are typically electrical circuits consisting of any combination of constant voltage sources, constant current sources, and resistors. The voltages across these circuits and the currents passing through them are constant over time. The fundamental laws of electrokinetics can be used to solve any DC circuit, i.e. to find the intensities of the currents flowing through the branches of the circuit and the voltage between two arbitrary points in the network if the values of the electromotive voltages of the sources and all the resistances in the circuit are known.

3.1 Electromotive force and the induced electric field

In electromagnetism and electronics, the electromotive force (EMF) represents the energy transferred by a source to an electric circuit per unit of electric charge; it is measured in volts. For example, batteries generate an EMF by converting chemical energy into electrical energy.[1] This energy conversion can only be achieved if, somewhere in the circuit, there is a force able to perform work on the electric charges. This force is associated with the presence of an induced electric field in an inhomogeneous conducting medium. A few examples of induced electric fields, together with the definition of the related electromotive force, will be given below.

3.1.1 The induced electric field ($\vec{E_i}$)

In the field of electrostatics [1], researchers studying conductors in electrostatic equilibrium have found that when a state of electrostatic equilibrium is reached, the intensity of the electric field inside a homogeneous conductor becomes zero.

[1] Electrical generators produce an EMF by converting mechanical energy into electrical energy. Similarly, electrical transducers provide an EMF by converting other forms of energy into electrical energy.

In inhomogeneous or accelerated conductors, when electrostatic equilibrium is reached, the intensity of the electric field has certain nonzero values independent of the external field into which the conductor is inserted. The field intensity usually depends on the local physicochemical state and the nature of the conductor. When a conductor reaches the electrostatic equilibrium state, the intensity of the electric field at a point within it is a property of the respective material that depends on the local nonelectrical physicochemical conditions (i.e. temperature, concentration, etc). This property is characterized by a vector quantity called the ***intensity of the induced electric field*** (or electromotive field), *defined in a macroscopic way as the value with the opposite sign to the intensity of the electric field that is established in the conductor upon reaching the electrostatic equilibrium state:*

$$\overrightarrow{E_i} = -(\overrightarrow{E})_{\text{electrostatic equilibrium}} \tag{3.1}$$

Under these conditions, it is clear that in homogeneous non-accelerated conductors, $\overrightarrow{E_i} = 0$. Conversely, in inhomogeneous or accelerated conductors, $\overrightarrow{E_i} \neq 0$, and its value depends on local physicochemical inhomogeneities (such as the chemical constitution, temperature, etc.) or on the acceleration of the body. From a microscopic point of view, the presence of this induced field can be explained in terms of the structure of the conducting medium. Conductors contain free particles that carry an electric charge (i.e. electrons in metals or ions in electrolytes and plasmas). An electrical force ($\overrightarrow{F_{\text{el}}} = q\overrightarrow{E}$) and a nonelectrical force ($\overrightarrow{F_{\text{nonel}}}$) are exerted on such a particle with a charge q due to local inhomogeneities (e.g. due to uncompensated collisions between the particle of interest and other particles) or due to the acceleration of the body (i.e. inertial forces). The macroscopic condition of electrostatic equilibrium, i.e. the absence of electric current, is expressed microscopically by the static condition of the absence of an oriented movement of charged particles. In other words, electrostatic equilibrium results from the cancellation of the mean value of the acceleration or the resulting force exerted on a particle:

$$\overrightarrow{F_{\text{el}}} + \overrightarrow{F_{\text{nonel}}} = 0 \Rightarrow q\overrightarrow{E} + \overrightarrow{F_{\text{nonel}}} = 0 \tag{3.2}$$

Dividing (3.2) by q gives:

$$\overrightarrow{E} + \frac{\overrightarrow{F_{\text{nonel}}}}{q} = 0 \text{ or } \overrightarrow{E} + \overrightarrow{E_i} = 0 \tag{3.3}$$

where:

$$\overrightarrow{E_i} = \frac{\overrightarrow{F_{\text{nonel}}}}{q} \tag{3.4}$$

Hence, at a microscopic scale, the intensity of the induced electric field is the ratio between the average nonelectric force exerted on a charged particle in a conductor and its charge. Depending on the nature of the nonelectric force, there are various induced electric fields. Some examples are presented below.

(a) ***Electric field induced by acceleration***

If a metallic disc rotates around its longitudinal axis with an angular velocity $\vec{\omega}$, its electrons are subject to centrifugal force. As a result, they accumulate on the periphery of the disc, which is therefore negatively charged, whereas the center of the disc becomes positively charged. In this case, the nonelectric force acting on an electron is the centrifugal force oriented outwards from the center of the disc:

$$\overrightarrow{F_{\text{nonel}}} = m\omega^2\vec{r} \tag{3.5}$$

As a result, the intensity of the induced electric field is, according to the definition:

$$\overrightarrow{E_i} = \frac{\overrightarrow{F_{\text{nonel}}}}{q} = \frac{m\omega^2\vec{r}}{-q} = -\frac{m\omega^2}{q}\vec{r} \tag{3.6}$$

According to (3.6), the intensity of the induced electric field is oriented toward the center of rotation (i.e. electrons always move in the opposite direction to that of an electric field, even if it is induced); see figure 3.1.

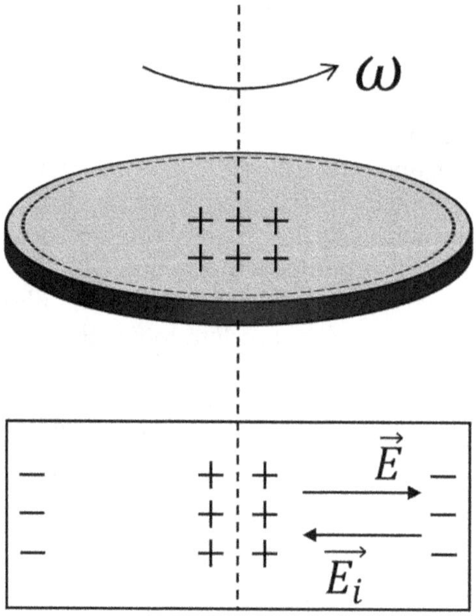

Figure 3.1. Electric field induced by acceleration.

When the force with which the electric field ($\vec{E}$), due to the polarization of the disc produced by centrifuging electrons, compensates for the action of the nonelectric force, equilibrium is achieved:

$$\overrightarrow{F_{\text{el}}} + \overrightarrow{F_{\text{nonel}}} = 0 \;\Rightarrow\; \vec{E} = -\overrightarrow{E_i} = \frac{m\omega^2}{q}\vec{r} \tag{3.7}$$

Consequently, between the beginning of disc rotation (i.e. when the free particles are set into motion) and the establishment of electrostatic equilibrium, a transient current of very short duration appears in the disc.

(b) ***Electric field induced by a concentration gradient (diffusion)***

A field may be induced by a concentration gradient if there is an inhomogeneous concentration of an electrolyte inside a vessel. To exemplify this, the porous wall (P) shown in figure 3.2 separates two solutions with

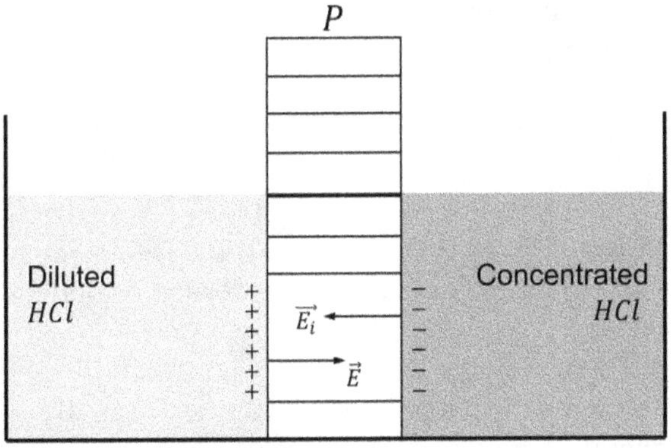

Figure 3.2. Electric field induced by a concentration gradient.

different concentrations of HCl. The concentrated electrolyte has a greater number of H^+ ions than the dilute electrolyte. H^+ ions, having higher mobility than Cl^- ions, easily cross the porous wall, positively charging the diluted electrolyte, while the more concentrated solution remains negatively charged. In this case, an electric field $\vec{E} = -\vec{E_i}$, appears in the opposite direction such that $\vec{E_i} = \frac{\vec{F_{\text{nonel}}}}{q}$. This field opposes the further diffusion of H^+ ions, and at equilibrium, the current ceases.

(c) ***Electric field induced by a temperature gradient***

As found in the study of the Thomson effect, the presence of a temperature gradient along a conductor causes a diffusion of electrons from the warm side to the cold side, ultimately leading to the appearance of an induced field $\left(\vec{E_i} = \frac{\vec{F_{\text{nonel}}}}{-q}\right)$; see figure 3.3(a). However, the field $\vec{E} = -\vec{E_i}$

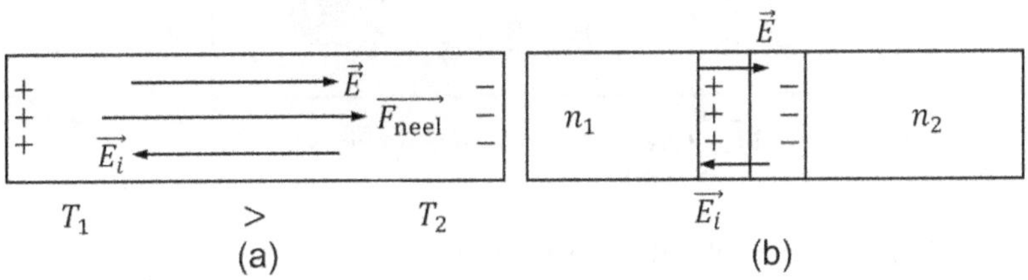

Figure 3.3. (a) Induced field due to temperature gradient. (b) Induced electric field by the contact potential difference between two different metals.

appears at equilibrium; it acts in a direction that opposes further diffusion. This type of field is extremely weak and can only be detected via some of its effects (for example, the Thomson effect, etc).

(d) ***Electric field induced by a contact potential difference between two conductors***

The induced contact field is located in the very thin inhomogeneous layer that separates two conductors brought into contact. In the case of contact between two different conductors, see figure 3.3(b), it has been found that a difference in contact potential occurs consisting of an extrinsic $\Delta V_{e_{12}}$ component and an intrinsic one, $\Delta V_{i_{12}}$. The difference in contact potential associated with the induced electric field is:

$$\Delta V_{12} = \frac{w_{e_2} - w_{e_1}}{e} + \frac{kT}{e}\ln\frac{n_1}{n_2} = \int_1^2 \overrightarrow{E_i}\overrightarrow{dl} \tag{3.8}$$

At equilibrium, $\vec{E} + \vec{E_i} = 0$, where $\vec{E}$ is the field given by the polarization of the contact. In the case of contact between two conductors, two cases can be highlighted:

(i) *Induced voltaic electric fields* are produced by contact between two metals with different work functions that are kept at the same temperature and are independent of any external factors.

(ii) *Induced galvanic electric fields* are present at the interface between a conductor of type I (e.g. a metal) and another of type II (e.g. a semiconductor, electrolyte, etc). An example is the electrode–electrolyte interface in a galvanic cell (see figure 3.4).

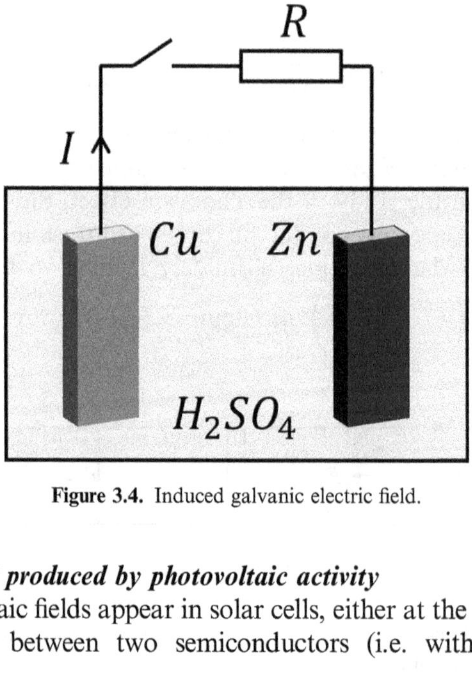

Figure 3.4. Induced galvanic electric field.

(e) ***Electric field produced by photovoltaic activity***

Photovoltaic fields appear in solar cells, either at the metal–semiconductor interface or between two semiconductors (i.e. within *p–n* junctions or heterojunctions). In such cases, the absorbed photons create electron–hole pairs, which are separated by the internal field and sent to the neutral regions n and p, respectively, thus leading to the polarization of the structure.

3.1.2 The electromotive force

According to the definition of the electrokinetic state, when conductors carry an electric conduction current, the relationship $\vec{E} + \vec{E_i} = 0$ valid at electrostatic equilibrium (resulting from the statistical compensation of the movements of the charged elementary particles), is no longer valid for a conductor in an electrokinetic state. In the latter case, statistical compensation no longer occurs, and as a result there is an oriented movement of charge carriers that generates an electric current. This oriented movement is maintained precisely as a result of the induced electric field, the electrokinetic state being characterized by the relationship:

$$\vec{E} + \vec{E_i} \neq 0 \tag{3.9}$$

Based on the sum of the intensity of the electric field $\vec{E}$ and the intensity of the induced electric field $\vec{E_i}$, *the* ***contour electromotive force*** *(CEMF) is defined as the integral of the sum of the electric field intensity and the induced electric field intensity along a closed path of integration,* Γ:

$$U_c = E = \oint_{\Gamma} (\vec{E} + \vec{E_i})\vec{dl} \tag{3.10}$$

U_c is measured in units of electric voltage. The energetic interpretation of this value can be directly derived from the definition of the electrokinetic state $(\vec{F_{el}} + \overrightarrow{F_{nonel}} \neq 0)$, namely:

$$E = \frac{1}{q}\oint_{\Gamma} (\vec{F_{el}} + \overrightarrow{F_{nonel}})\vec{dl} = \frac{L_{\Gamma}}{q} \tag{3.11}$$

In other words, *the contour electromotive force is the physical quantity numerically equal to the mechanical work performed by the resultant force (electric and non-electric) exerted on a particle with unit charge when the particle moves on a closed path.* In many works, the CEMF is also called the EMF. At electrostatic equilibrium (i.e. where $\vec{E} + \vec{E_i} = 0$), it is obvious that $E = 0$. If, however, the intensity of the induced electric field takes nonzero values ($\vec{E_i} \neq 0$) along a contour that passes through the conductor, electrostatic equilibrium is no longer maintained; therefore, in a closed circuit, an electric current appears.

Let us consider a system of three conductors, i.e. two metals and one electrolyte, constituting a galvanic cell as discussed above. If the electrodes are connected by an external conducting wire through an outer circuit having a resistance R (see figure 3.4), an electric current flows. If the external circuit is discontinued, the electrostatic equilibrium is restored again.

(a) If an electrokinetic electric state is stationary, the electric charges in the system have an unvarying distribution over time, and the electric field ($\vec{E}$) is a coulombic field. As the theorem of the stationary electrical potential is valid in this case (i.e. $\oint_\Gamma \vec{E}\vec{dl} = 0$, as also given for the stationary electrokinetic state), we can write:

$$E = \int_\Gamma (\vec{E} + \vec{E_i})\vec{dl} = \int_\Gamma \vec{E_i}\vec{dl} \tag{3.12}$$

that is, the EMF is given only by the circulation of the induced electric field. **Definition:** *the integral of the intensity of an electric field induced along an open or closed line of integration is called an EMF.*

Generally, as in the case of electric voltage, the calculation of the electromotive force requires us to specify the curve Γ along which the integration takes place. The EMF is often referred to as the *broad electric field strength*, which is the sum of the electric field intensity and the induced electric field intensity:

$$\overrightarrow{E_{\text{large}}} = \vec{E} + \vec{E_i} \tag{3.13}$$

In contrast to this name, the intensity of the coulombic electric field ($\vec{E}$) is also called *the intensity of the* ***electric field in the narrow sense.***

(b) If the electric state is nonstationary, the electric field in the narrow sense can be split into a coulombic field of intensity $\vec{E_c}$ and an electric field $\vec{E_s}$ produced by the variation over time of a magnetic field, as we will demonstrate later when discussing electromagnetic induction. This field is a solenoidal field (i.e. *a solenoidal field is a vector field with zero divergence*— the coulombic field is not a solenoidal field, since $\nabla \vec{E_c} = \frac{\rho}{\varepsilon}$). In this case, more generally, the definition of the CEMF becomes:

$$E = \oint_\Gamma (\vec{E} + \vec{E_i})\vec{dl} = \oint_\Gamma (\vec{E_c} + \vec{E_s} + \vec{E_i})\vec{dl} = \oint_\Gamma (\vec{E_s} + \vec{E_i})\vec{dl} \tag{3.14}$$

because the line integral of a coulombic field is always zero according to the theory of electrical potential ($\oint_\Gamma \vec{E_c}\vec{dl} = 0$). If the energy sources are characterized by the presence of induced electric fields and $\vec{E_s} = 0$, then the EMF is reduced to the voltage of the induced electric field (i.e. $E = \oint_\Gamma \vec{E_i}\vec{dl}$), which is nonzero only in the zones of inhomogeneity inside the sources. In this case, the EMF of the induced field is located exclusively along the curve on which there are inhomogeneous states that generate the induced electric field. Using decomposition of the electric field into a solenoidal component and a coulombic component (i.e. $\vec{E} = \vec{E_s} + \vec{E_c}$), we obtain an expression for the EMF in a portion of the circuit:

$$E_{AB} = \int_A^B (\overrightarrow{E_s} + \overrightarrow{E_i})\overrightarrow{dl} \tag{3.15}$$

Definition: *The EMF is the line integral of the non-coulombic part of the* ***electric field in the broad sense*** *(i.e.* $\overrightarrow{E} = \overrightarrow{E_s} + \overrightarrow{E_i}$*).*
The EMF generally depends on the integration path and is reduced to the line integral of the induced electric field in the case of the stationary regime ($\overrightarrow{E_s} = 0$).

3.2 Ohm's law for inhomogeneous conducting media

In this section, we determine the local and integral forms of Ohm's law for inhomogeneous but isotropic conducting media.

3.2.1 Local form of Ohm's law for inhomogeneous conducting media

In the classical treatment of the process of electric conduction in metals, the local form of Ohm's law was established ($\vec{j} = \sigma\vec{E}$), which only considers the intensity of the electric field in the narrow sense $\vec{E}$. In the case of a conducting medium in which there is also an induced electric field, the intensity of the resulting electric field in the broad sense ($\overrightarrow{E_{\text{large}}} = \vec{E} + \overrightarrow{E_i}$) must obviously be taken into account. In this case, the local form of Ohm's law becomes:

$$\vec{j} = \sigma(\vec{E} + \overrightarrow{E_i}) \tag{3.16}$$

This states that *at any point, the density of the conduction current density is proportional to the vector sum of the intensity of the electric field ($\vec{E}$) and the intensity of the electric field ($\overrightarrow{E_i}$) induced inside an isotropic inhomogeneous conductor present at that point.* The proportionality factor is a parameter of the material (σ) called the electrical conductivity of the conducting medium, which depends on its nature, temperature, etc.

3.2.2 Integral form of Ohm's law for non-homogeneous conducting media

Just as the integral form of Ohm's law $I = \dfrac{U}{R}$ was formulated using the local form of Ohm's law valid for homogeneous and isotropic conducting media ($\vec{j} = \sigma\vec{E}$), in the case of an inhomogeneous but isotropic medium, we can start from the local form of Ohm's law, equation (3.16), and establish the integral form of Ohm's law.

Let us consider a portion of a filiform circuit in which an induced electric field is present. An external generator maintains a potential difference ($V_1 - V_2 = \text{constant}$) between the ends of the circuit portion, see figure 3.5. The circuit is considered filiform if it is sufficiently thin that the current is uniformly distributed over the section. In this case, we can say that $j = \frac{I}{dA}$ (where dA represents its cross-sectional area) holds everywhere. In this case, the line integral of the electric field in a broad sense leads to the CEMF, which is:

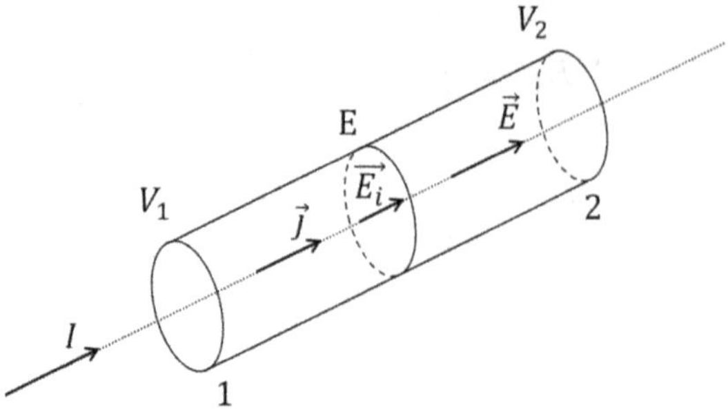

Figure 3.5. Portion of an inhomogeneous conducting medium containing the induced electric field ($\vec{E_i}$).

$$\int_1^2 (\vec{E} + \vec{E_i})\vec{dl} = \int_1^2 \frac{\vec{j}}{\sigma}\vec{dl} \tag{3.17}$$

or:

$$\int_1^2 \vec{E}\vec{dl} + \int_1^2 \vec{E_i}\vec{dl} = I\int_1^2 \rho\frac{dl}{dA} \tag{3.18}$$

Hence:

$$U_{12} + E_{12} = IR_{12} \tag{3.19}$$

where $U_{12} = \int_1^2 \vec{E}\vec{dl}$ is the electric voltage across the wire, $E_{12} = \int_1^2 \vec{E_i}\vec{dl}$ is the corresponding electromotive force, and $R_{12} = \int_1^2 \rho\frac{dl}{dA}$ is the electrical resistance of the conductor between points 1 and 2.

Relation (3.19) represents the integral form of the law of electrical conduction in inhomogeneous conducting media (the integral form of Ohm's law for inhomogeneous conducting media). It states that *for a certain portion of the filiform circuit, the sum of the local voltage (U_{12}) along the wire and the electromotive forces (E_{12}) of sources that are found in that portion of the circuit is equal to the product of the intensity of the electric current and a physical quantity specific to the circuit, called the electrical resistance.*

This law is valid if the conductor represents either just a portion of a closed circuit or the entire closed circuit. In the latter case, $U_{12} = 0$, then the sum $U_{12} + E_{12}$ is reduced to E_{12} (the total EMF along the circuit):

$$E_{12} = RI \tag{3.20}$$

In (3.20), R denotes the total resistance of the inhomogeneous conducting medium. If the EMF is zero along the inhomogeneous conducting medium (i.e. no source of electromotive force is present), the solution reduces to Ohm's law for a homogeneous conducting medium (a portion of the circuit without sources):

$$U_{12} = R_{12}I \tag{3.21}$$

The law of electrical conduction is valid both in direct current and in time-dependent current (i/e. alternating current—AC).

3.3 Joule–Lenz law for inhomogeneous conducting media

In this section, we derive the local and integral forms of the law of energy transformation (i.e. the Joule–Lenz law) for inhomogeneous and isotropic conducting circuits with energy sources.

3.3.1 Local form of the Joule–Lenz law for inhomogeneous conducting media

In the microscopic analysis of the conduction process, the local Joule–Lenz law was established, which states that *the power delivered per unit volume of a conductor by the electromagnetic field in the electrical conduction process is equal to the scalar product of the electric field intensity and the density of the conduction current* (i.e. $w_c = \vec{j}\,\vec{E}$). In this relation, the *power density* (w_c) was introduced, being defined as *the energy transferred by the electromagnetic field per unit time and per unit volume of the conducting medium.* In this case, relation $w_c = \vec{j}\,\vec{E}$ was written for homogeneous conductors, where $\vec{E}$ represents a coulombic field $\vec{E_c}$, the local form of Ohm's law is $\vec{j} = \sigma\vec{E_c}$, and, as a result, $w = \sigma\vec{E_c}^2 = \rho j^2 = \vec{E_c}\vec{j}$.

In the more general case of inhomogeneous conductors, i.e. in the presence of an induced electric field ($\vec{E_i}$), the above relation becomes:

$$w_c = \vec{j}\,\vec{E_c} = \vec{j}\,(\rho\vec{j} - \vec{E_i}) = \rho j^2 - \vec{j}\,\vec{E_i} = \rho j^2 - w_i \tag{3.22}$$

in which it was considered that $\vec{j} = \sigma(\vec{E_c} + \vec{E_i})$, therefore $\vec{E} = \rho\vec{j} - \vec{E_i}$. Relationship (3.22) comprises two terms, namely:

- The first term (ρj^2), which is always positive, representing the energy irreversibly dissipated by the electromagnetic field (produced by an external generator, regardless of the sense of the current flow) and transformed into heat according to the Joule–Lenz effect;
- The second term ($\vec{j}\,\vec{E_i}$) can be positive or negative; it represents the power density exchanged between the induced electric field sources and the electromagnetic field (external generator). The exact interpretation of the energy exchange between the source and the field is made according to the signs of these expressions:
  - If the vectors $\vec{j}$ and $\vec{E_i}$ are homoparallel, then $w_i > 0$ and thus the source in the inhomogeneous conducting medium gives up energy to the circuit, like the external generator. This occurs in any cell that generates an electric current and simultaneously produces electromagnetic energy by transforming its inner chemical energy into electrical energy, like the external generator;
  - If $\vec{j}$ and $\vec{E_i}$ are antiparallel, then $w_i < 0$ and the power w_i is received by the source in the inhomogeneous conducting medium from the

electromagnetic field (i.e. external sources). This occurs in charging an electric rechargeable battery, when the electromagnetic energy taken from the outside source is converted into chemical energy inside the inner source.

3.3.2 Integral form of the Joule–Lenz law for inhomogeneous conducting media

If the relationship (3.22), expressing the local form of the Joule–Lenz law, is integrated, the integral form of the law is obtained for a portion of the filiform conductor. At the ends 1 and 2 of the portion of interest, the voltage $U_{12} = V_1 - V_2$ is kept constant by an external generator:

$$P = \int_V w_c dv = \int_V \vec{j}\,\vec{E}\,dv = \int_V jA\vec{E}\,\vec{dl} = I\int_1^2 \vec{E}\,\vec{dl} = IU_{12} \tag{3.23}$$

This equation gives the electrical power transferred to the circuit by the external generator through its terminals 1 and 2 (see figure 3.6).

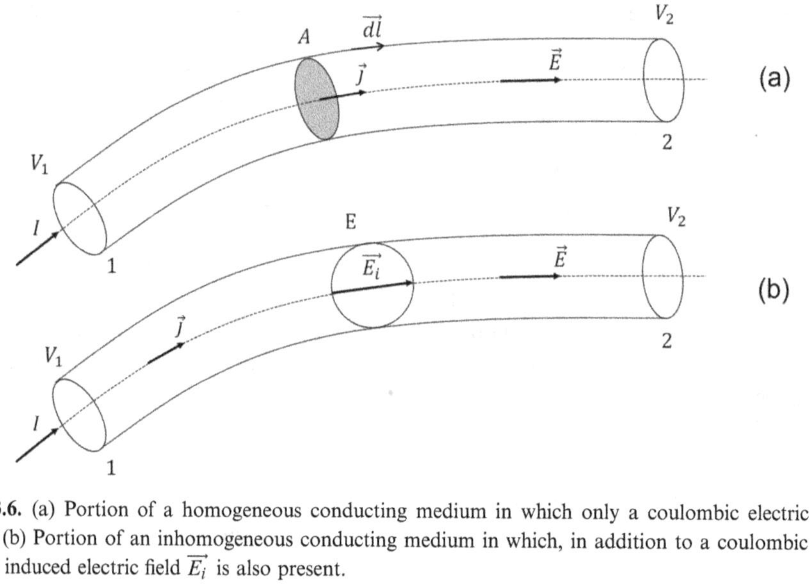

Figure 3.6. (a) Portion of a homogeneous conducting medium in which only a coulombic electric field is present. (b) Portion of an inhomogeneous conducting medium in which, in addition to a coulombic electric field, an induced electric field $\vec{E_i}$ is also present.

Definition: *the total power transferred by an electromagnetic field (produced by external generators) to a portion of a filiform conductor in the process of electrical conduction is equal to the product of the voltage applied between its ends and the intensity of the electric current flowing through it.*

If an EMF (E_{12}) is also present in the conducting medium, then taking into account that $U_{12} + E_{12} = IR_{12} \Rightarrow U_{12} = IR_{12} - E_{12}$, we obtain:

$$P = R_{12}I^2 - IE_{12} = IU_{12} \tag{3.24}$$

In other words, the power P transferred by the electromagnetic field to a portion of the filiform conductor consists of the power irreversibly given up in the form of heat in the conductor ($R_{12}I^2 > 0$) and a second term with a changed sign $P_i = E_{12}I$. The latter is the power exchanged with the electromagnetic field (external generators) by the source of the EMF (E_{12}); it is equal to the product of the current intensity I and its EMF (i.e. E_{12}). In this product, E_{12} can take the following signs:

- positive when it has the same sense as the current; or
- negative when it has the opposite sense to that of the current.

In the first case, the source produces energy, like the external generator (e.g. an electric battery powering a circuit without other sources, which behaves as an *active element of the circuit*), while in the second case the internal source of EMF (i.e. E_{12}) absorbs energy from the electromagnetic field—behaving like a receiver or a *passive element of a circuit*, such as when charging a rechargeable battery.

3.4 Linear and filiform DC circuits

In the following, we analyze filiform DC circuits (i.e. thin enough for the current intensity to be considered uniformly distributed over their cross-section) and linear circuits (i.e. circuits that have sides or branches of constant resistances, independent of the current values). Under these conditions, the fundamental relationships of the electrokinetic state are:

(1) The stationary electrical potential theorem: $U_\Gamma = \oint_\Gamma \vec{E}\,\vec{dl} = 0$;
(2) The continuity theorem for current lines (a particular form of the law of charge conservation): $i_\Sigma = I_\Sigma = \iint_\Sigma \vec{j}\,\vec{ds} = 0$;
(3) The law of electrical conduction (or Ohm's law) for an inhomogeneous circuit with energy sources: $U_{12} + E_{12} = R_{12}I$;
(4) The law of energy transformation for inhomogeneous conducting media (or the Joule–Lenz law): $P = RI^2 - E_{12}I = IU_{12}$.

The first three relationships allow the direct determination of the currents that are established through the branches of a network when the electromotive forces of the sources and the resistances of the passive and active elements in the circuit are known. As a rule, this problem is solved using the consequences of these theorems and laws, which are known as ***Kirchhoff's theorems***. The formulation and application of these theorems, in which the involved quantities (I, U, E_{12}) are algebraic quantities (positive, zero, or negative), require the unambiguous specification of the reference sense of these physical quantities.

3.4.1 Kirchhoff's theorems

Let us consider a complete electrical network with n nodes (side ends), l sides (representing unbranched portions bordered by two nodes), and m meshes (closed contours consisting of a sequence of sides of the grid without overlaps).[2] The sides that contain energy sources are called active branches, and those that do not contain sources are called passive branches. We consider network portions to be active or passive. A network portion is called passive if it derived from an active network portion by canceling the electromotive forces of its sources but without removing their internal resistances. A network is complete or isolated if it has no access terminals to the outside. An incomplete network with two access terminals is called a dipole with four quadrupoles. The network calculation consists of determining the intensities of the electric currents carried by the network branches. This is done using Kirchhoff's theorems, which lead to the required independent equations.

3.4.1.1 Kirchhoff's first theorem

Based on the law of the conservation of electric charge, the continuity theorem has been demonstrated in the stationary regime: $i_\Sigma = 0$. If the enclosed surface, Σ (see figure 3.7), surrounds node n of a network, it results in:

$$\int_\Sigma \vec{j}\,\vec{ds} = 0 \Rightarrow \int_{\Sigma_1} \vec{j_1}\,\vec{ds} + \int_{\Sigma_2} \vec{j_2}\,\vec{ds} + \int_{\Sigma_3} \vec{j_3}\,\vec{ds} + \int_{\Sigma_4} \vec{j_4}\,\vec{ds}$$
$$= -I_1 - I_2 + I_3 + I_4 = 0 \tag{3.25}$$

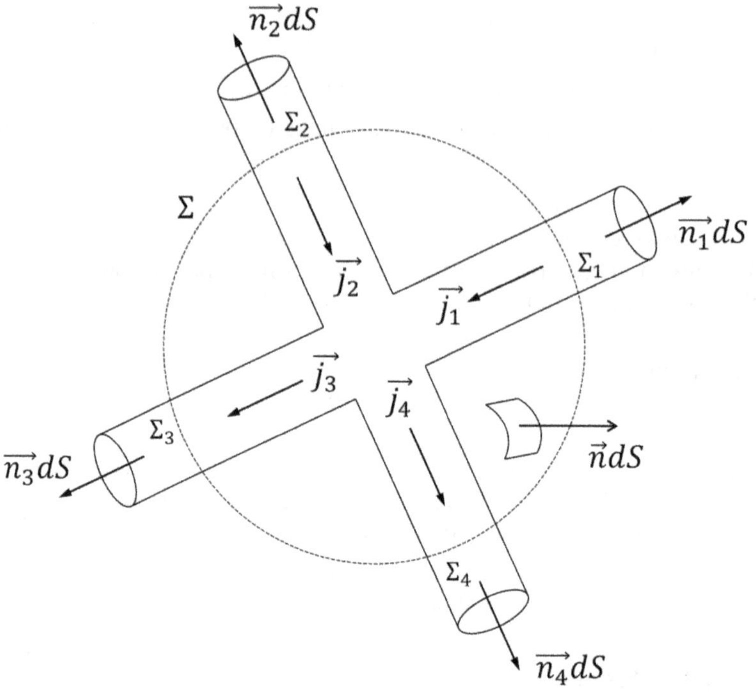

Figure 3.7. Demonstration of Kirchhoff's first theorem.

[2] In electronic circuits, a ***'loop'*** represents any closed path in which no node is encountered more than once, while a ***'mesh'*** is a loop that has no other loops inside it.

For a node in which k sides are connected, we can write the equation:

$$\sum_{(k)} I_k = 0 \tag{3.26}$$

Using Kirchhoff's first law, we obtain a number $n - 1$ of independent equations, in which n is the number of nodes.

3.4.1.2 Kirchhoff's second theorem

Writing the CEMF, i.e. $\int_\Gamma (\vec{E} + \vec{E}_i)\vec{dl}$ for the closed path Γ passing through the axis of the conductors forming the mesh (see figure 3.8) gives:

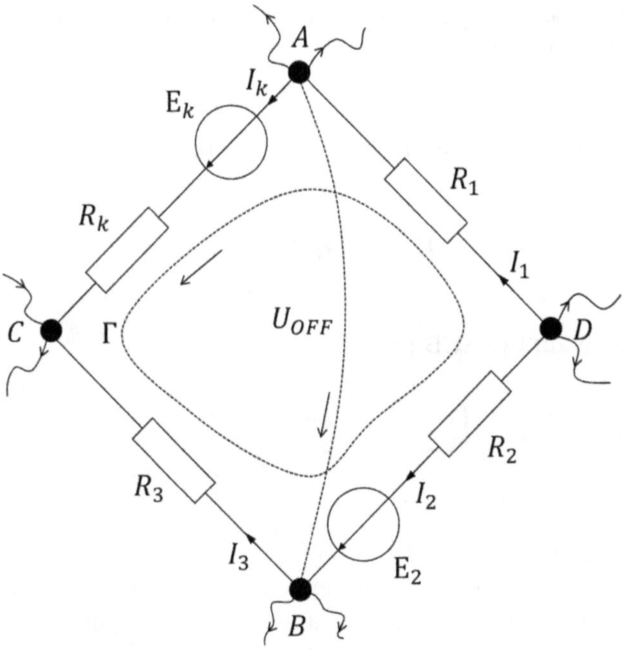

Figure 3.8. Illustration of Kirchhoff's second theorem.

$$\oint_\Gamma (\vec{E} + \vec{E}_i)\vec{dl} = \oint_\Gamma \rho\vec{j}\,\vec{dl} \tag{3.27}$$

However, in the stationary electrokinetic state, $\oint_\Gamma \vec{E}\vec{dl} = 0$ and:

$$\oint \vec{E}_i\vec{dl} = \sum_{k=1}^{n_p} E_k \tag{3.28}$$

where E_k is the EMF of branch k and the other branches are further numbered $k = 1, 2, \ldots, n_p$. On the other hand:

$$\oint \rho\vec{j}\, d\vec{l} = \sum_{k=1}^{n_m} I_k R_k \tag{3.29}$$

Therefore, for a single mesh of the network, one can write:

$$\sum_{k=1}^{n_p} E_k = \sum_{k=1}^{n_m} I_k R_k \tag{3.30}$$

For the whole network, we obtain $m = l - n + 1$ independent equations, in which $E_k > 0$ if its $\overrightarrow{E_i}$ has the same sense as the direction of current I_k. In equation (3.30), n_p is the number of sources present in the loop of the network, and n_m is the number of resistances present in it.

3.4.1.3 Calculation of the voltage between two nodes

For any two nodes A and B (see figure 3.8), if we apply Kirchhoff's second theorem, we obtain:

$$\sum_{(k)\,A\to B} E_k = \sum_{k\;A\to B} I_k R_k - U_{AB} \tag{3.31}$$

from which we obtain the ACB path:

$$U_{AB} = \left(\sum_k I_k R_k - \sum_k E_k\right)_{A\to B(\text{via } ACB)} \tag{3.32}$$

3.4.1.4 Balance of powers in a DC circuit

According to the energy conservation theorem, in a DC current circuit that has no connections with the outside, the energy and the power supplied by the sources are equal to the power received by the receivers:

$$\sum EI = \sum RI^2 \tag{3.33}$$

This equation allows us to verify that the correct solution is obtained when a DC circuit is analyzed.

3.4.1.5 Applications of Kirchhoff's theorems

1. **Combinations of resistors**
   Resistors connected by wires of negligible resistance form a passive network. If there are n external connections, it is said to be an n-terminal network. A two-terminal passive network has an **equivalent resistance** defined by $\frac{U}{I}$, where U is the voltage between the terminals when the current I enters via

one terminal and leaves via the other. The latter is used in conjunction with Kirchhoff's theorems to determine some useful formulae.

Series connection

For resistors in series (see figure 3.9(a)), the current passing through each resistor is the same, while the potential difference across each resistor is summed to give a total potential difference:

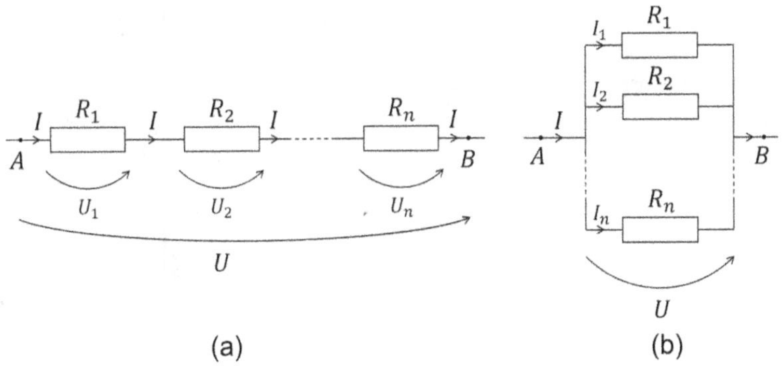

Figure 3.9. Connection of resistors: (a) in series and (b) in parallel.

$$U = U_1 + U_2 + \ldots + U_n = I(R_1 + R_2 + \ldots + R_n) \tag{3.34}$$

Hence, because the equivalent resistance is $R = \frac{U}{I}$, the result is:

$$R = R_1 + R_2 + \ldots + R_n$$

or, in general:

$$R_{es} = \sum_{k=1}^{n} R_k \tag{3.35}$$

Equation (3.35) shows that the equivalent resistance of a series combination is always larger than the resistance of each component resistor.

Parallel connections

For resistors in parallel (see figure 3.9(b)), the voltage drop across each resistor is common, having the same value (U) for all resistors, while the currents sum according to Kirchhoff's first law, so that:

$$I = I_1 + I_2 + \ldots + I_n = U(G_1 + G_2 + \ldots + G_n) \tag{3.36}$$

Hence, because the equivalent conductance is $G = \frac{I}{U}$, we obtain:

$$G = G_1 + G_2 + \ldots + G_n \text{ or } \frac{1}{R} = \frac{1}{R_1} + \frac{1}{R_2} + \ldots + \frac{1}{R_n}$$

or, in general:

$$\frac{1}{R_{ep}} = \sum_{i=1}^{n} \frac{1}{R_i} \tag{3.37}$$

Equation (3.37) shows that the equivalent resistance of a parallel combination is always smaller than the resistance of each component resistor.

2. **Combinations of generators**

   Using Kirchhoff's laws, we can obtain equivalent parameters for series or parallel connections of many generators, as described below.

Series connection

For generators with electromotive forces E_1, E_2, E_3, … , E_n and internal resistances r_1, r_2, r_3, … , r_n in series, see figure 3.10(a), the current is

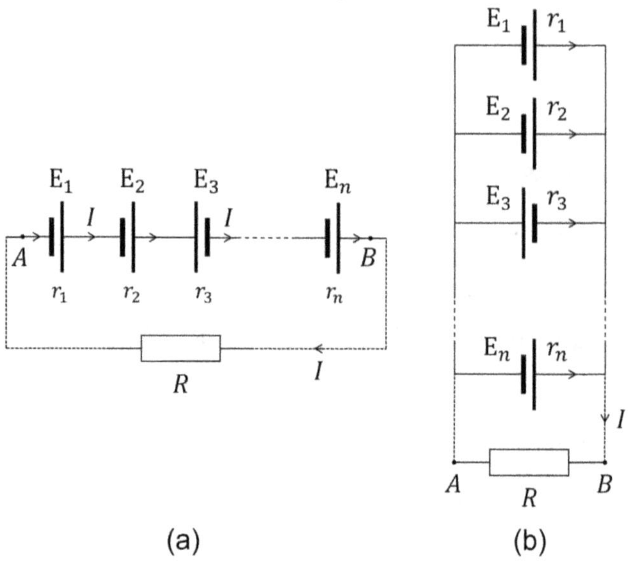

Figure 3.10. Connection of generators: (a) in series and (b) in parallel.

common and the final voltage between terminals A and B is the algebraic sum of the electromotive forces. The equivalent parameters are:

$$E_{es} = E_1 + E_2 - E_3 + \ldots + E_n = \sum_{k=1}^{n} E_k \tag{3.38}$$

and, respectively:

$$r_{es} = r_1 + r_2 + r_3 + \ldots + r_n = \sum_{k=1}^{n} r_k \tag{3.39}$$

Parallel connection

In this case (see figure 3.10(b)), the current passing through a load resistor (R) is, according to Kirchhoff's first law, the algebraic sum of the currents injected through it by each generator, so that the equivalent parameters are:

$$E_{ep} = \frac{\sum_{k=1}^{n} E_k g_k}{\sum_{k=1}^{n} g_k} \tag{3.40}$$

and, respectively:

$$g_{ep} = \sum_{k=1}^{n} g_k \tag{3.41}$$

where the EMF and internal conductance of the generator k are E_k and g_k, respectively.

3.4.1.6 *Theorem of maximum power transfer from a DC source to an external resistance*

Consider a circuit consisting of a source with an EMF E and an internal resistance r. It gives rise to a current I through a load resistance R that can be varied as shown in figure 3.11. The question arises of establishing the conditions under which the

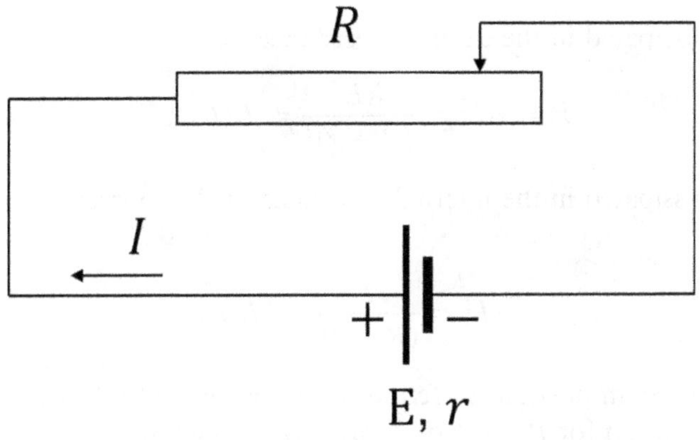

Figure 3.11. Simple DC circuit, with a variable load resistance.

maximum electric power is transferred from the source to the circuit. The form of Ohm's law for a closed, inhomogeneous circuit (with an energy source) is:

$$E = IR + Ir \tag{3.42}$$

Multiplying equation (3.42) by I yields:

$$EI = RI^2 + rI^2 \tag{3.43}$$

Hence, the total electrical power of the energy dissipated by the source in the circuit ($P = EI$) is consumed by the Joule effect in the circuit, being equal to the power $P_e = RI^2$ dissipated in the external resistance plus the power $P_i = rI^2$ dissipated in the internal resistance of the source. The ratio of the external power to the total power defines the yield:

$$\eta = \frac{P_e}{P} = \frac{R}{R + r} = \eta(R) \tag{3.44}$$

The current intensity in the circuit is:

$$I = \frac{E}{R + r} \tag{3.45}$$

Using this expression, we can obtain the following electrical quantities for the circuit:

- The voltage drop in the external circuit:

$$U_R = \frac{RE}{R + r} = f(R) \tag{3.46}$$

- Total power:

$$P = EI = \frac{E^2}{R + r} = P(R) \tag{3.47}$$

- Power dissipated in the external resistance:

$$P_e = RI^2 = \frac{RE^2}{(R + r)^2} = P_e(R) \tag{3.48}$$

- Power dissipated in the internal resistance of the source:

$$P_i = rI^2 = \frac{rE^2}{(R + r)^2} = P_i(R) \tag{3.49}$$

To find the maximum power transferred from the source to the external resistance, the limiting condition for P_e must be analyzed, as follows:

$$\begin{aligned} \frac{dP_e}{dR} &= E^2\left[\frac{(R + r)^2 - 2(R + r)R}{(R + r)^4}\right] = 0 \Rightarrow \frac{r - R}{(R + r)^3} = 0 \Rightarrow R = r \\ \frac{d^2P_e}{dR^2} &= E^2\left[\frac{r - R}{(R + r)^3}\right]'_R = E^2\frac{-(R + r)^3 - 3(R + r)^2(r - R)}{(R + r)^6} \\ &= -2E^2\frac{(2r - R)}{(R + r)^4} \end{aligned} \tag{3.50}$$

When $R = r$, $\frac{d^2P_e}{dR^2} < 0$, the maximum transfer of power takes place from the source to the circuit. The maximum value of the transferred power is:

$$P_{\max} = \frac{E^2}{4r} \tag{3.51}$$

Figure 3.12 shows the dependencies: $I(R)$, $U(R)$, $P(R)$, $P_e(R)$, $P_i(R)$, and $\eta(R)$.

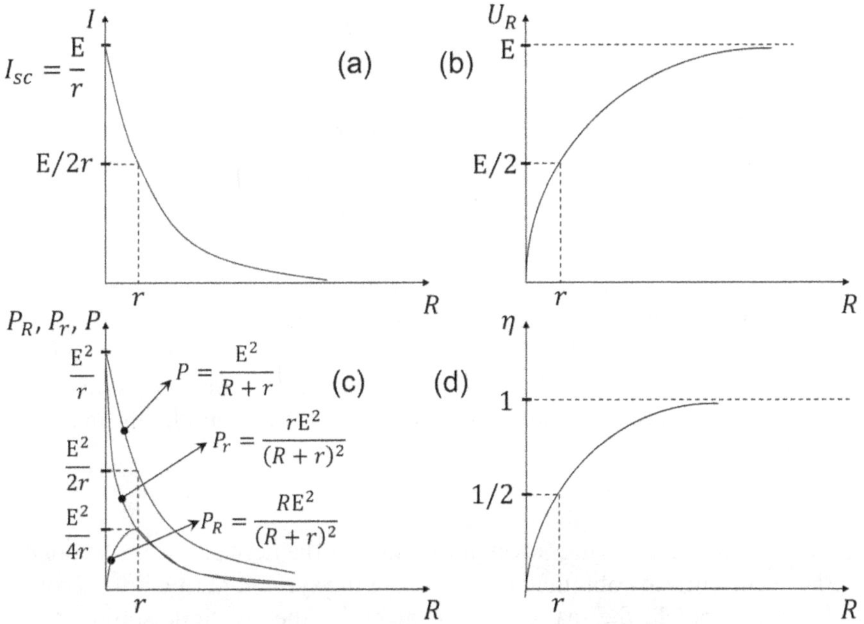

Figure 3.12. Dependencies of various values on the external resistance (R): (a) the current intensity; (b) the voltage drop across R; (c) the power dissipated by R and r as well as the total power consumed by the source; (d) the yield of the transfer.

3.4.2 Methods for solving DC circuits

These methods are based on Kirchhoff's theorems. We only use mathematical approaches and systematizations that simplify computation by introducing auxiliary unknowns or by performing a calculation step by step. Our approach does not require grouping equations into systems with a large number of equations and unknowns (as is the case with Kirchhoff's equations). In addition to these methods, a series of theorems can be stated that solve particular problems. 'Solving' a DC electric circuit means determining the currents flowing through its branches and obtaining (and checking) that its power is balanced.

1. **Cyclic currents method**

In this method, m fictive currents are introduced as unknowns, called cyclic currents, which are considered to be forced to flow through these m fundamental meshes (non-overlapping loops) of the network. The cyclic currents are forced to flow in the fundamental meshes, fulfilling the condition that their algebraic sum in each branch of the network is equal to the real current flowing through it. Using Kirchhoff's first theorem, the real currents through all sides of the network are obtained. This method reduces the system of l equations with l unknowns given by Kirchhoff's theorems to a system of m equations with m unknowns. If the cyclic currents are denoted by $I_{c1},\ I_{c2},\ I_{c3},\ \dots,\ I_{cm}$, the system given by the cyclic current method is:

$$\begin{cases} R_{11}I_{c1} + R_{12}I_{c2} + \dots + R_{1m}I_{cm} = E_{c1} \\ R_{21}I_{c1} + R_{22}I_{c2} + \dots + R_{2m}I_{cm} = E_{c2} \\ \vdots \\ R_{i1}I_{c1} + R_{ii}I_{ci} + R_{ij}I_{cj} + \dots + R_{im}I_{cm} = E_{ci} \\ \vdots \\ R_{m1}I_{c1} + R_{m2}I_{c2} + \dots + R_{mm}I_{cm} = E_{cm} \end{cases} \tag{3.52}$$

in which R_{ii} is the total resistance of loop i of the network; R_{ij} is the total resistance of the common branch of the i and j meshes (it is positive if the neighboring cyclic currents, I_{ci} and I_{cj} flow through it in the same sense, and negative if I_{ci} and I_{cj} flow through it in opposite senses); E_{ci} is the algebraic sum of the electromotive forces present in the fundamental mesh i. The method involves completing the following steps:

- a sense is chosen for each cyclic current;
- write and solve the system (3.52);
- the real currents flowing through the sides of the network are calculated using the cyclic currents obtained from solving the system, taking into account that for each branch, the real current is given by the algebraic sum of the cyclic currents.

Example 1. *Consider the circuit shown in figure* 3.13, *a Wheatstone bridge (a device used to measure electrical resistance) in the DC regime. Find the intensity of the currents flowing through the branches of the network (I_1 to I_6). The resistances R_1 to R_6 and the electromotive force (E) of the battery are known. Supposing that R_5 is a galvanometer (a null indicator), find the equilibrium condition of the device.*

Solution

Defining the cyclic currents I_{c1}, I_{c2}, and I_{c3} as shown in figure 3.13, the system given by the cyclic current method (3.52) becomes:

$$\begin{cases} I_{c1}(R_1 + R_4 + R_5) - I_{c_2}R_5 - I_{c_3}R_4 = 0 \\ -I_{c1}R_5 + I_{c_2}(R_2 + R_3 + R_5) - I_{c_3}R_3 = 0 \\ -I_{c1}R_4 + I_{c_2}R_3 + I_{c_3}(R_3 + R_4 + R_6) = E \end{cases} \tag{3.53}$$

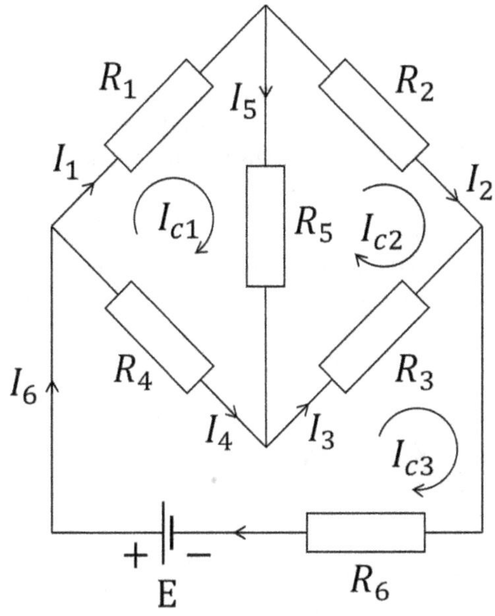

Figure 3.13. *Schematic of a Wheatstone bridge.*

Solving the system gives the cyclic currents; now, the real currents are obtained:

$$I_1 = I_{c1};\ I_2 = I_{c2};\ I_6 = I_{c3} \Rightarrow I_5 = I_{c1} - I_{c2};\ I_4 = I_{c3} - I_{c1};\ I_3 = I_{c3} - I_{c2} \quad (3.54)$$

The equilibrium condition of a Wheatstone bridge occurs when the current passing through the null indicator is zero; in this case, $I_5 = 0 \Rightarrow I_{c1} = I_{c2}$, *leading to*:

$$\begin{cases} I_{c1}(R_1 + R_4) = I_{c3}R_4 \\ I_{c2}(R_2 + R_3) = I_{c3}R_3 \end{cases} \quad (3.55)$$

Taking the ratio of the above equations, the equilibrium condition of the device is obtained as follows:

$$\frac{R_1 + R_4}{R_2 + R_3} = \frac{R_4}{R_3} \Rightarrow R_1R_3 = R_2R_4 \quad (3.56)$$

2. **The superposition method**

The use of this method is recommended when we are interested in finding the current flowing through one side of a network containing a large number of passive elements but a relatively small number of active elements (sources). *The current flowing through each branch of a linear circuit is equal to the algebraic sum of the currents generated in that branch by each source of EMF (E) on its own in the circuit (i.e. if the others were removed):*

$$I_j = \sum_{k=1}^{l} I_{jk} = \sum_{k=1}^{l} G_{jk}E_k \quad (3.57)$$

in which: G_{jk} is the transfer conductance between branches k and j, and I_{jk} is the current flowing through one of the sides (e.g. j), when just the EMF of a single source k is nonzero; the other sources are considered to introduce no EMF but keep their internal resistances. The theorem is a consequence of the linearity of the circuit equations for circuits with constant resistance (independent of current or voltage).

Example 2. *Let us consider the circuit shown in figure 3.14(a), in which the electromotive forces of the sources have the values:* $E_1 = E_3 = 10$ V and $E_2 = 20$ V, *their internal resistances are considered zero, and the resistances have*

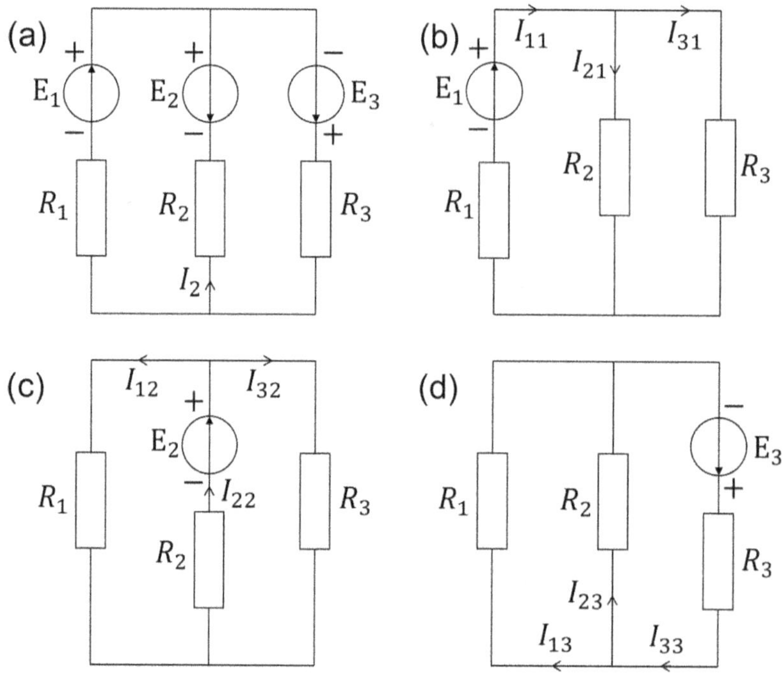

Figure 3.14. *Circuit diagrams illustrating the superposition method.*

the values: $R_1 = R_2 = 5\ \Omega$ and $R_3 = 10\ \Omega$. *Find the currents* I_2 *passing through the resistor* R_2 *using the superposition method.*

Solution

According to this method, we first compute the intensity of the partial currents I_{21}, I_{22}, and I_{23} *separately generated in the resistance* R_2 *by each source in the circuit acting alone. Finally, by taking their algebraic sum, we obtain the current* I_2. *These steps are presented below.*

By keeping just E_1 *in the circuit and removing* E_2 *and* E_3, figure 3.14*(a) can be transformed into* figure 3.14*(b). Analyzing the latter schematic gives the partial current* I_{21}:

$$I_{21} = I_{11} - I_{31} = I_{11} - \frac{I_{21}R_2}{R_3} \Rightarrow I_{21} = \frac{\frac{E_1}{R_1 + \frac{R_2 R_3}{R_2 + R_3}}}{1 + \frac{R_2}{R_3}} = \frac{\frac{10}{5 + \frac{10}{3}}}{\frac{3}{2}} = 0.8 \text{ A} \tag{3.58}$$

Keeping just E_2 in the circuit and removing E_1 and E_3 (see figure 3.14(c)) gives:

$$I_{22} = \frac{E_2}{R_2 + \frac{R_1 R_3}{R_1 + R_3}} = \frac{20}{5 + \frac{10}{3}} = \frac{60}{25} = 2.4 \text{ A} \tag{3.59}$$

Keeping just E_3 in the circuit and removing E_1 and E_2 (see figure 3.14(d)) gives:

$$I_{23} = I_{33} - \frac{I_{23}R_2}{R_1} \Rightarrow I_{23} = \frac{I_{33}}{1 + \frac{R_2}{R_1}} = \frac{\frac{E_3}{R_3 + \frac{R_1 R_2}{R_1 + R_2}}}{1 + \frac{R_2}{R_1}} = \frac{\frac{10}{10 + \frac{5}{2}}}{2} = 0.4 \text{ A} \tag{3.60}$$

The current I_2 passing through R_2, with the sense shown in figure 3.14(a), is the algebraic sum of the partial currents:

$$I_2 = -I_{21} + I_{22} + I_{23} = 2 \text{ A} \tag{3.61}$$

3. **The node potential method**

This method transforms the system of equations given by Kirchhoff's theorems (where the currents flowing through the sides of the network are unknown) from a system with l equations and l unknowns into a system with $n - 1$ equations where the potentials of the network nodes are unknown. The method is recommended when $n - 1 < m$ (the number of fundamental meshes). Considering a side of the network placed between nodes α and β, from a network with n nodes and l sides (see figure 3.15) and using the generalized form of Ohm's law, $\int_\alpha^\beta \vec{j}\,\rho \overrightarrow{dl} = \int_\alpha^\beta I\rho\frac{dl}{S} = \int_\alpha^\beta \overrightarrow{E_c}\overrightarrow{dl} + \int_\alpha^\beta \overrightarrow{E_i}\overrightarrow{dl}$, we obtain:

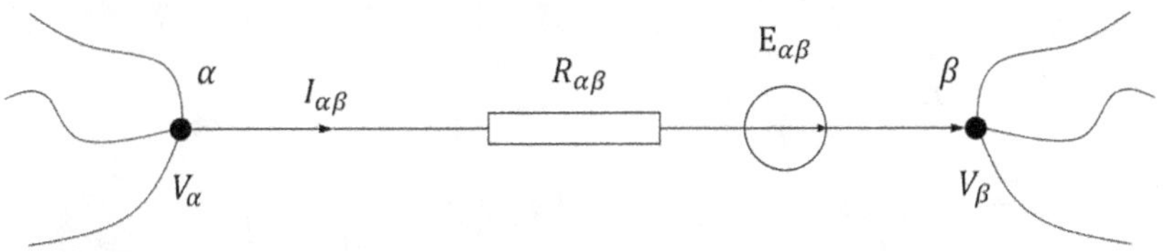

Figure 3.15. The network branch used to illustrate the node potential method.

$$I_{\alpha\beta}R_{\alpha\beta} = U_{\alpha\beta} + E_{\alpha\beta} \Longrightarrow I_{\alpha\beta} = \frac{V_\alpha - V_\beta}{R_{\alpha\beta}} + \frac{E_{\alpha\beta}}{R_{\alpha\beta}} \tag{3.62}$$

If the conductance of the branch $\alpha\beta$ is $\frac{1}{R_{\alpha\beta}} = G_{\alpha\beta}$, the intensity of the current flowing through it is:

$$I_{\alpha\beta} = V_{\alpha\beta}G_{\alpha\beta} - V_\beta G_{\alpha\beta} + E_{\alpha\beta}G_{\alpha\beta} \tag{3.63}$$

Applying Kirchhoff's first theorem to the node α, i.e. $\sum_{\beta} I_{\alpha\beta} = 0$ for each node, we obtain the equation:

$$V_\alpha \sum_{\beta} G_{\alpha\beta} - \sum_{\beta} V_\beta G_{\alpha\beta} + \sum_{\beta} E_{\alpha\beta} G_{\alpha\beta} = 0 \tag{3.64}$$

in which: $V_\alpha \sum_{\beta} G_{\alpha\beta}$ is the product of the node potential α and the sum of the transfer conductances of node α and all other nodes of the network; $\sum_{\beta} V_\beta G_{\alpha\beta}$ is the sum of the products of the potential of another node β that is connected to α and the transfer conductance between nodes α and β; $\sum_{\beta} E_{\alpha\beta} G_{\alpha\beta}$ is the algebraic sum of the products of the EMF and the conductance of each branch between nodes α and β. In short, we choose a zero potential as the reference potential for a particular node in the network and thus write $n - 1$ equations of the form (3.64). From these, the potentials V_α are determined; then, using the potentials, the actual currents flowing through the branches of the network are calculated.

Example 3. *Consider the circuit shown in figure 3.16, in which the electromotive forces of the sources have the values:* $E_1 = 10$ *V* and $E_3 = 30$ V, *and the resistances have the values* $R_1 = R_2 = R_3 = R_4 = R_5 = 1\ \Omega$. *Find the currents* ($I_1$ *to* I_5) *that flow through the branches of the network using the node potential method.*

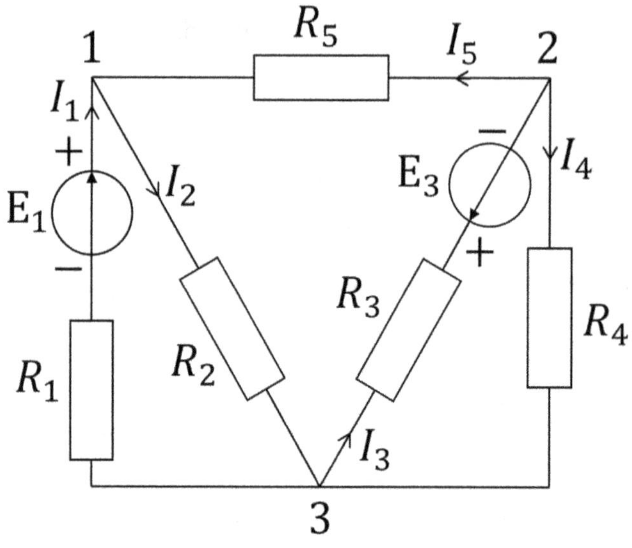

Figure 3.16. *Schematic of a circuit illustrating the node potential method.*

Solution

Let us assume that node 3 *is the reference point and is kept at a potential of* $V_3 = 0$. *According to the node potential method, for nodes* 1 *and* 2 *that have potentials of* V_1 *and* V_2, *respectively, we can write the following set of equations:*

$$\begin{cases} V_1\left(\dfrac{1}{R_1} + \dfrac{1}{R_2} + \dfrac{1}{R_3}\right) - \dfrac{V_2}{R_5} - E_1 R_1 = 0 \\ V_2\left(\dfrac{1}{R_3} + \dfrac{1}{R_4} + \dfrac{1}{R_5}\right) - \dfrac{V_1}{R_5} + E_3 R_3 = 0 \end{cases} \tag{3.65}$$

Upon replacing the values of the EMF and the resistances and solving the system, we obtain the potentials V_1, V_2, *and* V_3:

$$\begin{cases} 3V_1 - V_2 - 10 = 0 \\ 3V_2 - V_1 + 30 = 0 \end{cases} \Longrightarrow V_1 = 0;\ V_2 = -10V;\ V_3 = 0 \tag{3.66}$$

Using equation (3.62), *we can now obtain the currents passing through the sides of the network:*

$$\begin{cases} I_1 = \dfrac{V_3 - V_1 + E_1}{R_1} = 10\text{ A} \\ I_2 = \dfrac{V_1 - V_3}{R_2} = 0 \\ I_3 = \dfrac{V_3 - V_2 - E_3}{R_3} = \dfrac{10 - 30}{1} = -20\text{ A} \\ I_4 = \dfrac{V_2 - V_3}{R_4} = -10\text{ A} \\ I_5 = \dfrac{V_2 - V_1}{R_5} = -10\text{ A} \end{cases} \tag{3.67}$$

4. **Method of equivalence with a voltage generator (Thévenin's theorem)**
This method makes it possible to determine the current I flowing through a passive circuit element connected between nodes A and B of a complex network (see figure 3.17(a)) without the need to calculate the currents in all branches of the network. The complex network, with the exception of side AB, is equivalent to a voltage generator; its EMF U_{AB_0} is equal to the voltage between the two terminals A and B under open-circuit conditions, and its internal resistance R_{AB_0} is equal to the equivalent resistance nodes A and B of the passivated complex network (a network in which the EMFs are removed but the internal resistances remain). According to Thévenin's theorem, the current through resistor R is equal to the ratio between U_{AB_0} and the sum of the resistors R and R_{AB_0}:

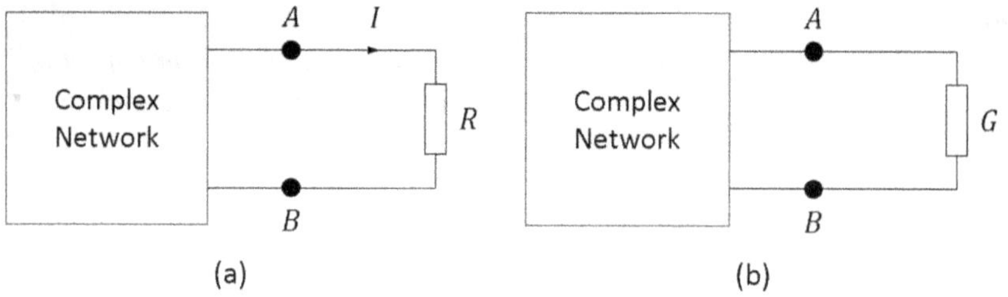

Figure 3.17. (a) Complex network with a passive element of resistance R connected between nodes A and B used to illustrate Thévenin's theorem; (b) Complex network with a passive element of conductance G connected between nodes A and B used to illustrate Norton's theorem.

$$I = \frac{U_{AB_0}}{R + R_{AB_0}} \tag{3.68}$$

Example 4. *Consider the circuit shown in figure 3.18(a), in which the electromotive forces of the sources have the values* $E_1 = E_3 = 10$ V and $E_2 = 20$ V, *their internal*

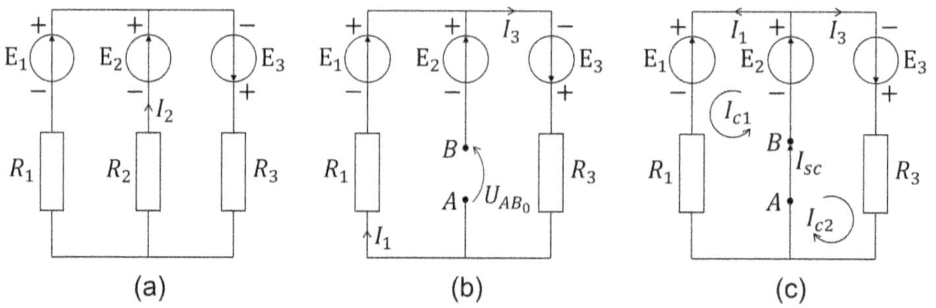

Figure 3.18. *Circuit schematics used to illustrate Thévenin's theorem and Norton's theorem.*

resistances are considered zero, and the resistances have the values $R_1 = R_2 = 5\ \Omega$ and $R_3 = 10\ \Omega$. *Find the current* I_2 *flowing through the resistor* R_2 *using Thévenin's theorem.*

Solution

According to Thévenin's method, we must first compute the open-circuit voltage indicated in figure 3.18(b):

$$U_{AB_0} = IR_1 - E_1 + E_2 = \frac{E_1 + E_3}{R_1 + R_3}R_1 + (E_2 - E_1) = \frac{10 + 10}{15}5 + 10 = \frac{50}{3}\ \text{V} \tag{3.69}$$

According to the scheme shown in figure 3.18(b), the equivalent resistance between nodes A and B under open-circuit conditions is:

$$R_{AB_0} = \frac{R_1 R_3}{R_1 + R_3} = \frac{50}{15} = \frac{10}{3}\ \Omega \tag{3.70}$$

Therefore, the current through side 2 is:

$$I_2 = \frac{U_{AB_0}}{R_2 + R_{AB_0}} = \frac{\frac{50}{3}}{5 + \frac{10}{3}} = \frac{50}{25} = 2\ \text{A} \tag{3.71}$$

5. **Method of equivalence with a current generator (Norton's theorem)**
A branched electrical circuit (complex network) is equivalent to a *current generator* that generates an electric current in one side of the circuit, considered to be a receiver. If, in figure 3.17(b), we note that $U_{AB} = \text{IR}$, then according to equation (3.68), we obtain:

$$U_{AB} = \frac{RU_{AB_0}}{R + R_{AB_0}} = \frac{U_{AB_0}}{1 + \frac{R_{AB_0}}{R}} \tag{3.72}$$

Introducing the external conductance $G = \frac{1}{R}$, the internal conductance $G_{AB_0} = \frac{1}{R_{AB_0}}$, and relationship (3.69), we can also write:

$$U_{AB} = \frac{U_{AB_0}}{1 + \frac{G}{G_{AB_0}}} = \frac{G_{AB_0} U_{AB_0}}{G + G_{AB_0}} = \frac{\frac{U_{AB_0}}{R_{AB_0}}}{G + G_{AB_0}} = \frac{I_{sc}}{G + G_{AB_0}} \tag{3.73}$$

Thus, Norton's theorem, in mathematical form, is:

$$U_{AB} = \frac{I_{sc}}{G + G_{AB_0}} \tag{3.74}$$

This describes the voltage across a passive element of conductance G in a complex network. The voltage across a passive element with conductance G equals the ratio of the short-circuit current between its terminals (A and B) and the sum of two conductances. These conductances are the element's own conductance G and the equivalent conductance G_{AB_0} *of the passivated network between terminals A and B.*

Example 5. *Consider the circuit shown in figure* 3.18*(a), in which the electromotive forces of the sources have the values* $E_1 = E_3 = 10$ V and $E_2 = 20$ V, *their internal resistances are considered zero, and the external resistances have the values* $R_1 = R_2 = 5\ \Omega$ and $R_3 = 10\ \Omega$. *Find the current* I_2 *passing through resistor* R_2 *using Norton's theorem.*

Solution
According to Norton's method, we must first compute the short-circuit current indicated in figure 3.18(c):

$$I_{sc} = I_{c1} + I_{c2} \tag{3.75}$$

According to the cyclic current method:

$$\begin{cases} I_{c1}R_1 = E_2 - E_1 \\ I_{c2}R_3 = E_3 + E_2 \end{cases} \Rightarrow I_{c1} = 2\text{ A};\ I_{c2} = 3\text{ A};\ I_{sc} = 5\text{ A} \tag{3.76}$$

According to (3.74), we obtain:

$$U_{AB} = \frac{I_{sc}}{G_2 + G_{AB_0}} = \frac{I_{sc}}{\frac{1}{R_2} + \frac{1}{R_{AB_0}}} = \frac{5}{\frac{1}{5} + \frac{3}{10}} = \frac{50}{5} = 10\text{ V} \tag{3.77}$$

We can now obtain the current passing through R_2, as follows:

$$I_2 = \frac{U_{AB}}{R_2} = \frac{10}{5} = 2\text{ A} \tag{3.78}$$

6. **'Delta-to-star' ('Δ − Y') conversions (Kennelly's method)**
A portion of a network in which three branches of the network are connected in a 'Y'-like configuration can be converted into a 'Δ'-like configuration and vice versa; see figure 3.19. If R_1, R_2, and R_3 are the resistances on the branches of the network in

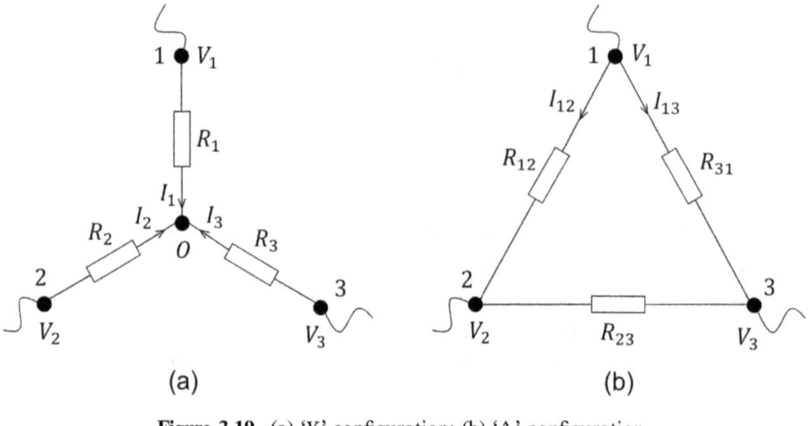

Figure 3.19. (a) 'Y' configuration; (b) 'Δ' configuration.

'Y' configuration and R_{12}, R_{23}, and R_{31} are those on the branches of the network in the 'Δ' configuration, the conversion relations are obtained by imposing the condition that after the transformation, the rest of the network must not have changed compared to its state before the transformation. This means that the potentials of points 1, 2, and 3 must remain the same after the conversion, and the net current from the rest of the network absorbed by each node must remain the same after the conversion has occurred. Taking these requirements into account, as an example, we can calculate the current absorbed through node 1 from the rest of the network in the 'Y' configuration and the 'Δ' configuration. Having equated these currents, we can obtain the relationships that give the resistances on the sides of the 'Δ' configuration as a function of the resistances on the sides of the 'Y' configuration.

In the 'Y' configuration (see figure 3.19(a)), applying first Kirchhoff's theorem to node O, we obtain:

$$I_1 + I_2 + I_3 = 0 \Rightarrow \frac{V_1}{R_1} + \frac{V_2}{R_2} + \frac{V_3}{R_3} = V_O\left(\frac{1}{R_1} + \frac{1}{R_2} + \frac{1}{R_3}\right) \Rightarrow$$

$$\Rightarrow V_O = \frac{\frac{V_1}{R_1} + \frac{V_2}{R_2} + \frac{V_3}{R_3}}{\frac{1}{R_1} + \frac{1}{R_2} + \frac{1}{R_3}} \tag{3.79}$$

Using this result, we can obtain the current intensity I_1 which flows from the rest of the network through resistor R_1 in the 'Y' configuration (see figure 3.19(a)):

$$I_1 = \frac{V_1 - V_O}{R_1} = \frac{V_1 - V_2}{R_1 + R_2 + \frac{R_1 R_2}{R_3}} + \frac{V_1 - V_3}{R_1 + R_3 + \frac{R_1 R_3}{R_2}} \tag{3.80}$$

In the 'Δ' configuration (see figure 3.19(b)), the intensity of the current entering node 1 from the rest of the network is:

$$I_1 = \frac{V_1 - V_2}{R_{12}} + \frac{V_1 - V_3}{R_{13}} \tag{3.81}$$

By equating (3.80) and (3.81), the resistances on the sides of the 'Δ' configuration can be obtained as a function of the resistances of the branches of the 'Y' configuration, as follows:

$$\begin{cases} R_{12} = R_1 + R_2 + \dfrac{R_1 R_2}{R_3} \\ R_{23} = R_2 + R_3 + \dfrac{R_2 R_3}{R_1} \\ R_{31} = R_3 + R_1 + \dfrac{R_3 R_1}{R_2} \end{cases} \tag{3.82}$$

Solving this reverse system (whose unknowns are the resistances R_1, R_2, and R_3 on the network sides in the 'Y' configuration, and knowing the resistances R_{12}, R_{23}, and R_{31}, on the network sides in the 'Δ' configuration), we obtain:

$$\begin{cases} R_1 = \dfrac{R_{12}R_{13}}{R_{12} + R_{23} + R_{31}} \\ R_2 = \dfrac{R_{23}R_{12}}{R_{12} + R_{23} + R_{31}} \\ R_3 = \dfrac{R_{31}R_{23}}{R_{12} + R_{23} + R_{31}} \end{cases} \tag{3.83}$$

References and further reading

[1] Antohe Ş and Antohe V-A 2023 *Electrostatics. Formalism of the Electrostatic Field in Vacuum and Matter* 1st edn (Bristol: IOP Publishing) 266 pages

[2] Barr J R 2023 *Principles of Direct-Current Electrical Engineering* (London: Palala Press) 592 pages

[3] Nasar S 1988 *Solved Problems in Electrical Circuits* 1st edn (New York: McGraw Hill) 768 pages

[4] Davis C 2016 *DC Circuits* 1st edn (Norman, OK: University of Oklahoma Libraries) 137 pages

[5] Fiore J M 2020 *DC Electrical Circuit Analysis: A Practical Approach* (New York: Independently published) 374 pages

[6] Fiore J M 2016 *DC Electrical Circuits: Laboratory Manual* (New York: Independently published) 90 pages

[7] Fiore J M 2017 *DC Electrical Circuits: Workbook* (New York: Independently published) 128 pages

[8] O'Malley J 2011 *Schaum's Outline of Basic Circuit Analysis* 2nd edn (New York: McGraw Hill) 432 pages

[9] Feynman R P, Leighton R B and Sands M 2011 *Feynman Lectures on Physics, Vol II: The New Millennium Edition: Mainly Electromagnetism and Matter* (New York: Basic Books) (Feynman Lectures on Physics) 589 pages

[10] Purcell E M and Morin D J 2013 *Electricity and Magnetism* 3rd edn (Cambridge: Cambridge University Press) 853 pages

[11] Breithaupt J 2002 *Teach Yourself Physics* 1st edn (New York: McGraw-Hill) 240 pages

IOP Publishing

Ştefan Antohe and Vlad-Andrei Antohe

Chapter 4

Magnetic field of the electrokinetic state

Magnetostatics deals with the study of magnetic fields generated by direct current (DC) established in different circuits. We will first discuss the magnetic field established around a DC current circuit in vacuum; then, the study of magnetic fields in any medium will be the subject of chapter 7, describing the magnetic properties of various substances.

4.1 Magnetic fields. Definition and characterization

4.1.1 Historical experiments. Magnetic forces

Since antiquity, it has been found that there are rocks between which forces of attraction or rejection are exerted, forces that depend on the relative position between them. The first such rock, discovered in a region of ancient Greece called *Magnesia*, gave rise to the phenomenon called ***magnetism***. Later on, in 1820, Oersted[1] observed that a conductor carrying an electric current acts on a permanent magnetic compass needle (see figure 4.1(a)), and based on the action that magnetite rock exerted on the compass needle, it was concluded that electric current creates a field similar to that created by magnetite rocks. Less than two weeks after Oersted's discovery, Ampère[2] showed that forces of attraction or rejection are exerted between

[1] **Hans C Oersted** (August 14, 1777–March 9, 1851), also known as **Ørsted,** was a Danish physicist and chemist who made the first connection between electricity and magnetism. Oersted's law and the *oersted* measuring unit (*Oe*) are named after him.

[2] **André–Marie Ampère** (January 20, 1775–June 10, 1836) was a French physicist and mathematician who was one of the founders of the science of classical electromagnetism, which he referred to as *electrodynamics*. The International System (SI) measuring unit for the intensity of the electric current (the *ampere—A*) is named after him.

doi:10.1088/978-0-7503-5854-5ch4

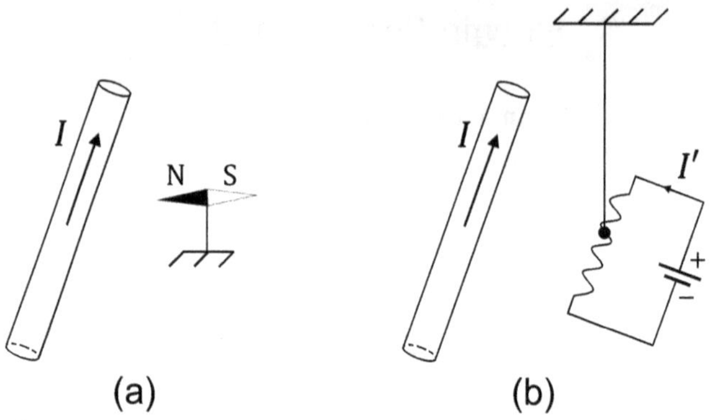

Figure 4.1. (a) The interaction between a conductor carrying a current and a magnetic needle. (b) The interaction between a conductor carrying electric current and a coil carrying electric current.

two conductors carrying electric currents if the two current flows have the same direction and sense or if they have opposite senses, respectively. He also noted that these forces are of a completely different nature than those exerted between two conductors that are electrically charged and have a linear charge density of λ. A first proof of the latter remark is the fact that a metal screen interposed between the two conductors does not change the force of interaction between them, as it would if the conductors were charged with electric charge. This observation led him to assume that this force must be of the same nature as the force exerted between a conductor carrying a current and the magnetic needle, so it must be a magnetic force. However, it is difficult to find the correspondence between the magnetic needle and a conductor carrying electric current. It was Ampère who found this correspondence. He assumed that the magnetic needle contained a very large number of microscopic circular currents, which he called *molecular currents*. Thus, in Oersted's experiment, if the magnetic needle was removed and replaced by a coil (see figure 4.1(b)), it rotated around the suspension wire like the magnetic needle, with the direction of rotation depending on the direction and sense of the electric current through both the coil and the conductor. It was therefore necessary to establish the role of conductors in the production of these forces.

In this context, the following questions arise: *'Are these forces due to the conductor's property of possessing quasi-free electrons that move chaotically in the conductors?'* or *'Are they due to the passage of electric current through the conductors?'*, that is, to the presence of electric charges in motion at a nonzero drift velocity. It has been found that this process of oriented movement of the charge carriers is the cause of the presence of magnetic forces. Thus, such magnetic forces are exerted between any two electric charges that move, as Maxwell later stated. Experiments with cathode rays have shown that in the magnetic field of a permanent magnet or in the magnetic field created by a conductor carrying a current, there are magnetic forces acting on the moving electric charges. The beam of electrons in a cathode ray tube, for example, is deflected if a conductor carrying current is brought nearby, and the direction of deflection depends on the sense of the flow of electric current. These magnetic forces that are exerted between moving electric charges can be attributed to the existence around them of a physical field called a ***'magnetic field'***. Consequently, an electric current can be associated with a magnetic field that manifests itself around it. Any other electrically charged particle that moves around a conductor carrying an electric current experiences a force proportional to the intensity of the magnetic field in that place. It has been experimentally verified that this force is always perpendicular to the speed of the charge's movement and to the magnetic field, so it can be stated that the magnetic field ($\vec{B}$) is a vector field that determines a force acting on a charged particle in motion, a force proportional to the speed of the particle and perpendicular to its direction of movement.

Next, we provide a thorough characterization of the magnetic field created by DC current, and on this occasion, we formulate the basic laws of magnetostatics. Typically, such studies presume, as in the case of electrostatics, the specification of a study model and a *'probe body'*.

4.1.2 The current loop

The current loop used to explore magnetic fields is a very thin circular metallic wire or a bundle of circular wires (a small coil) carrying an electric current. The wires connecting the current loop to a DC source are twisted (see figure 4.2) so that they do not perturb the mechanical actions induced by the magnetic field into which the current loop is introduced. In practice, such a current loop is often called an *exploration coil.* Under these conditions, the ponderomotive actions exerted by the

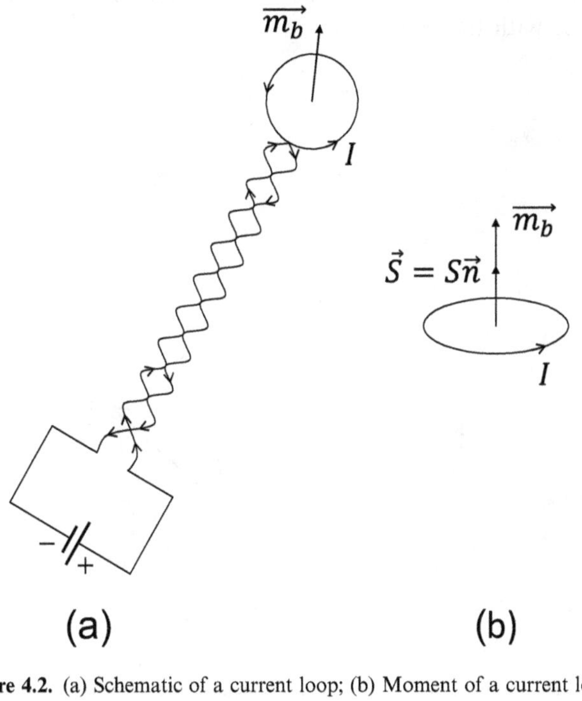

Figure 4.2. (a) Schematic of a current loop; (b) Moment of a current loop.

magnetic field into which it is inserted act only on the loop and not on the connecting conductors. The current loop is characterized by a vector quantity, $\overrightarrow{m_b}$, called *the magnetic moment of the loop*, defined by the relation:

$$\overrightarrow{m}_b = IS\vec{n} = I\vec{S} \tag{4.1}$$

where I is the intensity of the current that flows through the loop and $\vec{S} = S\vec{n}$ is the area of the surface bounded by the loop, $\vec{n}$ being the normal at the surface of the loop. The normal at the surface of the loop has an orientation related to the direction of the electric current flowing through it. The orientation of the vector $\vec{n}$ is determined by the ***right-handed screw rule***, i.e. *it is given by the direction of advance of a screw placed perpendicular to the plane of the loop when it is rotated in the direction of the current passing through the loop.*

From relation (4.1), it can be seen that if $S \to 0$, for the expression of the moment to be finite, the current intensity must tend to infinity ($I \to \infty$). If the curve Γ forming the contour of the loop is flat, the surface vector has as its magnitude the area of the plane curve. If the curve is not flat, as shown in figure 4.3, the area vector is defined by the relation:

$$\vec{S} = \frac{1}{2}\oint_{\Gamma} \vec{r} \times \vec{dl} \tag{4.2}$$

in which $\vec{r}$ is the positional vector of an element of length $\vec{dl}$ lying on the contour relative to the origin O of the chosen reference system. Relation (4.2) is inferred by the integration along the Γ contour of the $\vec{ds}$ elementary areas that form the area of the conical surface with the peak in O:

$$\vec{ds} = \frac{1}{2}\vec{r} \times \vec{dl} \tag{4.3}$$

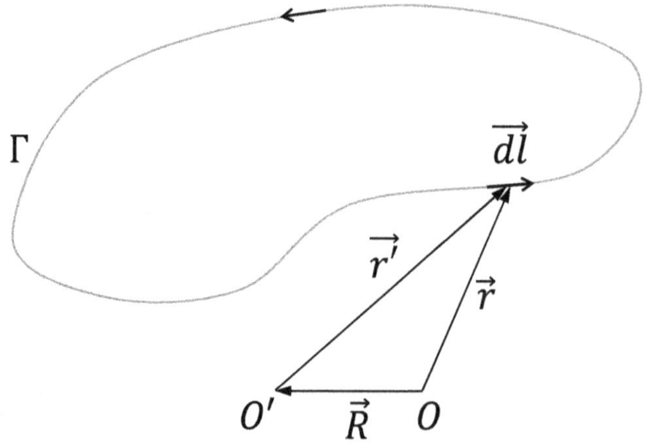

Figure 4.3. The moment of the loop is independent of the shape of the loop's surface.

The vector $\vec{S}$ does not depend on the chosen origin. For the demonstration of this property, another origin O' is considered, against which the area vector is assumed to have a different value $\vec{S'}$, which results in:

$$\vec{S'} = \frac{1}{2}\oint_{\Gamma} \vec{r'} \times \vec{dl} = \frac{1}{2}\oint_{\Gamma} (\vec{r} - \vec{R}) \times \vec{dl} = \frac{1}{2}\oint_{\Gamma} \vec{r} \times \vec{dl} - \frac{1}{2}\oint_{\Gamma} \vec{R} \times \vec{dl}$$
$$= \frac{1}{2}\oint_{\Gamma} \vec{r} \times \vec{dl} \tag{4.4}$$

In equation (4.4), $\frac{1}{2}\oint_{\Gamma} \vec{R} \times \vec{dl} = 0$. The reason for this is that since $\vec{R}$ (the distance between the two origins O and O′) is a constant size, when $\vec{dl}$ travels through the whole loop, the vector product $\vec{R} \times \vec{dl}$ changes sign such that $\oint_{\Gamma} \vec{R} \times \vec{dl} = 0$. If the magnitude of the area vector $\vec{S}$ of the loop is independent of the shape of the area bounded by the loop, its moment $\overrightarrow{m_b}$ retains the same property.

4.1.3 Magnetic induction in vacuum

4.1.3.1 The torque exerted on a current loop in a magnetic field

It has been found that when a current loop is brought into a magnetic field, ponderomotive actions (forces and torque of forces) are exerted on it from the side of the field. An introduction to the characteristic quantities of magnetic fields (intensity $\vec{H}$ or magnetic induction $\vec{B}$) involves, first of all, knowledge about the ponderomotive actions exerted by the respective field in its own source (here, a conductor carrying current or a magnetic moment associated with a permanent magnet or a current loop). The static moment (the torque moment) of the forces exerted on the current loop in relation to its center of mass is denoted by $\overrightarrow{M_c}$, and experimentally, the following observations can be made:

(1) At each point in the field, there is a preferential direction, characterized in figure 4.4 through the vector $\overrightarrow{e_B}(\vec{r})$, such that if the magnetic moment $\overrightarrow{m_b}$ is oriented in this direction, the torque $\overrightarrow{M_c}$ is zero.

(2) In some position of the loop, for which $\overrightarrow{m_b}$ makes the angle α with $\overrightarrow{e_B}(\vec{r})$, the torque tends to bring the magnetic moment of the loop ($\overrightarrow{m_b}$) after its orientation $\overrightarrow{e_B}(\vec{r})$ and is proportional to $\sin\alpha$, having as a proportionality factor the maximum value of the torque. So, we can write:

$$M_c = M_{C_{max}} \sin\alpha \Longleftrightarrow \overrightarrow{M_c} = M_{C_{max}} \vec{n} \times \overrightarrow{e_B} \tag{4.5}$$

(3) The maximum torque $M_{C_{max}}$ is proportional to the current intensity I and to the area of the loop S; otherwise, it depends on the point at which the loop is placed in the field by $\vec{B}(\vec{r})$:

$$M_{C_{max}} = ISB(\vec{r}) \tag{4.6}$$

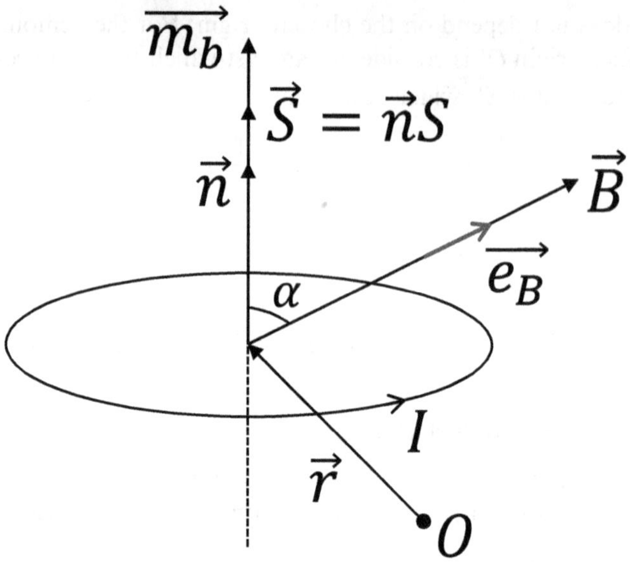

Figure 4.4. The torque moment exerted by the magnetic field onto a current loop.

The three previous experimental observations allow us to obtain an expression for the torque:

$$\overrightarrow{M_c} = IS\vec{n} \times B(\vec{r}) \cdot \overrightarrow{e_B} \tag{4.7}$$

in which it can be seen that the direction $\overrightarrow{e_B}(\vec{r})$ depends only on the point at which the loop is placed and not on the current or the surface area of the loop. Or, taking into account the definition of the moment of the loop, the torque can also be written as:

$$\overrightarrow{M_c} = \overrightarrow{m_b} \times \vec{B}(\vec{r}) \tag{4.8}$$

in which $\vec{B}(\vec{r}) = B(\vec{r})\overrightarrow{e_B}(\vec{r})$ is a quantity that characterizes the magnetic field at each point. The torque $(\overrightarrow{M_c})$ exerted on a current loop in vacuum is determined by the magnetic field into which it is inserted. This torque is equal to the vector product of two factors: the magnetic moment of the loop $\overrightarrow{m_b}$ and a vector quantity representing the state of the field $\vec{B}(\vec{r})$, called the *magnetic field induction* or *magnetic flux density*. Magnetic induction in vacuum characterizes the local state of the magnetic field in vacuum. The torque tends to rotate the loop in such a way that its magnetic moment becomes parallel to the induction of the magnetic field. Based on this experiment, magnetic field induction $(\vec{B}(\vec{r}))$ was introduced as a vector quantity that characterizes the local state of the field and is responsible for the presence of the magnetic torque exerted on a current loop introduced at that point in the field.

4.1.3.2 The force exerted by the magnetic field on a conductor carrying an electric current (electromagnetic force or Laplace force)

In order to introduce the magnetic induction produced by a magnetic field, we can also use other ponderomotive actions with which the magnetic field acts on the electric current, such as some specific forces. By experimentally measuring $\overrightarrow{dF}$, the force exerted on an element of length $\overrightarrow{dl}$ that is part of a circuit carrying a current of intensity I and placed in an external magnetic field of induction $\vec{B}$ (see figure 4.5), it is found that it can be written as:

$$\overrightarrow{dF} = I\overrightarrow{dl} \times \vec{B}(\vec{r}) \tag{4.9}$$

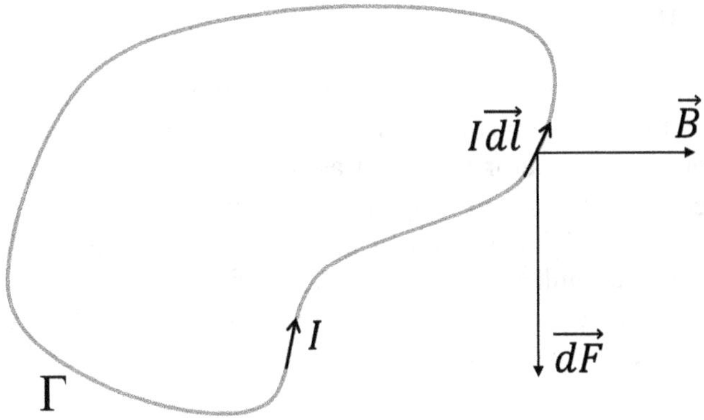

Figure 4.5. Electromagnetic force (Laplace).

This expression can be demonstrated with the help of the relationship (4.7) of the torque's moment. It will be used in the particular case of a rectangular frame of dimensions a and $\overrightarrow{dl}$ (see figure 4.6) carrying a current of intensity I and placed in a uniform magnetic field of induction $\vec{B}$. According to equation (4.9), the forces exerted on the sides of the frame are perpendicular to the plane formed by $\overrightarrow{dl}$ and $\vec{B}$. The frame can rotate around the dotted OO' axis. Consequently, only the parallel

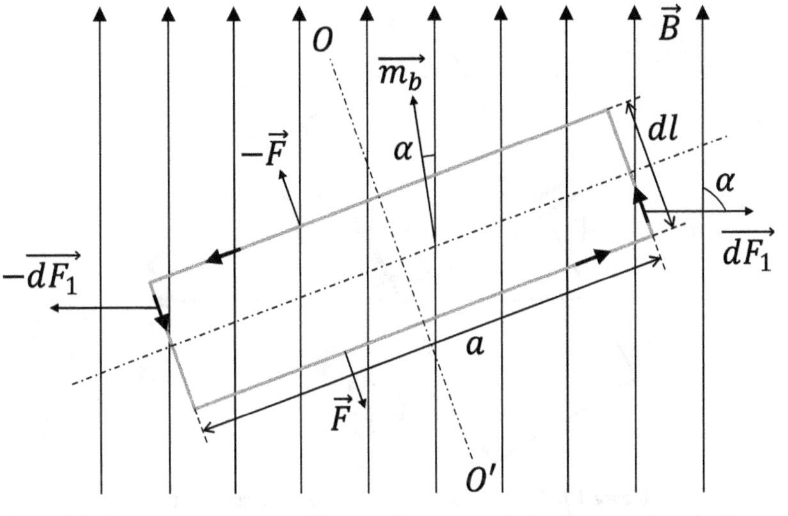

Figure 4.6. Torque moment exerted by a uniform magnetic field on a rectangular frame.

forces that are exerted on the long sides $\overrightarrow{dl}$ of the frame produce nonzero moments, because those that are exerted on the sides of length a are parallel to the rotation's axis. The torque moment is:

$$\overrightarrow{M_c} = \vec{a} \times \overrightarrow{dF_1} \tag{4.10}$$

Substituting the force $\overrightarrow{dF_1}$ from (4.9) into (4.10) results in:

$$\overrightarrow{M_c} = \vec{a} \times I\overrightarrow{dl} \times B = Iadl \cdot \vec{n} \times \vec{B} = \overrightarrow{m_b} \times \vec{B} \tag{4.11}$$

$\vec{a} \times \overrightarrow{dl} = adl \cdot \vec{n}$ because $\vec{a} \perp \overrightarrow{dl}$ is oriented in the normal direction to the plane of the wire; that is, it has the same orientation as $\overrightarrow{m_b}$. Assuming that $\overrightarrow{dF} = I\overrightarrow{dl} \times \vec{B}$, the force exerted on the conductor is correct as defined, since we find that the torque moment exerted on this current loop has the same expression $\overrightarrow{m_b} \times \vec{B}$ that was established as a result of the experiment described above. Any current loop can be decomposed into rectangular loops like the one in figure 4.6. As a result, we can say that the expression for the electromagnetic force with which the magnetic field acts on a conductor carrying a current of intensity I, placed in the field, is given by the Laplace force:

$$\vec{F} = \int I(\overrightarrow{dl} \times \vec{B}) \tag{4.12}$$

In the particular case of a linear conductor of length $\vec{l}$ carrying a current of intensity I, placed in a uniform magnetic field of induction $\vec{B}$ (see figure 4.7), the Laplace force takes the form:

$$\vec{F} = I(\vec{l} \times \vec{B}) \tag{4.13}$$

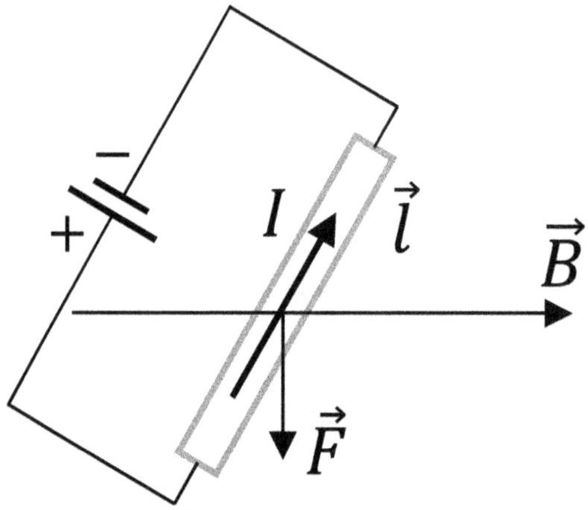

Figure 4.7. Laplace force in the case of a linear conductor.

Its magnitude is $F = IlB \sin\alpha$, its direction is $\vec{F} \perp (\vec{l}, \vec{B})$ and its sense is given by the right-hand screw rule (RSR). If $\vec{l} \perp \vec{B}$, the force F has a maximum value of:

$$F_{\max} = IlB \Longrightarrow B = \frac{F_{\max}}{Il} \tag{4.14}$$

On the basis of the electromagnetic force, *the magnetic induction or magnetic flux density ($\vec{B}$) can be defined as the vector quantity numerically equal to the force with which the respective magnetic field acts on a unit length of a conductor carrying a unit current when the conductor is placed perpendicular to the lines of the magnetic field.* The SI unit of measurement for the magnetic flux, is the Weber (Wb),[3] and the unit for the magnetic flux density is Wb m^{-2} in this system. However, the measuring unit of the magnetic flux density (magnetic induction, i.e. $\vec{B}$) is often given in the MKSA[4] system as *tesla* (T), or in the Gaussian system of units as *gauss* (Gs), the relation between them being:

$$\langle B \rangle = 1\ \frac{\text{Wb}}{\text{m}^2} = 1\ \text{T} = 10^4\ \text{Gs} \tag{4.15}$$

A magnetic field of 1 T is very intense. In order to understand this, the following examples are worth remembering: (i) the horizontal component of the Earth's magnetic field is 0.24 Gs $= 2.4 \times 10^{-5}$ T; (ii) the magnetic field between the poles of an electromagnet reaches 1–2 T, when a current of 50–100 A passes through its winding; and (iii) very intense magnetic fields (7–8 T) can be obtained with the help of special electromagnets, such as those constructed using superconductors. One *tesla* (T) denotes an induced magnetic field that exerts a torque of 1 Nm on a current loop with a moment of 1 A m^2. Nevertheless, the unit of Wb m^{-2} is also used for $\vec{B}$, which we will define later using the magnetic flux.

4.1.3.3 The density of the electromagnetic force

In the case of massive conductors, the density of the electromagnetic force comes into play. To introduce this topic, we consider a tube of current lines of intensity I placed in a magnetic field of induction $\vec{B}$. If the elementary volume in the tube $dV = dldS$ and an electromagnetic force of $\overrightarrow{dF}$ is exerted by the magnetic field on this element of volume by the conducting medium (see figure 4.8) we can write:

$$\overrightarrow{dF} = I(\overrightarrow{dl} \times \vec{B}) = \vec{j}\,\overrightarrow{ds}(\overrightarrow{dl} \times \vec{B}) = dldS(\vec{j} \times \vec{B}) = dV(\vec{j} \times \vec{B}) \tag{4.16}$$

[3] The *Weber* (Wb) is derived from the relationship 1 Wb = 1 V · s. It was named after the German physicist Wilhelm E Weber (October 24, 1804–June 23, 1891).

[4] The rationalized meter–kilogram–second (RMKS) system combines the MKS system with the rationalization of electromagnetic equations. The MKS units combined with the ampere as a fourth base unit are referred to as the MKSA system.

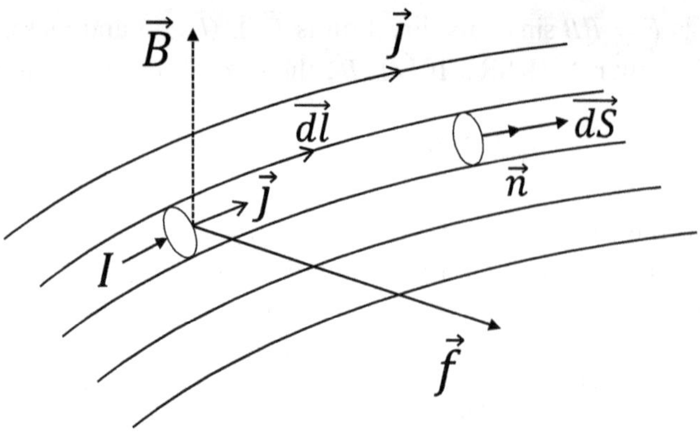

Figure 4.8. The density of the electromagnetic force.

As a result, the density of the electromagnetic force is:

$$\vec{f} = \lim_{dV \to 0} \frac{\overrightarrow{dF}}{dV} = \vec{j} \times \vec{B} \tag{4.17}$$

So, on a massive conductor carrying the conduction current I, a force is exerted whose volume density is equal to the vector product of the current density and the magnetic induction at the considered point ($\vec{j} \times \vec{B}$). In practice, there are many examples where this force makes itself felt. For example, the deformation of a circular wire made from elastic metal when a current of high intensity flows through it is a consequence of the presence of these transverse forces, which tend to widen the loop under the influence of its own magnetic field (see figure 4.9(a)). In induction furnaces, intense currents flow through the metal melt, creating their own magnetic field that acts on the melt with electromagnetic forces. Such forces create a *pinch* effect, an effect that gives rise to the interruption of the metal fluid column. In electromagnetic pumps (see figure 4.9(b)), a transverse current passes through the

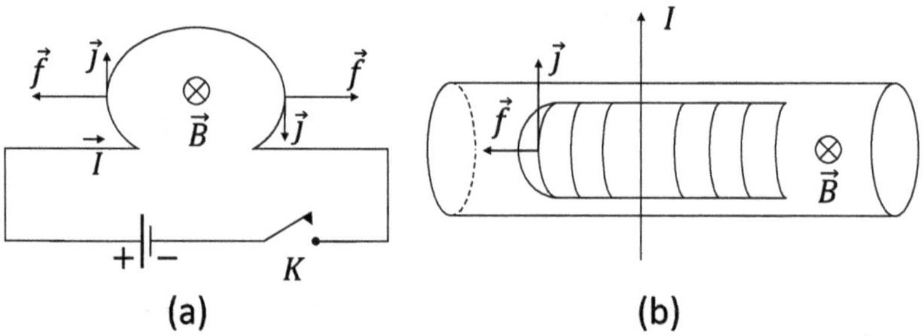

Figure 4.9. (a) A loop carrying an electric current is deformed as a result of the action of the electromagnetic force. (b) The operational principle of an electromagnetic pump.

conductive fluid to be pumped, located in the magnetic field $\vec{B}$. As a result, a force $\vec{f}$ is exerted that can set the liquid in motion.

4.1.3.4 Electrodynamic force between two parallel conductors carrying electric currents and Ampère's force

When analyzing a magnetic field, Ampère observed that between two filiform and very long parallel conductors carrying currents I_1 and I_2, an attractive force is exerted when the currents flow in the same direction, whereas a repulsive force is exerted when the currents circulate in opposite directions. From experimental study, it has been established that $\overrightarrow{F_{12}}$, the force with which conductor (1) acts on conductor (2) (see figure 4.10) is:

$$\overrightarrow{F_{12}} = -\frac{\mu_0}{4\pi}\frac{2I_1I_2l}{r_{12}}\overrightarrow{e_{12}} \tag{4.18}$$

where l is the conductor length for which the force is calculated, $\overrightarrow{r_{12}}$ is the relative position vector between the two conductors, and $\overrightarrow{e_{12}}$ is the versor of this vector, oriented from conductor (1) to conductor (2). The proportionality constant μ_0, called the *magnetic permeability of the vacuum*, is a universal constant that has the value $\mu_0 = 4\pi \times 10^{-7}$ N A^{-2}. Based on relation (4.18), Ampère defined the unit of measurement for the intensity of electric current, A, a unit named after him (ampere) and which is defined as *that value of the intensity of DC that, being maintained in two filiform conductors, parallel, rectilinear, infinitely long, placed in vacuum at a distance of* 1 m *one from each other, determines between them a force of* $2 \cdot 10^{-7}$ N *per meter*

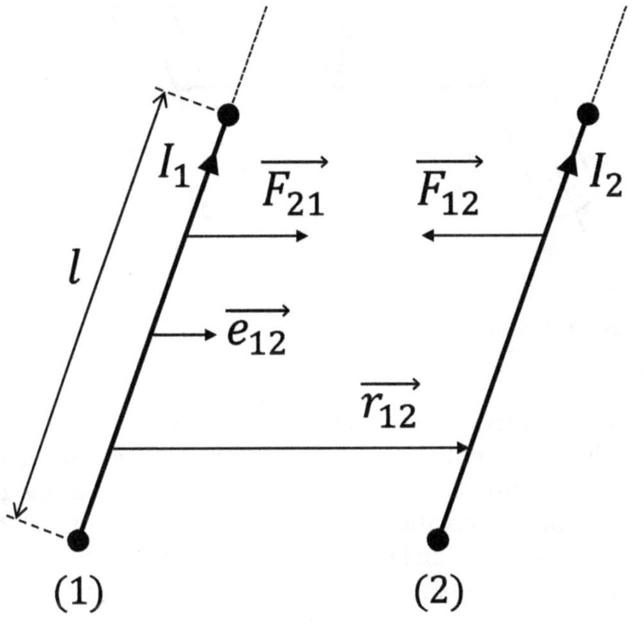

Figure 4.10. The electrodynamic force.

of length. With the help of the permeability of vacuum (μ_0) and magnetic induction in vacuum ($\vec{B}$), a vector quantity can be defined for the magnetic field that characterizes only the geometry of the current circuit that creates the field, without taking into account the properties of the medium in which the field propagates. This quantity is the intensity of the magnetic field ($\vec{H}$), defined by:

$$\vec{H} = \frac{\vec{B}}{\mu_0} \tag{4.19}$$

Observation. If one returns to the force of interaction between two parallel conductors carrying currents I_1 and I_2, $\overrightarrow{F_{12}} = -\frac{\mu_0}{4\pi}\frac{2I_1I_2l}{r_{12}}\overrightarrow{e_{12}}$, an analogy can be observed between it and the electrostatic force between two wires charged with electric charges described by linear charge densities λ_1 and λ_2 (see figure 4.11), a force given by:

$$\overrightarrow{F_{12}} = \frac{1}{4\pi\varepsilon_0}\frac{2\lambda_1\lambda_2 l}{r_{12}}\overrightarrow{e_{12}} \tag{4.20}$$

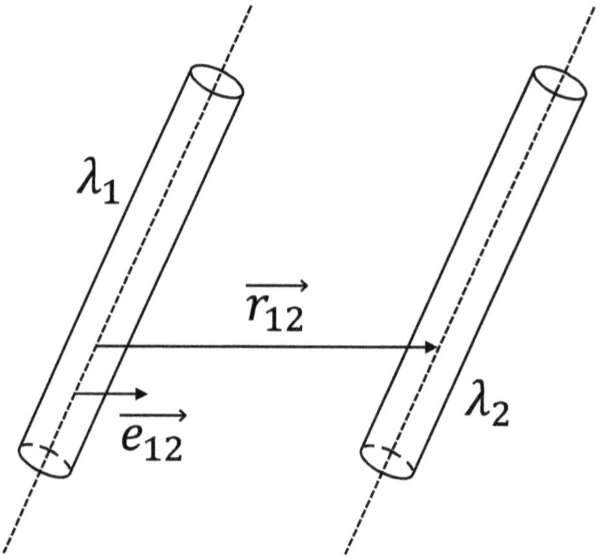

Figure 4.11. Force of interaction between two linear charge distributions.

However, in electrostatics, a repulsive force is exerted between the two linear distributions of charge when the two conductors are charged with charges that have the same sign, while an attractive force is exerted when the charges are of opposite signs, contrary to the observations regarding the interaction of currents. This analogy between electrostatics and magnetostatics allows the introduction of specific quantities that describe the magnetic field in a manner similar to that used for the introduction of the quantities characterizing the electrostatic field.

4.2 Magnetic fields of DC circuits in vacuum

4.2.1 Biot–Savart–Laplace formula

In electrostatics, the expression of the electric field $\vec{E}$ produced in vacuum by known charge distributions was established on a direct experimental basis using the relation:

$$\vec{F} = q\vec{E} \tag{4.21}$$

and Coulomb's formula:

$$\overrightarrow{F_{12}} = \frac{1}{4\pi\varepsilon_0}\frac{q_1 q_2}{r_{12}^3}\overrightarrow{r_{12}} \tag{4.22}$$

In the case of a point electric charge $\vec{E}$, the result is the relation:

$$\vec{E} = \frac{q}{4\pi\varepsilon_0 r^3}\vec{r} \tag{4.23}$$

and for a volume charge distribution, the result is the expression:

$$\vec{E} = \frac{1}{4\pi\varepsilon_0}\int_V \frac{\rho dv}{r^3}\vec{r} \tag{4.24}$$

In the study of the stationary magnetic field, an analogous approach can be used to experimentally establish an expression for the magnetic field produced by a DC current circuit; this expression is analogous expression to (4.24) and is known as the **Biot–Savart–Laplace formula**. The analogous relation for equation (4.21) is Laplace's formula for the electromagnetic force:

$$\overrightarrow{dF} = I\overrightarrow{dl} \times \vec{B} \tag{4.25}$$

and the analogous version of Coulomb's formula, i.e. equation (4.22), can be considered the electrodynamic force between two parallel conductors, $\overrightarrow{F_{12}} = -\frac{\mu_0}{4\pi}\frac{2I_1 I_2 l}{r_{12}}\overrightarrow{e_{12}}$. But this analogy has a particular character. In the magnetic field, there is no sample body as simple as the point charge; as a result, the reasoning in magnetostatics is more complicated. We present the reasoning in a simplified manner, as the experimental premise represented by the relation of the electrodynamic force (4.18) is basically insufficient. It was to be replaced and complemented by the experiences of Jean–Baptiste Biot and Félix Savart. As a result of these experiments, the following expression was established for the intensity of the magnetic field $\vec{H} = \frac{\vec{B}}{\mu_0}$, produced in a vacuum by a closed filiform circuit carrying the DC I:

$$\vec{H} = \frac{\vec{B}}{\mu_0} = \frac{I}{4\pi}\oint_\Gamma \frac{\overrightarrow{dl} \times \vec{r}}{r^3} \tag{4.26}$$

This is called the Biot–Savart–Laplace formula, in which:

- $\overrightarrow{dl}$ is the length element on the Γ contour of the circuit and the sense is given by the sense of the current, see figure 4.12.
- $\vec{r}$ is the position vector from $\overrightarrow{dl}$ at point P, where the field is calculated.

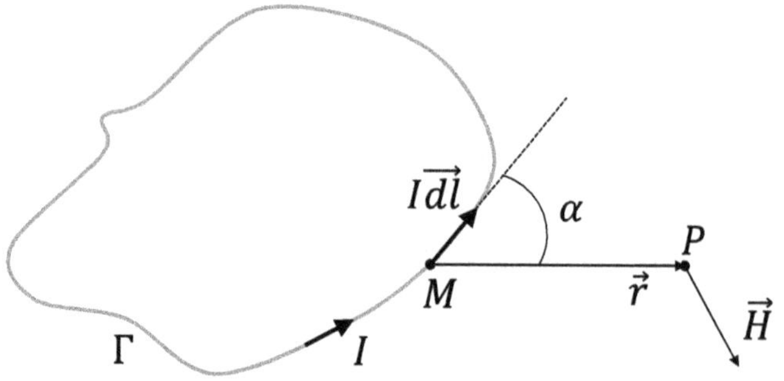

Figure 4.12. Illustration of elements appearing within the Biot–Savart–Laplace formula.

Observation. When operating with this formula, it should be emphasized that the integration should be carried out only on a closed curve, because the lines of current are always closed. The experiments verify this expression in stationary magnetic fields in vacuum, or, as we will see later, in homogeneous environments. Therefore, from the point of view of formulating the experimental basis of the theory of stationary magnetic fields, this formula plays same role that the expression of the coulombic electric field played $\vec{E} = \frac{1}{4\pi\varepsilon_0}\int_V \frac{\rho dv}{r^3}\vec{r}$ in formulating the experimental basis of the theory of electrostatics. A rigorous demonstration of the Biot–Savart–Laplace formula can be made using the general laws for magnetic fields, accepting the postulates of the macroscopic theory of electromagnetic fields, and it is given later. In the following, however, we present the deduction of this formula on an experimental basis, as conceived by Biot, Savart, and Laplace, a deduction that has more historical interest.

4.2.1.1 Biot–Savart's experiments (1821–80)

Using Ampère's electrodynamic force, $\overrightarrow{F_{12}} = -\frac{\mu_0}{4\pi}\frac{2I_1I_2l}{r_{12}}\overrightarrow{e_{12}}$, an experimentally measurable force, and trying to put it in the form of the Laplace force with which the magnetic field created by conductor (1) acts on conductor (2) of length $\vec{l}$, one can obtain an expression for the intensity of the magnetic field created by an infinitely long linear conductor (see figure 4.13). Thus:

$$\overrightarrow{F_{12}} = I_2(\vec{l} \times \overrightarrow{B_1}) = I_2(\vec{l} \times \mu_0\overrightarrow{H_1}) \tag{4.27}$$

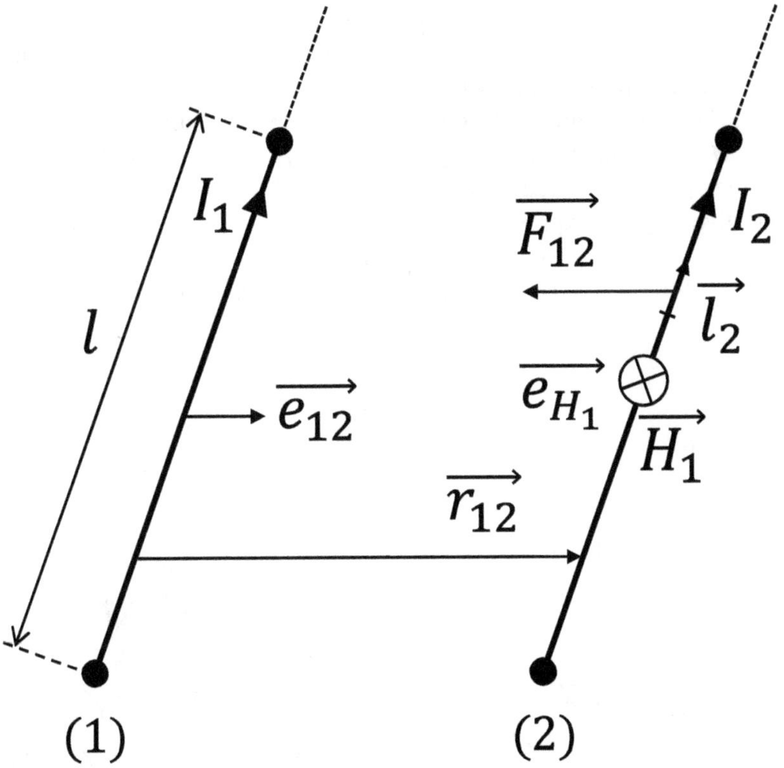

Figure 4.13. Electrodynamic force between two conductors carrying electric currents.

However:

$$\overrightarrow{F_{12}} = -\frac{\mu_0}{4\pi}\frac{2I_1I_2}{r_{12}}l\overrightarrow{e_{12}} = +I_2\left(\vec{l} \times \mu_0\frac{I_1}{2\pi r_{12}}\overrightarrow{e_H}\right) \tag{4.28}$$

In writing equation (4.28), a vector with the sense of I_2 was associated with the conductor $(\vec{l_2})$ carrying I_2, and it was considered that the eventual magnetic field created by conductor (1) at the point where conductor (2) is located can only be perpendicular to the plane of the conductors; in this way, the electromagnetic force acts in the plane of the conductors, as experimentally ascertained. Thus, applying relations (4.27) and (4.28), we obtain the following expression for the intensity of the magnetic field created by the first conductor at a distance of $\overrightarrow{r_{12}}$:

$$\overrightarrow{H_1} = \frac{I_1}{2\pi r_{12}}\overrightarrow{e_{H_1}} \tag{4.29}$$

The lines of the field $\vec{H}$ of a linear conductor are concentric circles located in transverse planes in relation to the current that produces it. Relation (4.29) was established independently of Ampère's formula; it was experimentally determined using a magnetic compass needle by Biot and Savart, but expression (4.29) is insufficient for the derivation of the magnetic field produced by a circuit of any form. Biot and Savart, however, extended their experiences to the field of an infinite conductor, bent at an angle of 2α (see figure 4.14). They measured the intensity of the magnetic field at a point P, placed on the bisector of the angle 2α, at a distance R from point M (the spike of the angle). The vector $\overrightarrow{H}$ was found to be perpendicular to the angular plane, oriented according to the RSR in relation to the sides of the angle, and given by the expression:

$$H = \frac{I}{2\pi R}\tan\frac{\alpha}{2} = \frac{I}{2\pi\frac{r}{\sin\alpha}}\tan\frac{\alpha}{2} = \frac{I}{2\pi r}\cdot 2\sin^2\frac{\alpha}{2} = \frac{I}{2\pi r}(1-\cos\alpha) \tag{4.30}$$

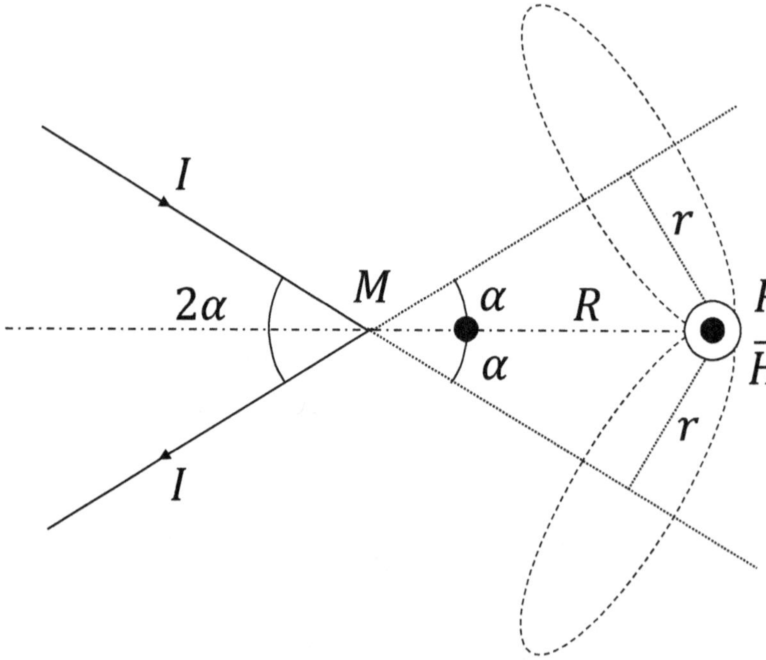

Figure 4.14. Infinitely long conductor carrying a current I and bent at an angle of 2α.

Relation (4.30) generalizes the previous result (4.29) because if $\alpha = \frac{\pi}{2}$, i.e. $2\alpha = \pi$, the result is $H = \frac{I}{2\pi R}$, obtained in the case of a straight infinite conductor carrying the current I.

In order to be able to deduce an expression for the intensity of the magnetic field produced by a circuit of any form from relation (4.30) (see figure 4.15), Laplace[5]

[5] **Pierre-Simon, Marquis de Laplace** (March 23, 1749–March 5, 1827) was a mathematician who made important contributions to the development of engineering, mathematics, statistics, physics, astronomy, and philosophy.

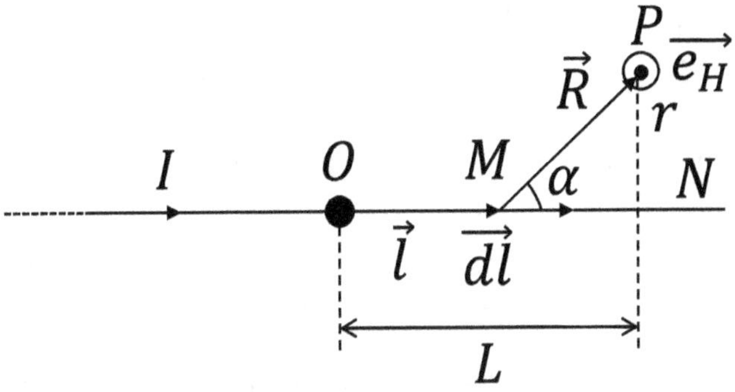

Figure 4.15. One side of the angular conductor carrying the electric current.

used the hypothesis that this field is the resultant obtained from the superposition of some elementary fields $\overrightarrow{dH}$ produced by each separate length element $\overrightarrow{dl}$ of a circuit:

$$\vec{H} = \oint_{\Gamma} d\vec{H}(I, d\vec{l}, \vec{r}) \tag{4.31}$$

Before examining the validity of this hypothesis (which physically proved inadequate), we will show how it can be used to infer the general expression for $\overrightarrow{H}$ from the expressions of Biot and Savart and in particular from (4.30).

For reasons of symmetry, when we apply the superposition principle given by relation (4.31) to the angle contour in figure 4.14, it can be said that each side of the angle contributes the following to the field:

$$\overrightarrow{H'} = \frac{\vec{H}}{2} = \frac{I}{4\pi r}(1 - \cos\alpha)\overrightarrow{e_H} \tag{4.32}$$

Taking into account one side of the semi-infinite conductor (see figure 4.15), the intensity of the magnetic field at a point P, located at a distance r = constant from the conductor, is: $\vec{H} = \frac{I}{4\pi r}(1 - \cos\alpha)\overrightarrow{e_H}$. On the other hand, according to the principle of superposition, this relation can be written as: $\vec{H} = \oint_{\Gamma} \overrightarrow{dH}$. In the particular case of the semi-infinite wire (it is accepted that the circuit is closed at infinity) and for a chosen reference frame whose origin is at the point O and oriented in the sense of I, this becomes:

$$\vec{H}(\vec{l}, \vec{r}, I) = \int_{-\infty}^{l} \overrightarrow{dH} = \frac{I}{4\pi r}(1 - \cos\alpha)\overrightarrow{e_H} \tag{4.33}$$

Differentiating this relation with respect to $\overrightarrow{dl}$ and assuming that the distance r (from point P to the conductor) is constant, we obtain:

$$\frac{\overrightarrow{dH}}{dl}dl = \overrightarrow{dH} = \frac{I}{4\pi r}d(1 - \cos\alpha)\overrightarrow{e_H} = \frac{I}{4\pi r}\sin\alpha\frac{d\alpha}{dl}dl \cdot \overrightarrow{e_H} \tag{4.34}$$

So:

$$\overrightarrow{dH} = \frac{I}{4\pi r}\sin\alpha\frac{d\alpha}{dl}dl \cdot \overrightarrow{e_H}$$

However, in the triangle MPN (with PN $= r =$ constant), we notice that MN $= L - l = r\cot\alpha$ and differentiating the relation with respect to $\overrightarrow{dl}$ gives:

$$-dl = -r\frac{d\alpha}{\sin^2\alpha} \quad \Rightarrow \quad \frac{d\alpha}{dl} = \frac{\sin^2\alpha}{r} \quad \Rightarrow$$

$$\Rightarrow \overrightarrow{dH} = \frac{I}{4\pi r^2}\sin^3\alpha \cdot dl \cdot \overrightarrow{e_H} = \frac{IdlR\sin\alpha}{4\pi R^3}\overrightarrow{e_H} = \frac{I \cdot \overrightarrow{dl} \times \overrightarrow{R}}{4\pi R^3} \tag{4.35}$$

This accounts for the fact that the field $\overrightarrow{dH}$ must be perpendicular to the plane containing the conductor and its sense must coincide with the sense of the vector product of the element of length $\overrightarrow{dl}$ and $\overrightarrow{R}$. Therefore, starting from the Biot–Savart expressions and considering Laplace's proposed principle of superposition to be valid, we obtain expression (4.35) for a hypothetical elementary field created by a circuit element $I\overrightarrow{dl}$ at a point P located at a distance $\overrightarrow{R}$ from it.

An expression for the intensity of a magnetic field $\overrightarrow{H}$ produced by the entire circuit Γ at point P (i.e. see figure 4.12) can now be obtained by integrating the Biot–Savart–Laplace formula:

$$\overrightarrow{H} = \frac{I}{4\pi}\oint_{\Gamma}\frac{\overrightarrow{dl} \times \overrightarrow{R}}{R^3} \tag{4.36}$$

Observation. Although the result obtained is correct, the hypothesis of the superposition of the elementary fields $\overrightarrow{dH}$ formulated by Pierre Laplace as expressed in relation (4.31) is incorrect. The reason for this is that no circuit elements can be broken by the circuit to which they belong (independent of it), since DC always circulates in closed paths. For this reason, relation (4.35) does not represent an elementary field (because $I\overrightarrow{dl}$ does not exist) but rather represents the integrant in relation (4.36), which gives the magnetic field created by a closed circuit Γ at a point P at a distance $\overrightarrow{R}$ from the chosen contour element $\overrightarrow{dl}$.

Obviously, for the induction of a magnetic field in vacuum, the following expression can be given:

$$\overrightarrow{B} = \frac{\mu_0 I}{4\pi}\oint_{\Gamma}\frac{\overrightarrow{dl} \times \overrightarrow{R}}{R^3} \tag{4.37}$$

The Biot–Savart–Laplace formula is often expressed in terms of the current density. Consider a volume element dV in a conductor with the volume $dV = dl \cdot dS$; then $I = \overrightarrow{j} \cdot \overrightarrow{ds} = j \cdot ds$. With this the relationship, (4.37) becomes:

$$\overrightarrow{B} = \frac{\mu_0 I}{4\pi}\oint_{\Gamma}\frac{\overrightarrow{dl} \times \overrightarrow{R}}{R^3} = \frac{\mu_0}{4\pi}\oint_{\Gamma}\frac{jds\overrightarrow{dl} \times \overrightarrow{R}}{R^3} = \frac{\mu_0}{4\pi}\int_{V_\Gamma}\frac{\overrightarrow{j} \times \overrightarrow{R}}{R^3}dv \tag{4.38}$$

where $\vec{B}$ is given by the integral over the volume of the circuit (carrying the conduction current density $\vec{j}$) of the relation $\frac{\vec{j} \times \vec{R}}{R^3}$. In principle, the Biot–Savart–Laplace formula in forms (4.37) or (4.38) allows one to calculate the magnetic induction in vacuum for any circuit geometry carrying a current. In the following, we give some examples showing the calculation of the quantities characterizing the magnetic fields generated by different circuits using the Biot–Savart–Laplace formula.

Example 1. *The calculation of the magnetic field created by a circular conducting wire that has a radius of a,* carrying *a current of intensity I, at a point P, located on the symmetry axis of the conducting wire, at distance h from its plane (see figure 4.16).*

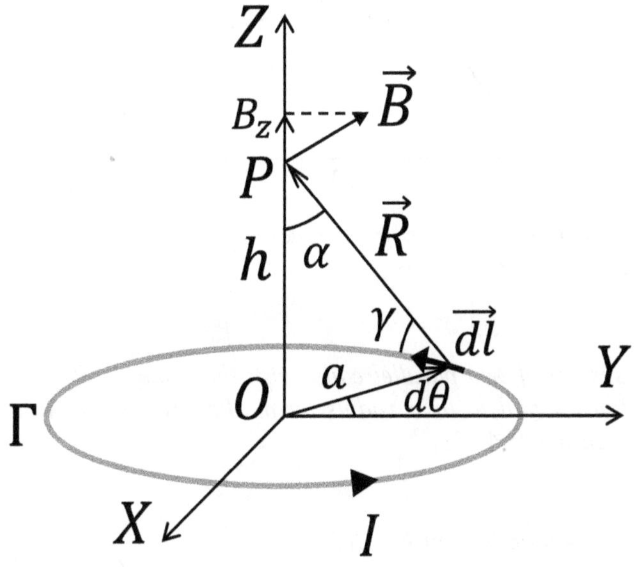

Figure 4.16. *Calculation of the magnetic field created by a circular conducting wire carrying a current I.*

Solution

According to the Biot–Savart–Laplace formula:

$$\vec{B} = \frac{\mu_0 I}{4\pi} \oint_\Gamma \frac{\vec{dl} \times \vec{R}}{R^3} \Longrightarrow$$

$$B_z = \frac{\mu_0 I}{4\pi} \oint_\Gamma \frac{dl \cdot R \sin \gamma}{R^3} \sin \alpha = \frac{\mu_0 I}{4\pi} \int_0^{2\pi} \frac{ad\theta R}{R^3} \frac{a}{R} = \frac{\mu_0 I a^2}{2(a^2 + h^2)^{3/2}} \tag{4.39}$$

Thus, $\vec{B}$ has only the normal component to the plane of the circular wire and is given by relation (4.39). For $h = 0$ (the center of the circular wire), $\vec{B} = \frac{\mu_0 I}{2a}$, while for $h \to \infty, B = 0$. A graph of B(h) on both sides of the plane of the circular wire is given in figure 4.17. If the field is evaluated at points sufficiently far from the center of the circular wire, it is found that the induction of the magnetic field can be expressed according to the magnetic moment of the loop, as follows:

$$\vec{B} = \frac{\mu_0 I a^2}{2R^3}\vec{k} = \frac{\mu_0 I S}{2\pi R^3}\vec{k} = \frac{\mu_0 \vec{m}}{2\pi R^3} \tag{4.40}$$

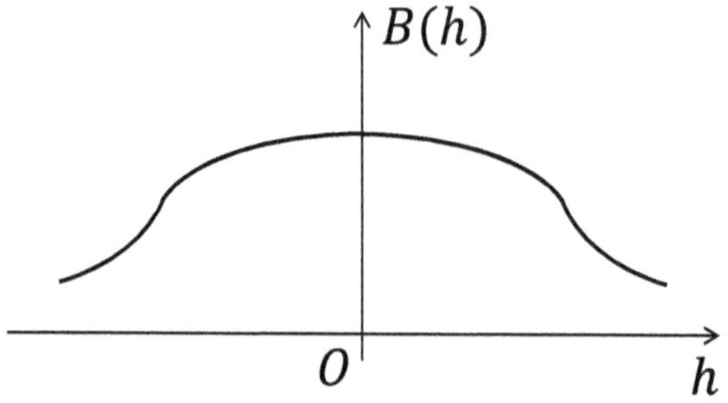

Figure 4.17. *Dependence of the magnetic induction created by a circular wire conducting a current on the distance h from the plane of its symmetry axis.*

Example 2. *A system of two parallel coils, with the same number of circular wires, placed at a distance equal to their radius forms the system known as the Gaugain–Helmholtz coil, see* figure 4.18.

Solution

The resulting magnetic field at $h = a/2$ is:

$$B = \frac{2\mu_0 I a^2}{2\left(a^2 + \frac{a^2}{4}\right)^{3/2}} = \frac{\mu_0 I}{a\left(\frac{5}{4}\right)^{3/2}} = 0.715\frac{\mu_0 I}{a}$$

The resulting magnetic field at the center of one of the coils is:

$$B = \frac{\mu_0 I}{2a} + \frac{\mu_0 I a^2}{2(a^2 + a^2)^{3/2}} = \frac{\mu_0 I}{2a}\left(1 + \frac{1}{2^{3/2}}\right) = 0.676\frac{\mu_0 I}{a}$$

Between the two coils, B ranges from $0.68\frac{\mu_0 I}{a}$ to $0.71\frac{\mu_0 I}{a}$, so it can be considered practically constant.

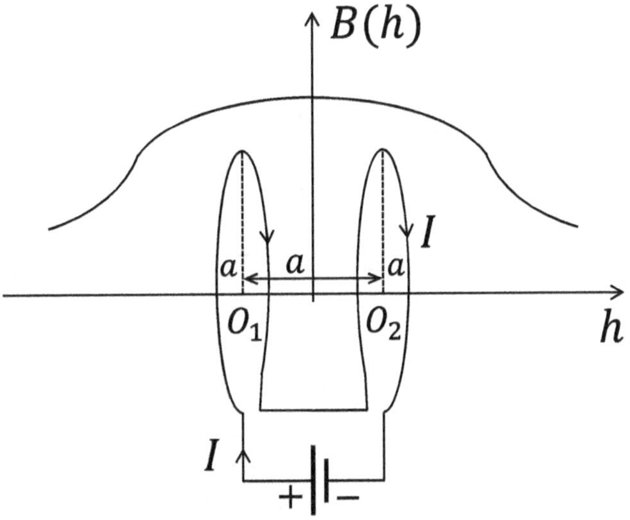

Figure 4.18. *System of Gaugain–Helmholtz coils.*

4.2.2 Scalar magnetic potential of a closed circuit carrying a current

In the case of the circular conducting wire shown in figure 4.16, for which the magnetic field induction was calculated using the Biot–Savart–Laplace formula, the magnetic field strength is:

$$H_x = H_y = 0; H_z = \frac{Ia^2}{2R^3} = \frac{Ia^2}{2(a^2 + z^2)^{3/2}} \tag{4.41}$$

Thus, there is only the normal component of H at the circuit plane, and this component is dependent on $1/R^3$. This behavior also produces the intensity of electric field created by a circular wire uniformly charged with the linear charge density λ. Analyzing this electrostatic–magnetostatic analogy, it is natural to ask ourselves the question, can a scalar magnetic potential, whose gradient leads to the intensity of the magnetic field, be introduced for the magnetic field? Under certain conditions, which will be discussed further, by analogy with the electrical potential defined by:

$$\vec{E} = -\nabla V, \ V(\vec{r}) = V(\vec{r_0}) - \int_{\vec{r_0}}^{\vec{r}} \vec{E}\vec{dl} \tag{4.42}$$

a scalar magnetic potential may also be introduced for the magnetic field, so that:

$$\vec{H} = -\nabla V_m \tag{4.43}$$

In this relation, V_m is the scalar magnetic potential, which can be obtained using the Biot–Savart–Laplace formula. It can be introduced, however, only where the relation $\nabla \times \vec{H} = 0$ is satisfied; that is, at points in the magnetic field far from any source of the magnetic field. This means that, at least locally at those points, the

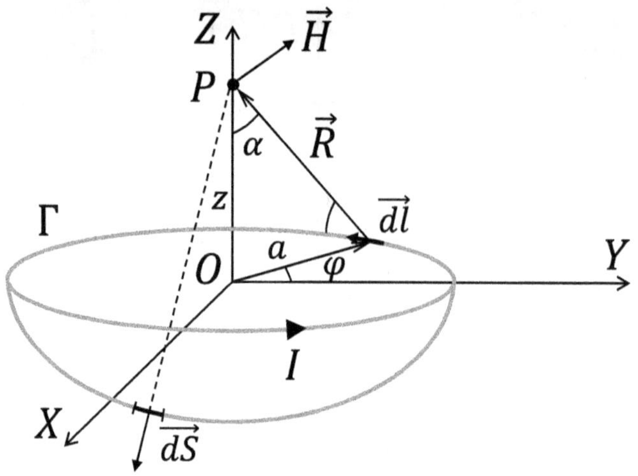

Figure 4.19. Calculation of the scalar magnetic potential in the case of a circular wire of radius a conducting a current of intensity I.

field can be considered to have open field lines. In the case of the conductive circular ring in figure 4.19, the scalar magnetic potential at point P is calculated by analogy with equation (4.42) as follows:

$$V_m(z) = V_m(z_0) - \int_{z_0}^{z} \vec{H}\vec{dz} \tag{4.44}$$

Taking into account, by convention, that $V_m \to \infty$ when $z_0 \to \infty$, the result is:

$$V_m(z) = -\int_{\infty}^{z} H_z dz = -\int_{\infty}^{z} \frac{Ia^2}{2(a^2 + z^2)^{3/2}} dz = \int_{z}^{\infty} \frac{Ia^2}{2R^3} dz \tag{4.45}$$

If the exchange of variables shown below is made:

$$z = a\cot\alpha \Rightarrow dz = -\frac{a}{\sin^2\alpha} d\alpha, \quad R = \frac{a}{\sin\alpha}$$

the result is:

$$V_m = \frac{I}{2}\int_{\alpha}^{0} \frac{a^2\left(-\frac{a}{\sin^2\alpha} d\alpha\right)}{\frac{a^3}{\sin^3\alpha}} = \frac{I}{2}\int_{\alpha}^{0} (-\sin\alpha) d\alpha = \frac{I}{2}(1 - \cos\alpha) \tag{4.46}$$

However, according to figure 4.19, the solid angle subtended by the circular circuit with the associated contour Γ at point P is:

$$\Omega_\Gamma = \int \frac{\vec{dS}\vec{R}}{R^3} = \int_0^{2\pi}\int_0^{\alpha} \frac{R^2 \sin\alpha}{R^2} d\alpha d\varphi = 2\pi(1 - \cos\alpha) \tag{4.47}$$

Thus, the scalar magnetic potential can be expressed according to the solid angle Ω_Γ subtended by the contour Γ traveled by the current I at the point where V_m is calculated. So, in the case of this circuit, we obtain:

$$V_m = \frac{I}{4\pi}\Omega_\Gamma \tag{4.48}$$

It can be demonstrated that formula (4.48) is valid regardless of the shape of the circuit and the position of the point in the magnetic field at which the scalar magnetic potential is computed. Therefore, the scalar magnetic potential (defined up to an additive arbitrary constant) is equal to the product of the intensity of the electric current that flows through the circuit and the solid angle subtended by the circuit at the considered point, divided by 4π. For the solid angle Ω_Γ, the following sign convention is used: $\Omega_\Gamma > 0$ if the circuit, as seen from point P, is traveled by the current in the trigonometric sense, and $\Omega_\Gamma < 0$ for the opposite sense (clockwise).
Observation. It will be shown a little later that the scalar magnetic potential can be defined only in the stationary regime and only in regions where the density of the electric current is zero, i.e. where $\nabla \times \vec{H} = 0$. Since these regions are generally connected, the scalar magnetic potential is non-unique; that is, it can have different sets of values at the same point. Of course, under the specified conditions of validity, one can easily calculate the scalar magnetic potential through the relation $V_m = \frac{I}{4\pi}\Omega_\Gamma$, and by calculating its gradient, one can find the intensity of the magnetic field, which is sometimes much easier than using the Biot–Savart–Laplace law; for example:

$$\vec{H} = -\nabla V_m = -\nabla\left(\frac{I\Omega_\Gamma}{4\pi}\right) \tag{4.49}$$

4.2.3 Ampère's theorem (magnetic circuit law)

In electrostatics, the general properties of the electrostatic field in vacuum are described by:

a) The theorem of electrostatic potential: $\oint_\Gamma \vec{E}\vec{dl} = 0$

b) Gauss's theorem: $\int_\Sigma \vec{E}\vec{ds} = \frac{q_i}{\varepsilon_0}$

In the context of the electrostatics–magnetostatics analogy, it is expected that the magnetic field created in vacuum by a stationary electric current should be characterized by generally analogous properties; one related to the circulation of the vector field $\vec{B}$ along a closed path and another to the flux of the magnetic induction $\vec{B}$ or to the flux of the intensity of the magnetic field $\vec{H}$ through a surface.

4.2.3.1 Integral form of the magnetic circuit law (Ampère's law)

As in the case of the electric field, we can calculate the circulation of the magnetic field on a closed path Γ. For example, we first consider the case of a field created by

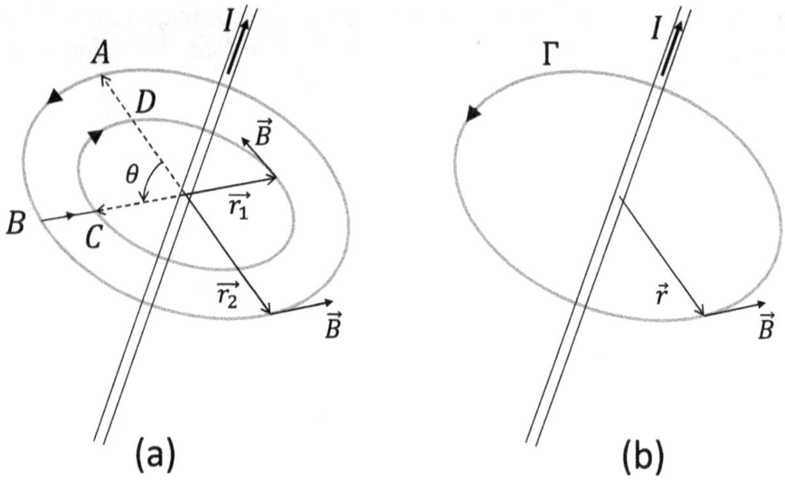

Figure 4.20. Demonstration of Ampère's law.

an infinitely long conductor for which we know the induced magnetic field at a distance r from it. The circulation of $\vec{B}$ on the contour $ABCDA$ that does not contain the conductor, see figure 4.20(a), is:

$$\begin{aligned}\int_{ABCDA} \vec{B}\vec{dl} &= \int_A^B \frac{\mu_0 I}{2\pi r_2} dl_2 + \int_B^C \vec{B}\vec{dl} + \int_C^D - \frac{\mu_0 I}{2\pi r_1} dl_1 + \int_D^A \vec{B}\vec{dl} \\ &= \frac{\mu_0 I}{2\pi r_2} r_2\theta + 0 - \frac{\mu_0 I}{2\pi r_1} r_1\theta + 0 = 0\end{aligned} \tag{4.50}$$

Observation. Because any closed path Γ that does not surround the current can be decomposed into paths of type $ABCDA$, it can be stated that the circulation of the magnetic field along a closed path that does not enclose the current I is zero.

We now consider a closed path Γ surrounding the current; it lies exactly on a field line (see figure 4.20(b)), and $\vec{B}$ is constant at the distance $\vec{r}$ from the conductor. This gives the result:

$$\oint_\Gamma \vec{B}\vec{dl} = \int_0^{2\pi} \frac{\mu_0 I}{2\pi r} \cdot r d\theta = \mu_0 I \tag{4.51}$$

For generalization, a second example will be considered: that of a magnetic field created by a flat, closed circuit Γ' traveled by a current I (see figure 4.21). Calculating the circulation of the magnetic field produced by this circuit along the contour $ACBDA$ that encircles the circuit carrying the current I, and whose points A and B are in the plane of the circuit, we obtain:

$$\oint_\Gamma \vec{B}\vec{dl} = \int_{ACB} \vec{B}\vec{dl} + \int_{BDA} \vec{B}\vec{dl} \tag{4.52}$$

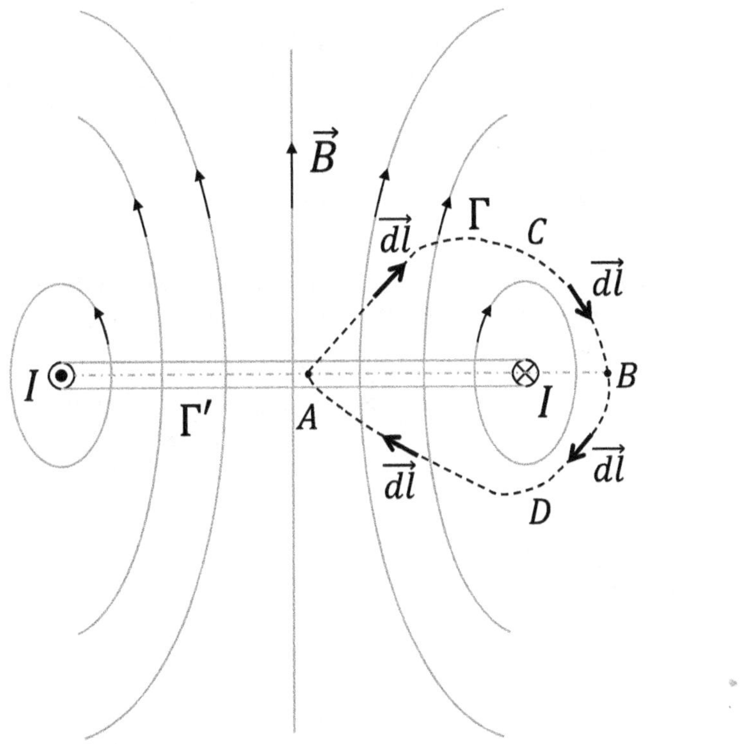

Figure 4.21. Circuit of arbitrary form carrying a current of intensity I, used in the demonstration of the Ampère's law.

However:

$$\oint_{ACB} \vec{B}\vec{dl} = \int_{ACB} (-\mu_0 \nabla V_m)\vec{dl} = -\mu_0 \int_{ACB} (\nabla V_m)\vec{dl} = -\mu_0 \int_{ACB} dV_m$$

$$= -\mu_0[V_m(B) - V_m(A)]$$

$$= \mu_0 \frac{I}{4\pi}[\Omega_c(A) - \Omega_c(B)] = \mu_0 \frac{I}{4\pi}(2\pi - 0) = \mu_0 \frac{I}{2} \tag{4.53}$$

Indeed, $\Omega_c(A) = 2\pi$ (half of the entire solid angle) and is positive, due to fact that the circuit is viewed from above and the sense of the current through it is the trigonometric one, while $\Omega_c(B) = 0$ because the closed plane circuit is viewed from an outer point located in its plane. In turn:

$$\oint_{BDA} \vec{B}\vec{dl} = \int_{BDA} (-\mu_0 \nabla V_m)\vec{dl} = -\mu_0 \int_{BDA} (\nabla V_m)\vec{dl} = -\mu_0 \int_{BDA} dV_m$$

$$= -\mu_0[V_m(A) - V_m(B)]$$

$$= \mu_0 \frac{I}{4\pi}[\Omega_c(B) - \Omega_c(A)] = \mu_0 \frac{I}{4\pi}(0 - (-2\pi)) = \mu_0 \frac{I}{2} \tag{4.54}$$

Here, $\Omega_c(B) = 0$, as in the previous case, while $\Omega_c(A) = -2\pi$ (half of the entire solid angle) and is negative, due to fact that the circuit is viewed from below and the sense of the current through is the reverse of the trigonometric sense. Returning to (4.53) and (4.54), we obtain:

$$\int_{\Gamma=ACBDA} \vec{B}\vec{dl} = \mu_0 I \tag{4.55}$$

This can be extended to the case of a closed path that surrounds several conductors carrying electric currents (see figure 4.22) by applying the principle of superposition, and then one can write:

$$\oint_{\Gamma} \vec{B}\vec{dl} = \mu_0 \sum_{k=1}^{n} I_k \tag{4.56}$$

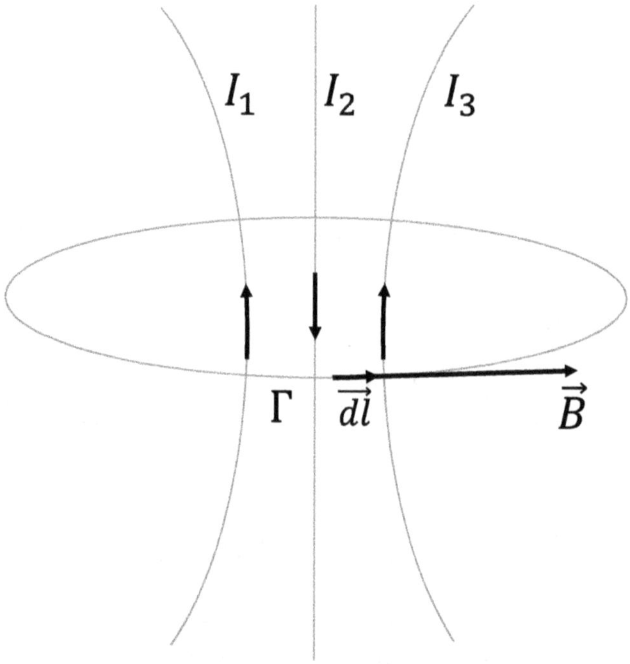

Figure 4.22. Ampère's law for a closed path Γ that surround several conductors carrying electric current.

Relation (4.56) expresses Ampère's theorem, which states: *the line integral of magnetic induction in vacuum is equal to the product of the vacuum magnetic permittivity (μ_0) and the algebraic sum of the chained current intensities.* Currents are considered positive when the sense in which they cross the contour Γ is associated, according to the RSR, with the sense in which the integration is performed. For example, in figure 4.22, I_3 and $I_1 > 0$, while $I_2 < 0$, so:

$$\oint_{\Gamma} \vec{B}\vec{dl} = \mu_0(I_1 - I_2 + I_3) \tag{4.57}$$

When the closed path surrounds the conducting wires of a coil with N loops, the relationship (4.56) becomes:

$$\oint_{\Gamma} \vec{B}\vec{dl} = \mu_0 NI \tag{4.58}$$

because I passes through the surface S_Γ N times. Generally, when the closed path Γ encircles several circuits, the relation (4.58) can be rewritten as:

$$\int_{\Gamma} \vec{B}\vec{dl} = \mu_0 \sum_k N_k I_k = \theta \tag{4.59}$$

where θ is called the magnetomotive voltage.

Definition: *the magnetomotive voltage along a closed curve, i.e. the line integral of the magnetic field, is proportional to the algebraic sum of the intensities of all currents encircled by that closed curve.*

Observation. The magnetic voltage between two points does not depend on the path unless the paths connecting the two points form a closed curve that does not encircle different circuits carrying currents. In the general case, the magnetic voltage between two points depends on the path of integration. This is due to the fact that in the presence of electric currents, the magnetic field is not irrotational. Only in fields where the current density is zero everywhere is the magnetic field irrotational, in which case it can be considered to be the gradient of the scalar magnetic potential. These observations can be expressed in a manner similar to that used in electrostatics by invoking the local form of the magnetic circuit law, that is, the local form of Ampère's law.

4.2.3.2 Local (differential) form of the magnetic circuit law (Ampère's law)

If a closed path Γ surrounds a tube of current lines of density $\vec{j}$ (see figure 4.23), then Ampère's law can be written as:

$$\oint_{\Gamma} \vec{B}\vec{dl} = \mu_0 \int_{S_\Gamma} \vec{j}\,\vec{ds} \tag{4.60}$$

where S_Γ is the area of the surface bounded by the closed curve Γ. Using Stokes's theorem, we get:

$$\oint_{\Gamma} \vec{B}\vec{dl} = \int_{S_\Gamma} (\nabla \times \vec{B})\vec{ds} = \mu_0 \int_{S_\Gamma} \vec{j}\,\vec{ds} \tag{4.61}$$

which results in the differential form of Ampère's law:

$$\nabla \times \vec{B} = \mu_0 \vec{j} \tag{4.62}$$

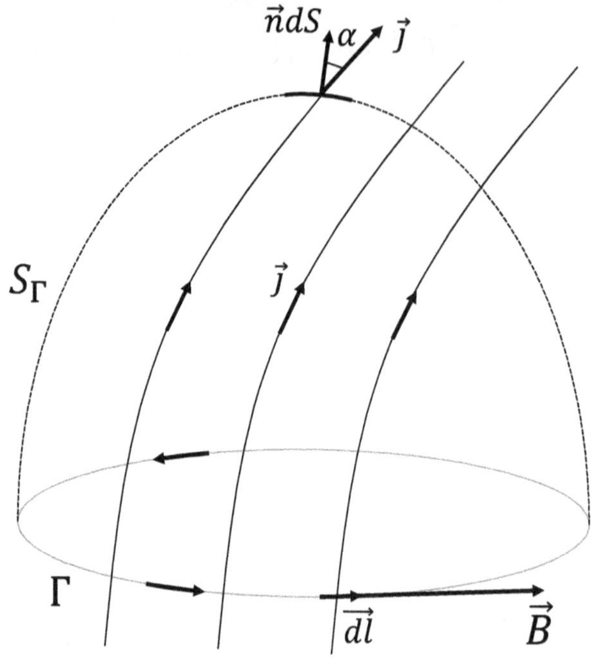

Figure 4.23. Ampère's law for the case of a curve Γ that surrounds a tube of current lines.

This equation is one of Maxwell's equations. It is valid only for magnetostatics, that is, the stationary electrokinetic state. In a regime that varies over time, e.g. alternating or transient current, equation (4.61) is no longer valid unless another term is added to the right-hand side of equation (4.61). This term was added by Maxwell and is known as *the displacement current density*, which was already introduced in the first chapter. So, in the case of a system of stationary bodies ($\vec{v} = 0$), the first of Maxwell equation takes the form:

$$\nabla \times \vec{B} = \mu_0\left(\vec{j} + \frac{\partial \vec{D}}{\partial t}\right) \tag{4.63}$$

in which the term $\frac{\partial \vec{D}}{\partial t}$ is the density of the displacement current.

In an even more general context, in the case of moving media where a surface $S - \Gamma$ moves together with the bodies in their movement, the local developed form of the magnetic circuit law becomes:

$$\nabla \times \vec{B} = \mu_0\left[\vec{j} + \frac{\partial \vec{D}}{\partial t} + \rho\vec{v} + \nabla \times (\vec{D} \times \vec{v})\right] \tag{4.64}$$

in which, in addition to the conduction current density $\vec{j}$, the displacement current density $\frac{\partial \vec{D}}{\partial t}$, and the density of the convection current $\rho\vec{v}$, the theoretical current density or Röentgen current density $\nabla \times (\vec{D} \times \vec{v})$ also appears. The theoretical current density or Röentgen current density $\vec{j_t} = \nabla \times (\vec{D} \times \vec{v})$ is produced by the movement of a polarized body. In one of Röentgen's experiments, the rotation of a dielectric disc between the armatures of a charged capacitor gave rise to a magnetic field associated with the theoretical current. All these terms arise if one calculates the substantial derivative of the flux of the displacement vector $\vec{D}$ (electrical induction) through a moving surface (see figure 4.24).

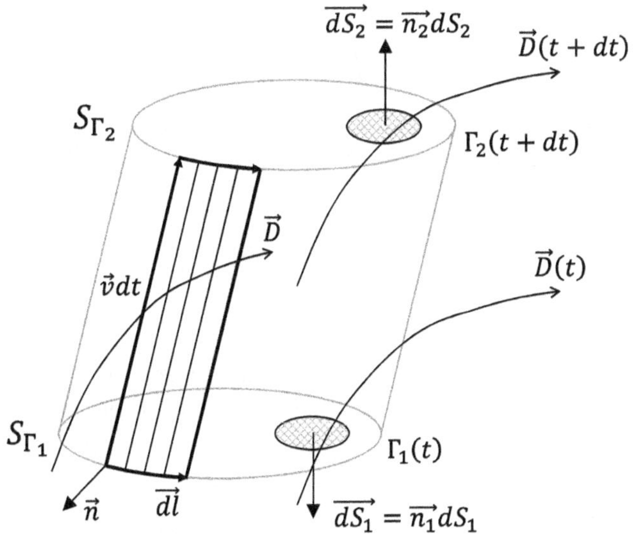

Figure 4.24. Figure illustrating the calculation of the substantial derivative of the electric induction's flux (i.e. the displacement vector's flux).

Exercise. *Demonstration of the substantial derivative of the electric induction's flux:*

$$\begin{aligned}\frac{d\Phi_D}{dt} &= \lim_{\Delta t\to 0}\frac{\Delta\Phi_D}{\Delta t} = \lim_{\Delta t\to 0}\frac{\int\limits_{S_{\Gamma(t+dt)}}\vec{D}(t+dt)\overrightarrow{ds_2} - \int\limits_{S_\Gamma}\vec{D}(t)\overrightarrow{ds_1}}{\Delta t}\\ &= \lim_{\Delta t\to 0}\frac{\int\limits_{S_{\Gamma_2}}\left[\vec{D}(t) + \frac{\partial\vec{D}}{\partial t}dt\right]\overrightarrow{ds_2} - \int\limits_{S_{\Gamma_1}}\vec{D}(t)\overrightarrow{ds_1}}{\Delta t}\\ &= \lim_{\Delta t\to 0}\frac{1}{\Delta t}\left[\int\limits_{S_{\Gamma_2}}\vec{D}(t)\overrightarrow{ds_2} - \int\limits_{S_{\Gamma_1}}\vec{D}(t)\overrightarrow{ds_1}\right] + \int\limits_{S_{\Gamma_2}}\frac{\partial\vec{D}}{\partial t}\overrightarrow{ds_2}\end{aligned} \tag{4.65}$$

However, according to the Gauss–Ostrogradsky theorem, if $\int_{S_{tot}}\vec{D}\vec{ds} = \int_{V_{tot}}\nabla\vec{D}dv$ *is applied to the closed total surface shown in figure* 4.24, *we can write:*

$$-\int_{S_{\Gamma(t)}} \vec{D}(t)\overrightarrow{ds_1} + \int_{S_{\Gamma_2}} \vec{D}(t)\overrightarrow{ds_2} - \int_{\Gamma} \vec{D}(\vec{v}dt \times \vec{dl}) = \int_{V} \nabla \vec{D}\ \vec{ds}\vec{v}dt \tag{4.66}$$

From which we obtain:

$$\int_{\Gamma_2} \vec{D}(t)\overrightarrow{ds_2} - \int_{S_{\Gamma_1}} \vec{D}(t)\overrightarrow{ds_1} = \int_{S_{\Gamma_1}} \nabla \vec{D}\ \vec{v}\vec{ds}dt + \int_{\Gamma} (\vec{D} \times \vec{v})\vec{dl}dt \tag{4.67}$$

Given these results, relation (4.65) *becomes:*

$$\begin{aligned}\frac{d\Phi_D}{dt} &= \int_{S_\Gamma} \frac{\partial \vec{D}}{\partial t}\vec{ds} + \int_{S_\Gamma} \rho\vec{v}\vec{ds} + \int_{S_\Gamma} \nabla \times (\vec{D} \times \vec{v})\vec{ds} \\ &= \int_{S_\Gamma} \left[\frac{\partial \vec{D}}{\partial t} + \rho\vec{v} + \nabla \times (\vec{D} \times \vec{v})\right]\vec{ds}\end{aligned} \tag{4.68}$$

So, returning to the law of magnetic circuits in the local form (4.63) and following surface integration over S_Γ bounded by the closed path Γ, we obtain:

$$\int_{S_\Gamma} (\nabla \times \vec{B})\vec{ds} = \mu_0 \int_{S_\Gamma} \vec{j}\,\vec{ds} + \mu_0 \int_{S_\Gamma} \left[\frac{\partial \vec{D}}{\partial t} + \rho\vec{v} + \nabla \times (\vec{D} \times \vec{v})\vec{ds}\right] \tag{4.69}$$

The developed integral form of the law of magnetic circuit becomes:

$$\int_{\Gamma} \vec{B}\vec{dl} = \mu_0 \int_{S_\Gamma} \vec{j}\,\vec{ds} + \mu_0 \frac{d\Phi_D}{dt} \tag{4.70}$$

The circulation of the magnetic field on a closed curve Γ *is equal to the enclosed conduction current and the substantial derivative of the electric flux through the mobile surface* S_Γ *bounded by the curve* Γ. The developed integral form of the magnetic circuit law can be used to calculate the magnetic field in the most diverse situations.

Exercise. *Calculate the magnetic field of a charge* q *in slow motion with speed* $\vec{v}$; *see figure 4.25.*

Solution: *a positive particle moving with a velocity* $\vec{v}$ *is equivalent to a filiform current; it gives rise to a magnetic field with field lines in the form of concentric circles that lie in a plane perpendicular to the particle's trajectory. If a closed path* Γ *of circular shape with a radius of* r *is considered, lying along the magnetic field line, then according to equation* (4.70), *we obtain:*

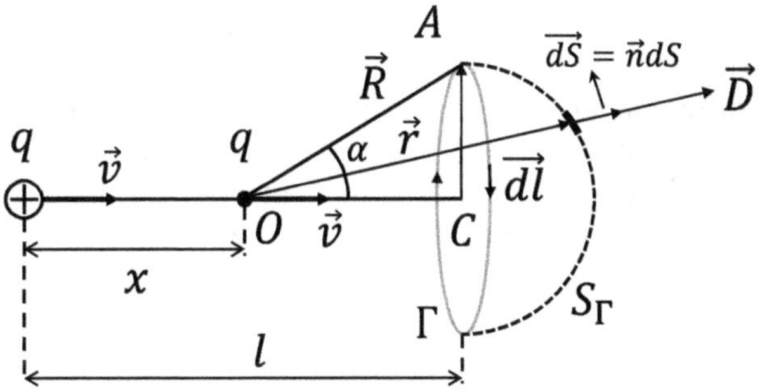

Figure 4.25. *Calculation of the magnetic field induced by an electric charge q moving at a low speed* $\vec{v}$.

$$\int_\Gamma \vec{B}\vec{dl} = \mu_0 \frac{d\Phi_D}{dt} = \mu_0 \frac{d}{dt}\int_{S_\Gamma} \vec{D}\vec{ds} = \mu_0 \frac{d}{dt}\int_{S_\Gamma} \frac{q\vec{R}\vec{ds}}{4\pi R^3} = \frac{q\mu_0}{4\pi}\frac{d}{dt}\int_{S_\Gamma} d\Omega$$

$$= \frac{q\mu_0}{4\pi}\frac{d}{dt}[2\pi(1 - \cos\alpha)] = \frac{\mu_0 q}{2}\sin\alpha\frac{d\alpha}{dt}$$

However, according to figure 4.25, we can write:

$$l - x = r\cot\alpha \Longrightarrow -\frac{dx}{dt} = -r\frac{1}{\sin^2\alpha}\frac{d\alpha}{dt} \Longrightarrow \frac{d\alpha}{dt} = \frac{v\sin^2\alpha}{r}$$

Thus:

$$B2\pi r = \frac{\mu_0 q}{2}\frac{v\sin^3\alpha}{r} \quad \Rightarrow \quad B = \frac{\mu_0 q v\sin^3\alpha}{4\pi r^2} = \frac{\mu_0 q v\sin\alpha}{4\pi R^2} \Longrightarrow$$

$$\vec{B} = \frac{\mu_0 q\cdot\vec{v}\times\vec{R}}{4\pi R^3}$$

Observation: we return to the case $\vec{v} = 0$ and the purely stationary regime and even more so, the regions of the magnetic field where the density of the conduction current is zero, for example in vacuum; under these conditions, according to equation (4.64), the curl of the magnetic field is zero ($\nabla \times \vec{H} = 0$) or ($\nabla \times \vec{B} = 0$). Therefore, in this case, according to the theorem used in the vector analysis (which states that the curl of the gradient of a scalar field is always zero) we can also define a scalar potential for the magnetic field. In general, however, the magnetic field is not a potential vector field (a field derived from the gradient of a scalar function), because $\nabla \times \vec{H} = 0$ is not always true but only holds in certain regions where there is no magnetic field source.

Ampère's law in integral form is very useful for calculating the intensity of the magnetic field for different geometries of current. Many examples can be given, such as the determination of magnetic field induction: (i) at a distance r from an infinitely long linear conductor carrying an electric current; (ii) on the axis of an infinitely long solenoid; (iii) in a toroidal coil, etc. The law of magnetic circuits also allows us derive an equation for the transmission of magnetic field intensity at an interface between two media with different properties, which will be presented below.

In the case of discontinuous surfaces between two different magnetic media (say, 1 and 2; see figure 4.26), applying the law of magnetic circuits along the small contour Γ, in the case where there are no currents on the surface, gives:

$$\int_{\Gamma} \overrightarrow{H}\overrightarrow{dl} = 0 \Rightarrow \overrightarrow{H_{2_t}}\overrightarrow{dl} - \overrightarrow{H_{1_t}}\overrightarrow{dl} = 0 \Longrightarrow = H_{1_t} = H_{2_t} \tag{4.71}$$

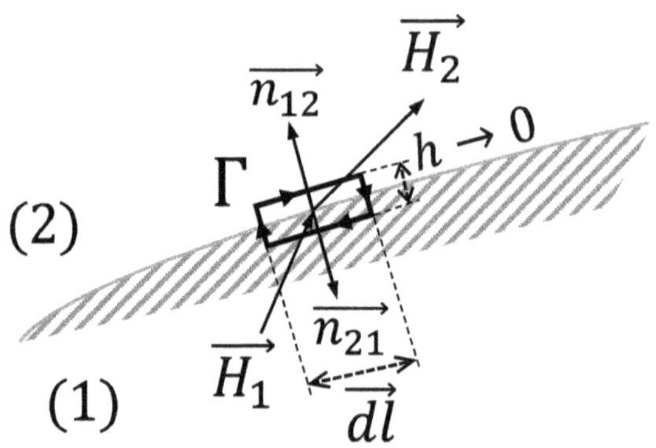

Figure 4.26. Interface between two media with different magnetic properties.

So, if there are no superficial currents, the tangential component of the magnetic field intensity is continuous when the magnetic field passes through the separation surface between two different media.

4.2.4 Gauss's theorem for magnetic fields

4.2.4.1 Integral form of Gauss's theorem for magnetic fields

The second important general property of the magnetic field refers to the flux of the magnetic induction vector ($\overrightarrow{B}$) through a closed surface (Σ). This is zero regardless of the nature of the closed surface ($\int \overrightarrow{B}\overrightarrow{ds} = 0$), as opposed to the situation in electrostatics, where the flux of an electric induction vector passing through a closed surface is equal to the charge located inside the surface. To verify this property, we consider the particular case of a magnetic field produce by an infinitely long conducting wire carrying a current of intensity I. We calculate its surface integral $\overrightarrow{B}$ for two areas Σ_1 and Σ_2, as shown in figure 4.27 (Σ_1 is a cylindrical surface containing the conducting wire, whereas Σ_2 does not contain the conducting wire carrying the current). Considering the fact that the field lines are concentric circles arranged in a plane perpendicular to the conductor, it is obvious that there is no flux through the bases of Σ_1 and also that the flux is zero ($\vec{n} \perp \overrightarrow{B}$) through the lateral surface in any position, such that $\int_{\Sigma_1} \overrightarrow{B}\overrightarrow{ds} = 0$.

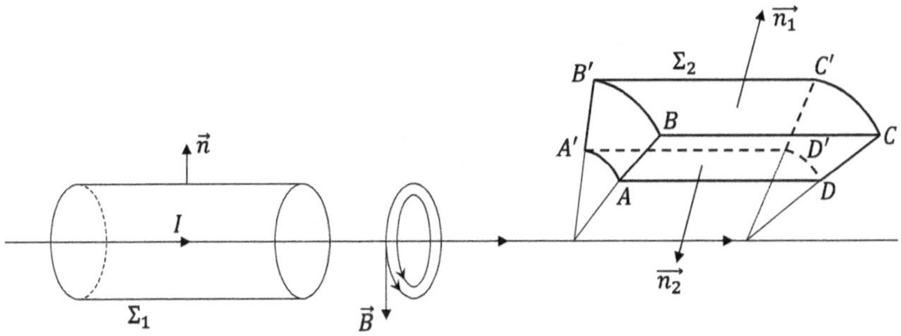

Figure 4.27. Surfaces used in the demonstration of Gauss's law for a magnetic field created by a linear conductor carrying a current of intensity I.

The flux of $\vec{B}$ through the $AA'BB'$ and $DD'CC'$ faces of the closed surface Σ_2 parallel to the field lines are zero, and the net flows through the $ABCD$ and $A'B'C'D'$ faces intersected normally by the field lines are again zero, given that the normal vectors $\vec{n_1}$ and $\vec{n_2}$ are antiparallel. So, for the closed surface Σ_2 which does not contain any current, we may also write $\int_{\Sigma_2} \vec{B}\,\vec{ds} = 0$.

Therefore, regardless of the shape of the closed surface and regardless of whether or not it contains a conductor carrying a current, the flux of a magnetic field passing through a closed surface is zero:

$$\int_{\Sigma} \vec{B}\,\vec{ds} = 0 \tag{4.72}$$

In fact, a demonstration of this theorem can be given in a general context if the Biot–Savart–Laplace formula used in the above reasoning is applied to the current geometry, namely:[6]

$$\begin{aligned}
\int_{\Sigma} \vec{B}\,\vec{ds} &= \int_{V_\Sigma} \nabla\,\vec{B}\,dv = \int_{V_\Sigma} \nabla\left(\frac{\mu_0 I}{4\pi}\int \frac{\vec{dl}\times\vec{R}}{R^3}\right) = \int_{V_\Sigma}\left(\frac{\mu_0 I}{4\pi}\int \nabla\,\frac{\vec{dl}\times\vec{R}}{R^3}\right)\\
&= \int_{V_\Sigma} \frac{\mu_0 I}{4\pi}\int\left[\frac{\nabla(\vec{dl}\times\vec{R})}{R^3} + (\vec{dl}\times\vec{R})\,\nabla\left(\frac{1}{R^3}\right)\right]\\
&= \int_{V_\Sigma} \frac{\mu_0 I}{4\pi}\int\left[\frac{\vec{R}(\nabla\times\vec{dl}) - \vec{dl}(\nabla\times\vec{R})}{R^3} - \frac{3(\vec{dl}\times R)\vec{R}}{R^5}\right] = 0
\end{aligned} \tag{4.73}$$

However, $\nabla \times \vec{dl} = 0$, because the operator ∇ acts on x, y, and z (see figure 4.28), and:

[6] In the equation below, the relation from vector analysis was used: $\nabla(\vec{a}\times\vec{b}) = \vec{b}\;\nabla\times\vec{a} - \vec{a}\;\nabla\times\vec{b}$.

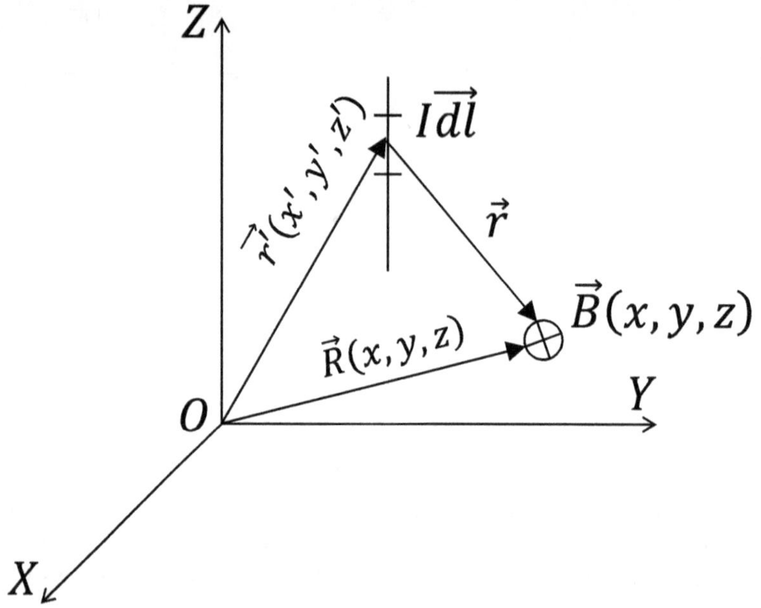

Figure 4.28. Figure justifying the mathematical operations performed to demonstrate Gauss's law for magnetic fields.

$$\begin{cases} \nabla \times \vec{R} = \begin{vmatrix} \vec{i} & \vec{j} & \vec{k} \\ \dfrac{\partial}{\partial x} & \dfrac{\partial}{\partial y} & \dfrac{\partial}{\partial z} \\ x & y & z \end{vmatrix} = 0 \\ \vec{R}(\vec{dl} \times \vec{R}) = \begin{vmatrix} x & y & z \\ dl_x & dl_y & dl_z \\ x & y & z \end{vmatrix} = 0 \end{cases} \tag{4.74}$$

Based on these demonstrations, Gauss's theorem for the magnetic field can be stated as follows: *the flux of a magnetic field passing through a closed surface is zero.* This property is due to the shape of the field lines of magnetic induction, which are always closed lines. The flux of magnetic induction is also constant along of a tube of field lines, i.e. if a tube of field lines is considered, as shown in figure 4.29, we obtain:

$$\int_{\Sigma} \vec{B}\vec{ds} = \int_{S_2} \vec{B}\vec{ds} - \int_{S_1} \vec{B}\vec{ds} = 0 \Longrightarrow \int_{S_2} \vec{B}\vec{ds} = \int_{S_1} \vec{B}\vec{ds} \tag{4.75}$$

In other words, the flux that enters is equal to the flux that leaves the volume bounded by the considered closed surface, and this property is valid for any magnetic field. Thus, Gauss's theorem for magnetic fields can be stated as follows: *the flux of a magnetic field passing through a closed surface is equal to zero, regardless of the shape of the field or the structure of the sources giving rise to it.*

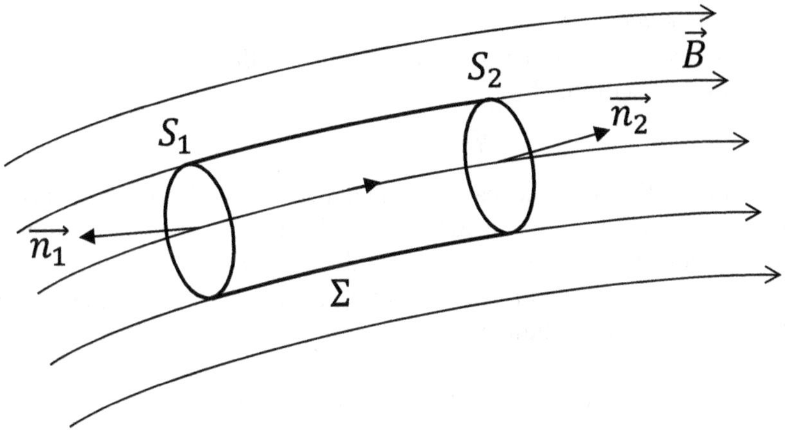

Figure 4.29. Conservation of flux of a magnetic field.

4.2.4.2 Local form of Gauss's theorem for magnetic fields

On the basis of the Gauss–Ostrogradsky theorem, we can write:

$$\int_S \vec{B}\vec{ds} = \int_V (\nabla\vec{B})dv = 0 \tag{4.76}$$

where S is the closed surface and V the volume limited by the surface S, resulting in:

$$\nabla\vec{B} = 0, \tag{4.77}$$

Equation (4.77) represents one of Maxwell's equations for magnetostatics. This can be demonstrated directly on the basis of the Biot–Savart–Laplace formula shown above. Relationships (4.76) and (4.77) represent the integral form and the differential form, respectively, of Gauss's law for magnetic fields. These relationships suggest that the lines of the magnetic field are closed. Unlike the electric field ($\vec{E}$), for which this conclusion is valid only in a space without sources (electric charges), for the magnetic field, this principle is valid everywhere. This corresponds to the fact that in the case of the magnetic field, there is no analog for the electric charges present in an electric field. In other words, in the case of the magnetic field, there are no magnetic charges (magnetic monopoles). The fundamental experimental argument that disproves the existence of magnetic charges lies in the fact that elementary particles brought into a uniform magnetic field with magnetic induction $\vec{B} = \mu_0\vec{H}$ do not experience parallel forces where the intensity of the magnetic field follows $\vec{F} = q_m\vec{H}$.

4.2.5 The vector magnetic potential

Previously, it was shown that one of Maxwell's equations ($\nabla\vec{B} = 0$) describes a property of magnetic fields, namely that they are not created by magnetic charges. However, taking into account the theorem from algebra and vector analysis (stating that: *if the divergence of a vector field is zero, then there is another vector field for which its curl is equal to a vector field that has zero divergence*), we find that for magnetic fields, we can use a vector field $\vec{A}$ called the *vector magnetic potential*, which satisfies the relation:

$$\vec{B} = \nabla \times \vec{A} \tag{4.78}$$

The vector magnetic potential can be calculated using the Biot–Savart–Laplace formula. Considering the case of a conductor of length l, as shown in figure 4.28, and expressing the intensity of the current I as the product of the conduction current density $\vec{j}\,(x', y', z')$ and the surface element $\vec{ds}$, in accordance with the Biot–Savart–Laplace law, the magnetic induction is:

$$\vec{B} = \frac{\mu_0}{4\pi} \int_{V_\Gamma} \frac{\vec{j}\,(x', y', z') \times \vec{r}}{r^3} dv \tag{4.79}$$

Expressing the position vector $\vec{r}$ and the conduction current density $\vec{j}$ in the Cartesian coordinate system, we obtain:

$$\vec{r} = \vec{R} - \vec{r}'(x - x')\vec{i} + (y - y')\vec{j} + (z - z')\vec{k};\ \vec{j} = j_x\vec{i} + j_y\vec{j} + j_z\vec{k} \Longrightarrow$$

$$\vec{j} \times \vec{r} = \begin{vmatrix} \vec{i} & \vec{j} & \vec{k} \\ j_x & j_y & j_z \\ x - x' & y - y' & z - z' \end{vmatrix}$$

Thus, the x-axis component of the magnetic flux density is:

$$B_x(x, y, z) = \frac{\mu_0}{4\pi}\left(\int_V \frac{j_y(z - z') - j_z(y - y')}{r^3} dv \right)$$

However,

$$\frac{z - z'}{r^3} = -\frac{\partial}{\partial z}\left(\frac{1}{r}\right) \quad \text{and} \quad \frac{y - y'}{r^3} = -\frac{\partial}{\partial y}\left(\frac{1}{r}\right)$$

Therefore,

$$B_x(x, y, z) = \frac{\mu_0}{4\pi} \int_V \left[j_z \frac{\partial}{\partial y}\left(\frac{1}{r}\right) - j_y \frac{\partial}{\partial z}\left(\frac{1}{r}\right) \right] dv \tag{4.80}$$

Taking into account that the partial derivatives are taken after y and z, and the density of conduction current, $\vec{j}\,(x', y', z')$, is a function of the coordinates (x', y', z'), the components of the conduction current density can be placed after the derivative operator, so that equation (4.80) becomes:

$$B_x(x, y, z) = \frac{\mu_0}{4\pi} \int_V \left[\frac{\partial}{\partial y}\left(\frac{j_z}{r}\right) - \frac{\partial}{\partial z}\left(\frac{j_y}{r}\right) \right] dv \tag{4.81}$$

The integration is performed on the volume $V(x', y', z')$ of the finished portion of the conductor, and the derivation operator acts on the coordinates of the point where the magnetic induction is calculated. As a result, the integration operator can be swapped with the derivative operator, and equation (4.81) becomes:

$$B_x(x, y, z) = \frac{\partial}{\partial y}\left(\int_V \frac{\mu_0 j_z}{4\pi r} dv\right) - \frac{\partial}{\partial z}\left(\int_V \frac{\mu_0 j_y}{4\pi r} dv\right) \tag{4.82}$$

It can be seen that if $\vec{A}$ denotes the vector field $\vec{A} = \frac{\mu_0}{4\pi}\left(\int_V \frac{\vec{j}(x', y', z')}{r} dv\right)$, on the right-hand side of equation (4.82) there is the component after the x direction that represents the curl of $\vec{A}$. Using the same procedure to compute the components of the magnetic flux density B_y and B_z, the following are obtained:

$$\begin{cases} B_x = \dfrac{\partial A_z}{\partial y} - \dfrac{\partial A_y}{\partial z} \\ B_y = \dfrac{\partial A_x}{\partial z} - \dfrac{\partial A_z}{\partial x} \\ B_z = \dfrac{\partial A_y}{\partial x} - \dfrac{\partial A_x}{\partial y} \end{cases} \tag{4.83}$$

The vector field defined by:

$$\vec{A} = \frac{\mu_0}{4\pi}\left(\int_V \frac{\vec{j}(x', y', z')}{r} dv\right) \tag{4.84}$$

is called the *vector magnetic potential* (or simply, the *vector potential*) and has the property: $\vec{B} = \nabla \times \vec{A}$. The expression for the vector magnetic potential can be determined in a simpler manner by starting from Biot–Savart–Laplace's law and considering the set of coordinates on which the derivation and integration operators act. Thus, for the current geometry used in figure 4.28, we obtain:

$$\begin{aligned} \frac{\mu_0 I}{4\pi} \oint_\Gamma \left(\frac{\vec{dl} \times \vec{R}}{R^3}\right) &= \frac{\mu_0 I}{4\pi} \oint_\Gamma \vec{dl} \times \left(-\nabla \frac{1}{R}\right) = \frac{\mu_0 I}{4\pi} \oint_\Gamma \nabla \frac{1}{R} \times \vec{dl} \\ &= \nabla \times \oint_\Gamma \frac{\mu_0 I \vec{dl}}{4\pi R} \Rightarrow \nabla \times \vec{A} \quad \Rightarrow \quad \vec{A} = \frac{\mu_0 I}{4\pi} \oint_\Gamma \frac{\vec{dl}}{R} \end{aligned} \tag{4.85}$$

The vector field $\vec{A}$ represents the second form of the vector potential of the magnetic field created by a filiform circuit Γ carrying a current of intensity I. This time, the integration is done according to the contour Γ of the circuit rather than the volume of the circuit carrying the conduction current density $\vec{j}$ as used in relation (4.85).

Both relationships are useful for solving problems because it is often more convenient to calculate the vector potential $\vec{A}$ and then the magnetic flux density $\vec{B}$ using the curl applied to $\vec{A}$.

Note: The systems of equations shown in (4.78) or (4.83) have an infinite number of solutions for the vector components $\vec{B}$. This result can be seen immediately if we observe that the vector $\vec{A'} = \vec{A} + \nabla\,\Psi$, where Ψ is some scalar, and has the same curl as the vector $\vec{A}$. Indeed, because the curl of the gradient of any scalar function is zero,

$$\nabla \times \vec{A'} = \nabla \times (\vec{A} + \nabla\,\Psi) = \nabla \times \vec{A} + (\nabla \times \nabla\,\Psi) = \nabla \times \vec{A}$$

Thus, the magnetic vector potential is not uniquely determined by this function. This unicity imposes an additional condition on the vector $\vec{A}$, namely $\nabla\vec{A} = 0$. Indeed, $\vec{A}$ is directly related to $I\vec{dl}$ (from a closed circuit) and there is no reason for it to have a nonzero divergence. *The vector $\vec{A}$ that simultaneously satisfies equations $\vec{B} = \nabla \times \vec{A}$ and $\nabla\vec{A} = 0$ is called the vector magnetic potential of the field $\vec{B}$.* In fact, a vector quantity is perfectly determined if its curl and divergence are known. The existence of this quantity is conditioned by the validity of the law of magnetic flux, respectively by Gauss's law for magnetism. With the help of the vector potential, Gauss's law becomes $\int_{S_\Gamma} \vec{B}\,\vec{ds} = \int_{S_\Gamma} \nabla \times \vec{A}\,ds = \oint_\Gamma \vec{A}\,\vec{dl}$, showing that the flux through any surface which is supported by the closed contour Γ is the same. Although the usefulness of introducing this form is not obvious, it will become apparent in several specific cases where it is used to obtain the magnetic flux density of the magnetic field created by a certain distribution of currents.

Applications

(1) *Demonstration of Rowland's theorem, which states that convection currents create a magnetic field. An electrically charged body whose volume has a charge density of ρ moves uniformly at a speed of $\vec{v}$ along the* OX *axis. Compute the magnetic induction $\vec{B}$ of the magnetic field created by this charged body moving at the speed $\vec{v}$.*

Solution:

The charged body moving at the speed $\vec{v}$ (see figure 4.30) is equivalent to a convection current whose density is $\vec{j_c} = \rho\vec{v}$. In vacuum, this electric current generates a magnetic field whose vector potential is:

$$\vec{A} = \frac{\mu_0}{4\pi}\int_V \frac{\vec{j_c}\,dv}{r} = \frac{\mu_0}{4\pi}\int_V \frac{\rho\vec{v}dv}{r} \tag{4.86}$$

at a distance $\vec{r}$ from the origin of the considered coordinate system. In accordance with the definition of the vector potential, the induced magnetic field at the same point is:

$$\vec{B} = \nabla \times \vec{A} = \mu_0\,\nabla \times \int_V \frac{\rho dv}{4\pi r}\vec{v} = \mu_0\,\nabla \times \varepsilon_0 V\vec{v} \tag{4.87}$$

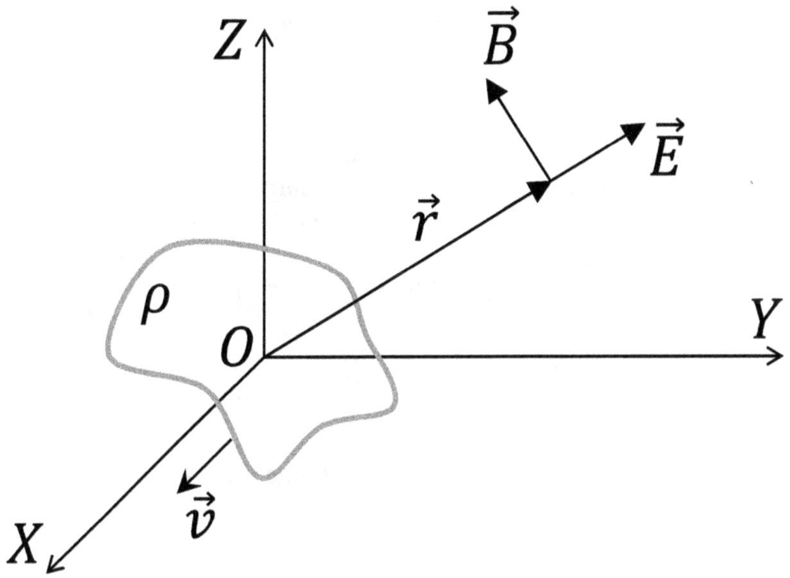

Figure 4.30. *Body charged with an electric charge and moving at a speed $\vec{v}$, creating an electric field of intensity $\vec{E}$ and a magnetic field of induction $\vec{B}$ at a distance $\vec{r}$ from its center.*

The last term of equation (4.87) considers the fact that at the point where the vector potential is calculated, the electric potential created by the charge distribution is $V = \int_V \frac{\rho dv}{4\pi\varepsilon_0 r}$. Since the velocity $\vec{v}$ has the components $(v_x, 0, 0)$, we can write the following expressions for the components of the magnetic flux density:

$$\begin{cases} B_x = \mu_0\varepsilon_0 V\left[\dfrac{\partial}{\partial y}(0) - \dfrac{\partial}{\partial z}(0)\right] = 0 \\ B_y = \mu_0\varepsilon_0\left[\dfrac{\partial}{\partial z}(Vv_x) - \dfrac{\partial}{\partial x}(0)\right] = \mu_0\varepsilon_0 v_x\dfrac{\partial V}{\partial z} = -\mu_0\varepsilon_0 v_x E_z \\ B_z = \mu_0\varepsilon_0\left[\dfrac{\partial}{\partial x}(0) - \dfrac{\partial}{\partial y}(Vv_x)\right] = -\mu_0\varepsilon_0 v_x\dfrac{\partial V}{\partial y} = \mu_0\varepsilon_0 v_x E_y \end{cases} \tag{4.88}$$

With these, the magnitude of magnetic induction is:

$$B = \sqrt{B_y^2 + B_z^2} = \mu_0\varepsilon_0 v\sqrt{E_z^2 + E_y^2} = \mu_0\varepsilon_0 vE = \mu_0 vD \tag{4.89}$$

Indeed, if we consider that a charge distribution which has the density ρ always creates an electric field at a distance $\vec{r}$ from its center, described by its vectors $\vec{E}$ or $\vec{D}$, then, at that point, there is clearly also a magnetic field whose induction is directly related to the vectors $\vec{E}$ or $\vec{D}$ of the electric field, as shown by equations (4.88) and (4.89).

(2) *The flux of the magnetic field is independent of the shape of any surface bounded by a closed contour* Γ.

Solution:

Let us consider two surfaces S_{Γ_1} *and* S_{Γ_2} *that lie on the curve* Γ, *(see figure 4.31). It can be demonstrated that the magnetic flux passing through any of these surfaces is the same. Indeed, using the vector potential* $\vec{A}$, *one can describe the magnetic flux passing through the two surfaces as follows:*

$$\begin{cases} \Phi = \int\limits_{S_{\Gamma_1}} \vec{B}\,\overrightarrow{ds_\Gamma} = \int\limits_{S_{\Gamma_1}} (\nabla \times \vec{A})\overrightarrow{ds_\Gamma} = \int\limits_{\Gamma} \vec{A}\,\vec{dl} \\ \Phi = \int\limits_{S_{\Gamma_2}} \vec{B}\,\overrightarrow{ds_\Gamma} = \int\limits_{S_{\Gamma_2}} (\nabla \times \vec{A})\overrightarrow{ds_\Gamma} = \int\limits_{\Gamma} \vec{A}\,\vec{dl} \end{cases} \tag{4.90}$$

It follows, then, that the flux passing through any surface S bounded by the contour Γ *is the same and depends only on the shape of the contour* Γ.

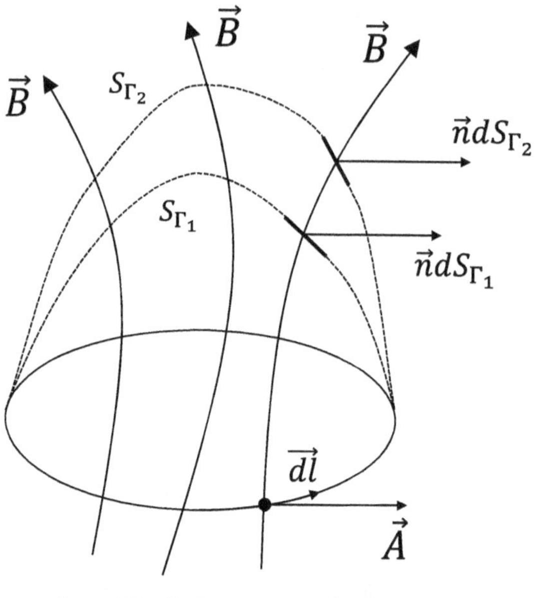

Figure 4.31. *Surfaces supported by the contour* Γ.

4.2.5.1 Multipolar development of the vector potential

In electrostatics, it has been demonstrated that the potential of an electric field generated by a static charge distribution, at a point sufficiently far from the center of the charge distribution, can be expressed (after a corresponding series development) in the form of the sum of the potential of a point charge (the monopole potential), the dipole potential, the quadrupole potential, and so on. Furthermore, by taking into account the relationship $\vec{E} = -\nabla V$, it is possible to express the intensity of an electric field of arbitrary charge distribution as the intensity of an electric point charge, an electric dipole, of an electric quadrupole, etc. Based on the analogy between electrostatics and magnetostatics, which has often been invoked, we will further present the development of the magnetic vector potential at a point far from the center of a current distribution, highlighting the similar terms, namely the potential due to a monopole, dipole, magnetic quadrupole, and so on.

Consider the distribution of currents inside the volume V', characterized at each point $\vec{r'}$ by a density of current $\vec{j}(\vec{r'})$ (see figure 4.32). The vector potential of this current distribution at point $P(\vec{r})$ is:

$$\vec{A}(\vec{r}) = \frac{\mu_0}{4\pi}\int_{V'}\frac{\vec{j}(\vec{r'})}{R}dv' \tag{4.91}$$

where:

$$\begin{cases} \vec{R} = \vec{r} - \vec{r'} = (x - x')\vec{i} + (y - y')\vec{j} + (z - z')\vec{k} \\ R = \sqrt{r^2 + r'^2 - 2rr'\cos\theta'} \end{cases}$$

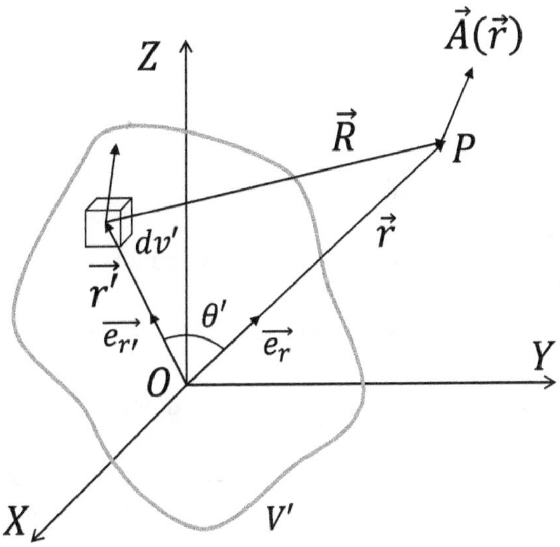

Figure 4.32. Multipolar expansion of the vector potential.

The condition r, $R \gg r'$ results in:

$$\begin{cases} R = r\left[1 + \left(\frac{r'}{r}\right)^2 - 2\frac{r'}{r}\cos\theta'\right]^{1/2} \\ \frac{1}{R} = \frac{1}{r}\left[1 + \left(\frac{r'}{r}\right)^2 - 2\frac{r'}{r}\cos\theta'\right]^{-1/2} \end{cases} \tag{4.92}$$

If $\frac{r'}{r} \ll 1$, the result is:

$$\begin{cases} \dfrac{1}{R} = \dfrac{1}{r}\left[1 - \dfrac{1}{2}\left(\dfrac{r'^2}{r^2} - 2\dfrac{r'}{r}\cos\theta'\right) + \dfrac{3}{8}\left(\dfrac{r'^2}{r^2} - 2\dfrac{r'}{r}\cos\theta'\right)^2 + \cdots = \right] \\ = \dfrac{1}{r}\left(1 - \dfrac{1}{2}\dfrac{r'^2}{r^2} + \dfrac{r'}{r}\cos\theta' + \dfrac{3}{8}\dfrac{4r'^2}{r^2}\cos^2\theta' + \ldots\right) = \\ = \dfrac{1}{r}\left(1 + \dfrac{r'}{r}\cos\theta' + \dfrac{r'^2}{r^2}\dfrac{3\cos^2\theta' - 1}{2} + \ldots\right) \end{cases} \tag{4.93}$$

Taking this into account, the vector potential of the electric current's distribution becomes:

$$\begin{cases} \vec{A}(\vec{r}) = \dfrac{\mu_0}{4\pi}\displaystyle\int_{V'} \dfrac{\vec{j}(\vec{r'})}{R} dv' = \dfrac{\mu_0}{4\pi}\displaystyle\int_{V'} \dfrac{\vec{j}(\vec{r'})}{r} dv' + \\ + \dfrac{\mu_0}{4\pi}\displaystyle\int_{V'} \dfrac{\vec{j}(\vec{r'})(\vec{r'}\vec{r})}{r^3} dv' + \dfrac{\mu_0}{4\pi}\displaystyle\int_{V'} \vec{j}(\vec{r'})\left[\dfrac{3(\vec{r'}\vec{r})^2}{2r^5} - \dfrac{r'^2}{2r^3}\right] dv' = \\ = \dfrac{\mu_0}{4\pi}\displaystyle\int_{V'} \dfrac{\vec{j}(\vec{r'})}{r} dv' + \dfrac{\mu_0}{4\pi}\displaystyle\int_{V'} \dfrac{\vec{j}(\vec{r'})(\vec{r'}\vec{e_r})}{r^2} dv' + \\ + \dfrac{\mu_0}{4\pi}\displaystyle\int_{V'} \vec{j}(\vec{r'})\dfrac{r'^2[3(\vec{e_r}'\vec{e_r}) - 1]}{2r^3} dv' \end{cases} \tag{4.94}$$

According to the last line of (4.94), the vector potential can be written as:

$$\vec{A}(\vec{r}) = \overrightarrow{A_m}(\vec{r}) + \overrightarrow{A_d}(\vec{r}) + \overrightarrow{A_q}(\vec{r})$$

where $\overrightarrow{A_m}(\vec{r})$, $\overrightarrow{A_d}(\vec{r})$ and $\overrightarrow{A_q}(\vec{r})$ are the monopole, dipole, and magnetic quadrupole contributions to the vector potential at point $P(\vec{r})$, respectively. It can be seen that, as in the case of the electric potential, the dependence of these terms on distance $\vec{r}$ takes the form of $\frac{1}{r}$ for the magnetic monopole term, $\frac{1}{r^2}$ for the magnetic dipole term, $\frac{1}{r^3}$ for the magnetic quadrupole term, etc. The further we depart from the center of the current distribution, the more the contribution of the magnetic potentials of higher degrees (octupoles, etc.) to the total magnetic potential at point $P(\vec{r})$ becomes increasingly weak. It can also be noted that in each term of the vector potential development (4.94), the integrant depends both on the coordinates of the current distribution (via $\vec{r'}$) and on the position of the observation point (via $(\vec{e_r}, \vec{r'}) = r'\cos\theta'$). Upon analyzing each term of the vector potential development in turn (4.94), one will understand the significance of the monopole, dipole, and magnetic quadrupole potentials.

(a) ***Magnetic monopole***

In the case of the distribution of electric charges, this term for the electrical potential, which depends on $\frac{1}{r}$, is associated with the potential of a point charge. Since it varies as $\frac{1}{r}$, it makes the most important contribution to the value of the electric potential of the charge distribution. In the case of the magnetic field, as stated earlier in the interpretation of Gauss's law of magnetism, the monopole potential is zero because there are no magnetic charges (magnetic monopoles).

Indeed: (i) stationary electric currents are associated only with closed circuits and (ii) the origin of these currents is the volume V' (currents do not enter through the surface S' that delimits the volume V'). Based on these observations, one can imagine the currents in volume V' circulating through closed current tubes that fill the entire volume without leaving it. If we consider the current intensity through the jth current tube (whose cross-section is a_j), denoted by $I_j = \int_{a_j} \vec{j_j}\,\overrightarrow{da_j}$, then:

$$\begin{aligned}\overrightarrow{A_m}(\vec{r}) &= \frac{\mu_0}{4\pi}\int_{V'}\frac{\vec{j}\,(\vec{r'})}{r}dv' = \frac{\mu_0}{4\pi}\sum\nolimits_j\int_{V'}\frac{\vec{j_j}\,(\vec{r'})\overrightarrow{da_j}\,\overrightarrow{dl_j}}{r} \\ &= \frac{\mu_0}{4\pi}\sum_j\oint_{\Gamma_j}\frac{I_j\overrightarrow{dl_j}}{r} = \frac{\mu_0}{4\pi}\sum_{j-1}^{n} I_j\left(\oint_{\Gamma_j}\frac{\overrightarrow{dl_j}}{r}\right) = 0\end{aligned} \tag{4.95}$$

Since each current tube is a closed circuit, $\oint_{\Gamma_j}\overrightarrow{dl_j} = 0$, equation (4.95) shows that the monopole term makes a zero contribution to the vector potential.

(b) ***Magnetic dipole***

The term corresponding to the magnetic dipole potential in (4.94) may be processed in such a way that it is expressed as a function of the magnetic moment of the dipole, after a long calculus [1], as follows:

$$\overrightarrow{A_d}(\vec{r}) = \frac{\mu_0}{4\pi}\int_{V'}\frac{\vec{j}\,(\vec{r'})(\vec{r'}\cdot\vec{e_r})}{r^2}dv' = \frac{\mu_0}{4\pi}\frac{\vec{m}\times\vec{r}}{r^3} \tag{4.96}$$

where:

$$\vec{m} = \frac{1}{2}\int_{V'}[\vec{r'}\times j(\vec{r'})]dv' \tag{4.97}$$

$\overrightarrow{m}$ is called the **magnetic moment** of the distribution of currents. It is given by the integral $\frac{1}{2}\int_{V'}[\vec{r'}\times\vec{j}\,(\vec{r'})]dv'$, which depends only on the distribution of some currents from volume V', and is a specific quantity for the distribution of currents from volume V'. Expression (4.96) for the magnetic vector potential of dipole is valid for any distribution of currents, since it has been deduced by considering a certain distribution of currents but without specifying the spatial shape of the currents in volume V'. The only restriction imposed is that the point at which $\overrightarrow{A_d}$ is calculated must be sufficiently far removed from the distribution of currents. From (4.96), it follows that the magnetic vector potential of a magnetic dipole is perpendicular to the plane $(\vec{m}, \vec{r})$, and its direction is given by the RSR for the cross product $\overrightarrow{m} \times \overrightarrow{r}$.

The distribution of current most commonly encountered in practice is a plane loop of any shape carrying a current of intensity I (see figure 4.33). If the conductor section is $\vec{a}$, then $I = \vec{a} \cdot \vec{j}$. The magnetic moment $\vec{m}$ of the loop is also defined by (4.97), as follows:

$$\vec{m} = \frac{1}{2}\int_{V'} [\vec{r'} \times j(\vec{r'})]dv' = \frac{1}{2}\int_{V'} (\vec{r'} \times \vec{j}\,)\vec{a}\vec{dl} = \frac{1}{2}\oint_{\Gamma} [\vec{r'} \times (\vec{j} \cdot \vec{a}\vec{dl})]$$
$$= \frac{1}{2} I \oint_{\Gamma} \vec{r'} \times \vec{dl} = I \cdot \frac{1}{2}\oint_{\Gamma} \vec{r'} \times \vec{dl} = I \cdot \vec{S} = IS\vec{n} \tag{4.98}$$

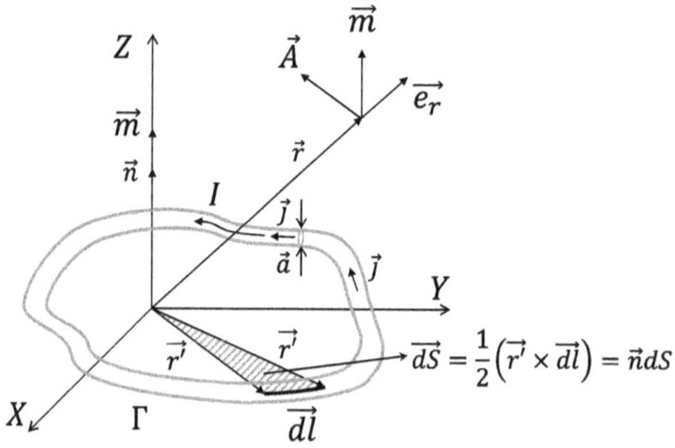

Figure 4.33. Calculation of the magnetic moment of a current loop.

It can be seen that $\frac{1}{2}(\vec{r'} \times \vec{dl})$ is the area of the surface element $\overrightarrow{ds}$ shown by a dashed area in figure 4.33; this area, integrated over the curve Γ, gives the total surface area of the loop oriented by the normal vector to surface $\vec{n}$. In the case of this plane current loop, which has a moment of $\vec{m} = I\vec{S} = IS\vec{n}$, the magnetic vector potential of the magnetic dipole is given by equation (4.96):

$$\overrightarrow{A_d} = \frac{\mu_0}{4\pi}\frac{\vec{m} \times \vec{r}}{r^3} = \frac{\mu_0}{4\pi}\frac{m \sin\theta}{r^2}(\vec{n} \times \vec{e_r}) \tag{4.99}$$

In figure 4.33, the vector potential of the current loop is present at any point $P(\vec{r})$ located at a distance $\vec{r}$ from the origin of the coordinate system, and it can be seen that it is parallel to the electric current passing through the loop.

(c) ***Magnetic quadrupole***

The third term in the equation (4.94) of the vector potential is:

$$\overrightarrow{A_q} = \frac{\mu_0}{4\pi}\int_{V'} \frac{\vec{j}(\vec{r'})r'^2[3(\vec{e_{r'}}\cdot\vec{e_r}) - 1]}{r^3}dv' \tag{4.100}$$

which is the magnetic potential of a quadrupole. It is dependent on $\frac{1}{r^3}$ and is even less of interest, so if the monopole potential is zero anyway and the quadruple potential $\overrightarrow{A_q}$ is very weak, then the vector potential of a certain current distribution is practically reduced to the dipole magnetic potential, i.e. the second term in equation (4.94); therefore, in general, it can be written as:

$$\vec{A} = \frac{\mu_0}{4\pi}\frac{\vec{m}\times\vec{r}}{r^3} \tag{4.101}$$

for any current distribution with the dipole moment $\vec{m}$.

Magnetic field of a current loop

Returning again to the current loop in figure 4.33, we aim to calculate the magnetic induction $\vec{B}$ at a point outside it at a distance $\vec{r}$ from the origin of the reference system (RS) using the vector potential. We want to express $\vec{B}$ according to its specific physical size ($\vec{m}$), which is the moment of the magnetic dipole:

$$\vec{B} = \nabla\times\vec{A};\ \vec{A} = \frac{\mu_0}{4\pi}\frac{\vec{m}\times\vec{r}}{r^3} \tag{4.102}$$

However,

$$\vec{m}\times\vec{r} = \begin{vmatrix} \vec{i} & \vec{j} & \vec{k} \\ m_x & m_y & m_z \\ x & y & z \end{vmatrix} = \begin{vmatrix} \vec{i} & \vec{j} & \vec{k} \\ 0 & 0 & m_z \\ x & y & z \end{vmatrix} = -\vec{i}m_z y + \vec{j}m_z x \tag{4.103}$$

With $m_z = m$, we obtain:

$$\vec{A} = -\frac{\mu_0 m y}{4\pi r^3}\vec{i} + \frac{\mu_0 m x}{4\pi r^3}\vec{j} = A_x\vec{i} + A_y\vec{j}$$

Then:

$$\vec{B} = \nabla\times\vec{A} = \begin{vmatrix} \vec{i} & \vec{j} & \vec{k} \\ \frac{\partial}{\partial x} & \frac{\partial}{\partial y} & \frac{\partial}{\partial z} \\ A_x & A_y & A_z \end{vmatrix} = \begin{vmatrix} \vec{i} & \vec{j} & \vec{k} \\ \frac{\partial}{\partial x} & \frac{\partial}{\partial y} & \frac{\partial}{\partial z} \\ A_x & A_y & 0 \end{vmatrix} \Longrightarrow$$

$$\Longrightarrow B_x = \vec{i}\left(-\frac{\partial A_y}{\partial z}\right) = -\vec{i}\frac{\mu_0 m}{4\pi}\frac{\partial}{\partial z}\left(\frac{x}{r^3}\right) = -\vec{i}\frac{\mu_0 mx}{4\pi}\frac{\partial}{\partial z}(x^2 + y^2 + z^2)^{-3/2}$$

$$= -\vec{i}\frac{\mu_0 m}{4\pi}x \cdot 2z\left(-\frac{3}{2}\right)\frac{1}{r^5} = \frac{\mu_0 m}{4\pi r^3}\left(\frac{3xz}{r^2}\right)\vec{i}$$

Following the same calculus:

$$B_y = \vec{j}\frac{\partial}{\partial z}A_x = -\vec{j}\frac{\partial}{\partial z}\left(\frac{\mu_0 my}{4\pi r^3}\right) = +\frac{\mu_0 m}{4\pi r^3}\left(\frac{3yz}{r^2}\right)\vec{j}$$

and:

$$B_z = \vec{k}\left[\frac{\partial A_y}{\partial x} - \frac{\partial A_x}{\partial y}\right] = \vec{k}\frac{\mu_0 m}{4\pi}\left[\frac{\partial}{\partial x}\left(\frac{x}{r^3}\right) + \frac{\partial}{\partial y}\left(\frac{y}{r^3}\right)\right]$$

$$= \frac{\mu_0 m}{4\pi}\left[\begin{array}{c}\dfrac{(x^2 + y^2 + z^2)^{\frac{3}{2}} - \frac{3}{2}x \cdot 2x((x^2 + y^2 + z^2)^{\frac{1}{2}})}{(x^2 + y^2 + z^2)^3} + \\ \dfrac{(x^2 + y^2 + z^2)^{\frac{3}{2}} - \frac{3}{2}y \cdot 2y(x^2 + y^2 + z^2)^{\frac{1}{2}}}{(x^2 + y^2 + z^2)^3}\end{array}\right]\vec{k}$$

$$= \frac{\mu_0 m}{4\pi}\frac{(x^2 + y^2 + z^2 - 3x^2) + (x^2 + y^2 + z^2 - 3y^2)}{(x^2 + y^2 + z^2)^{\frac{5}{2}}}\vec{k}$$

$$= \frac{\mu_0 m}{4\pi}\frac{(3z^2 - r^2)}{r^5}\vec{k}$$

Therefore,

$$\begin{gathered}\vec{B} = \frac{\mu_0 m}{4\pi r^5}[3(xz\vec{i} + yz\vec{j}\,) + (3z^2 - r^2)\vec{k}] = \\ \frac{\mu_0 m}{4\pi}\left[\frac{3z\vec{r}}{r^4 \cdot r} - \frac{r^2}{r^3 \cdot r^2}\vec{k}\right] = \frac{\mu_0}{4\pi}\left(\frac{3mz\vec{r}}{r^3 \cdot r^2} - \frac{m\vec{k}}{r^3}\right) \\ \frac{\mu_0}{4\pi}\left[\frac{3(\vec{m} \cdot \vec{r}) \cdot \vec{r}}{r^3 \cdot r^2} - \frac{\vec{m}}{r^3}\right] = \frac{\mu_0}{4\pi r^3}\left[\frac{3(\vec{m} \cdot \vec{r})\vec{r}}{r^2} - \vec{m}\right]\end{gathered} \tag{4.104}$$

Indeed, $\vec{m} = m_x\vec{i} + m_y\vec{j} + m_z\vec{k} = m_z\vec{k}$, and $\vec{m} \cdot \vec{r} = m_z z$ because $m_x = m_y = 0$. It follows that in the equation above, $3mz\vec{r} = 3(\vec{m} \cdot \vec{r})\vec{r}$, resulting in:

$$\vec{B} = \frac{\mu_0}{4\pi r^3}\left[\frac{3(\vec{m} \cdot \vec{r})\vec{r}}{r^2} - \vec{m}\right] \tag{4.105}$$

Using the spherical coordinate system r, θ and φ, we obtain:

$$x = r\sin\theta\cos\varphi, \quad y = r\sin\theta\sin\varphi, \quad z = r\cos\theta \Rightarrow$$

$$\Rightarrow \begin{cases} B_x = \dfrac{\mu_0 m 3xz}{4\pi r^5} = \dfrac{\mu_0 m}{4\pi r^3}(3\sin\theta\cos\theta\cos\varphi) \\ B_y = \dfrac{\mu_0 m 3yz}{4\pi r^5} = \dfrac{\mu_0 m}{4\pi r^3}(3\sin\theta\cos\theta\sin\varphi) \\ B_z = \dfrac{\mu_0 m}{4\pi}\dfrac{(3\cos^2\theta - 1)}{r^3} \end{cases} \tag{4.106}$$

Comparing relation (4.105), or the component (4.106) with the electric field of an electric dipole located at the origin of the coordinate system that has its moment $\vec{p}$ oriented in the same way as $\vec{m}$:

$$\vec{E} = \frac{1}{4\pi\varepsilon_0 r^3}\left[\frac{3(\vec{p}\cdot\vec{r})\vec{r}}{r^2} - \vec{p}\right] \tag{4.107}$$

we find that they are expressed by the same relationship. Therefore, the topography of the magnetic field created by a current loop that has a moment of $\vec{m}$ at long distances from the loop is identical to the topography of an electric field created by an electric dipole that has a moment of $\vec{p}$ placed at the origin of the coordinate system and oriented in the same way as $\vec{m}$. However, at points close to the electric dipole and the circular current loop, and especially between the charges of the electric dipole and inside the current loop, the electric field of the electric dipole is totally different from the magnetic field of the loop.

It should be noted that between the electrical charges of an electric dipole, the electric field is oriented in the opposite direction to $\vec{p}$, while the magnetic field inside the loop is oriented in the same direction as $\vec{m}$, although outwardly, the topographies of the fields are similar. The topography of the lines of the magnetic field is such that it ensures the validity of the equation $\nabla\vec{B} = 0$ at all points of the field and therefore also inside the source. The lines of the magnetic field are closed; they do not have a beginning and an end, as is the case for an electric field. Its relationship, $\vec{B}$ obtained for components starting from $\vec{B} = \nabla\times\vec{A}$, can be directly established by resorting to vector analysis as follows:

$$\vec{B} = \nabla\times\vec{A} = \frac{\mu_0}{4\pi}\nabla\times\left(\frac{\vec{m}\times\vec{r}}{r^3}\right) = \frac{\mu_0}{4\pi}\left[\frac{-3\vec{r}\times(\vec{m}\times\vec{r})}{r^5} + \frac{\nabla\times(\vec{m}\times\vec{r})}{r^3}\right]$$

$$= \frac{\mu_0}{4\pi}\left[\frac{\nabla\times(\vec{m}\times\vec{r})}{r^3} - \frac{3(\vec{m}\cdot\vec{r}\vec{r})}{r^5} + \frac{3\vec{r}(\vec{m}\cdot\vec{r})}{r^5}\right] =$$

$$= \frac{\mu_0}{4\pi}\left\{\frac{[(\nabla\cdot\vec{r})\vec{m} - (\nabla\cdot\vec{m})\vec{r} + (\vec{r}\cdot\nabla)\vec{m} - (\vec{m}\cdot\nabla)\vec{r}]}{r^3} - \frac{3\vec{m}}{r^3} + \frac{3(\vec{m}\cdot\vec{r})\vec{r}}{r^5}\right\}$$

$$= \frac{\mu_0}{4\pi}\left[\frac{(3\vec{m} - 0 + 0 - \vec{m})}{r^3} - \frac{3\vec{m}}{r^3} + \frac{3(\vec{m}\cdot\vec{r})\vec{r}}{r^5}\right]$$

$$= \frac{\mu_0}{4\pi r^3}\left[\frac{3(\vec{m}\cdot\vec{r})\vec{r}}{r^2} - \vec{m}\right]$$

Alternatively, the easiest way to compute $\vec{B}$ is to use the relationship between $\vec{B}$ and the scalar magnetic potential V_m, namely, $\vec{B} = -\mu_0 \nabla V_m$, where:

$$V_m = \frac{I\Omega}{4\pi} = \frac{I}{4\pi}\int_{S_\Gamma} \frac{\vec{ds}\cdot\vec{r}}{r^3} = \frac{1}{4\pi}\int_{S_\Gamma} \frac{I\vec{ds}\cdot\vec{r}}{r^3} = \frac{1}{4\pi}\int_{S_\Gamma} \frac{\vec{dm}\cdot\vec{r}}{r^3} = \frac{\vec{m}\cdot\vec{r}}{4\pi r^3} \tag{4.108}$$

Then,

$$\vec{B} = -\mu_0 \nabla \frac{\vec{m}\cdot\vec{r}}{4\pi r^3} = -\frac{\mu_0}{4\pi}\left[\frac{\nabla(\vec{m}\cdot\vec{r})}{r^3} + (\vec{m}\cdot\vec{r})\nabla\frac{1}{r^3}\right]$$

$$= -\frac{\mu_0}{4\pi}\left[\frac{\vec{m}}{r^3} - \frac{3(\vec{m}\cdot\vec{r})\vec{r}}{r^5}\right] = \frac{\mu_0}{4\pi r^3}\left[\frac{3(\vec{m}\cdot\vec{r})\vec{r}}{r^2} - \vec{m}\right] \tag{4.109}$$

4.2.6 Analogies with electrostatics

(1) In electrostatics, the physical phenomena are produced by charge distributions characterized by charge density (ρ) and are described by the vector fields $\vec{E}$ and $\vec{D} = \varepsilon_0\vec{E} + \vec{P}$, whose properties are given by the relations:

$$\nabla \times \vec{E} = 0,\ \vec{E} = -\nabla V,\ \nabla\vec{D} = \rho \tag{4.110}$$

In homogeneous and isotropic dielectrics,

$$\vec{D} = \varepsilon_0\vec{E} + \vec{P} = \varepsilon_0\vec{E} + \varepsilon_0\chi_E\vec{E} = \varepsilon_0\vec{E} + \varepsilon_0(\varepsilon_r - 1)\vec{E}$$

$$= \varepsilon_0\varepsilon_r\vec{E} = \varepsilon\vec{E} \tag{4.111}$$

finally resulting in the Poisson equation:

$$\nabla\vec{D} = \rho,\ \nabla\vec{E} = \frac{\rho}{\varepsilon},\ \vec{E} = -\nabla V \Rightarrow \Delta V = -\frac{\rho}{\varepsilon} \tag{4.112}$$

the solution of which is:

$$V = \frac{1}{4\pi\varepsilon}\int_V \frac{\rho dv}{r} \tag{4.113}$$

If $\rho = 0$, the Laplace equation ($\Delta V = 0$) is valid.

(2) In magnetism, the physical phenomena are produced by the current distributions described by the current density $\vec{j}$ (or by equivalent magnets) and are characterized by the vectorial physical quantities $\vec{B}$ and $\vec{H} = \frac{\vec{B}}{\mu_0} - \vec{M}$, for which the following equations are valid:

$$\nabla \times \vec{H} = \vec{j},\ \vec{B} = \nabla \times \vec{A},\ \nabla\vec{B} = 0 \tag{4.114}$$

In a perfectly homogeneous and isotropic medium, $\vec{M} = \chi_m \vec{H}$, and therefore:

$$\vec{B} = \mu_0(\vec{H} + \vec{M}) = \mu_0(\vec{H} + \chi_m \vec{H}) = \mu_0 \mu_r \vec{H} = \mu \vec{H} \tag{4.115}$$

Since the substance is homogeneous and isotropic, we obtain:

$$\nabla \times \vec{B} = \mu \vec{j}, \ \vec{B} = \nabla \times \vec{A}, \ \nabla \vec{B} = 0 \tag{4.116}$$

Therefore,

$$\nabla \times \vec{B} = \mu \vec{j} \quad \Rightarrow \quad \nabla \times (\nabla \times \vec{A}) = \mu \vec{j}$$

$$\Rightarrow \nabla (\nabla \vec{A}) - \nabla^2 \vec{A} = \mu \vec{j} \ \nabla (\nabla \vec{A}) - \Delta \vec{A} = \mu \vec{j} \tag{4.117}$$

Using the condition imposed on the magnetic vector potential, namely:

$$\nabla \vec{A} = 0 \tag{4.118}$$

we obtain the Poisson equation for the vector potential:

$$\Delta \vec{A} + \mu \vec{j} = 0 \tag{4.119}$$

which has the solution:

$$\vec{A} = \frac{\mu}{4\pi} \int_V \frac{\vec{j}\, dv}{r} \tag{4.120}$$

Alternatively, if $\vec{j} = 0$, the Laplace equation results:

$$\Delta \vec{A} = 0 \tag{4.121}$$

In other words, in the magnetostatic formalism, the magnetic vector potential $\vec{A}$ defined by $\vec{B} = \nabla \times \vec{A}$ and $\nabla \vec{A} = 0$ behaves similarly to the scalar potential in electrostatics, i.e. it satisfies the same Poisson and Laplace equations and is expressed by a similar relationship (4.120), depending on the sources of the magnetic field.

4.3 Lorentz force

The action of the magnetic field on the electric current is essentially the action of the magnetic field on the moving electric charges. The force with which the magnetic field acts on a microscopic scale on a moving electric charge is called the **Lorentz force**. The expression for the Lorentz force can be established using the expression for the electromagnetic force (Laplace) with which the magnetic field acts on an electric current. Thus, considering a portion of a conductor of length $\vec{dl}$ carrying a current of intensity I (see figure 4.34) on the current element $I\vec{dl}$, the magnetic field $\vec{B}$ acts with the force:

$$\vec{dF} = I\vec{dl} \times \vec{B} \tag{4.122}$$

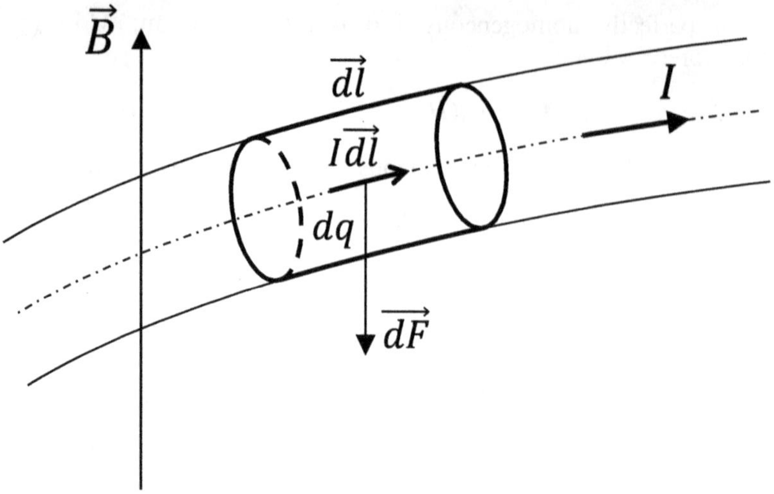

Figure 4.34. Element of length $\vec{dl}$ which is part of a conductor carrying a current in the magnetic induction field $\vec{B}$.

If dq is an elementary positive charge carried by a current of intensity I, in a time interval dt, through a portion of conductor of length $\vec{dl}$, then:

$$I = \frac{dq}{dt} \tag{4.123}$$

and the speed of movement of the charge is:

$$\vec{v} = \frac{\vec{dl}}{dt} \tag{4.124}$$

Thus, the element of length $\vec{dl}$ in the conductor can be expressed in terms of the speed of movement $\vec{v}$ of the elementary charge dq through the relation:

$$\vec{dl} = \vec{v}dt \tag{4.125}$$

Returning to the relationship of the Laplace force, upon combining (4.122) with (4.123) and (4.125), we get:

$$\vec{dF} = \frac{dq}{dt}(\vec{v}dt \times \vec{B}) = dq(\vec{v} \times \vec{B}) \tag{4.126}$$

where dt is a scalar. By integration over the entire charge, we obtain the force with which the magnetic field $\vec{B}$ acts on the charge q that flows through the conductor:

$$\vec{F} = \int_0^q dq(\vec{v} \times \vec{B}) = q(\vec{v} \times \vec{B}) \tag{4.127}$$

This is the formula for the Lorentz force. It can be seen that the force is perpendicular to the plane determined by the speed of the charge and the magnetic induction, while the direction and sense are given by the vector product, $\vec{v} \times \vec{B}$, modified to adopt the sign of the charge. For example, in the case of the electron, the Lorentz force is: $\vec{F} = -e(\vec{v} \times \vec{B})$, i.e. it acts in the opposite direction to the vectorial product $\vec{v} \times \vec{B}$. The Lorentz force is of a nonelectrical nature, being equivalent to the force produced by an induced electric field with the intensity:

$$\vec{E_i} = \frac{\overrightarrow{F_{\text{nonelectric}}}}{q} = \frac{-e(\vec{v} \times \vec{B})}{-e} = \vec{v} \times \vec{B} \tag{4.128}$$

4.3.1 Movement of an electron in a magnetic field $\vec{B}$

The motion of an electron in a magnetic field can easily be analyzed using the equation of motion:

$$\vec{F} = m\frac{d^2\vec{r}}{dt^2} \tag{4.129}$$

If the electron enters a magnetic field that has a magnetic induction of $\vec{B}$, travelling at a velocity of $\vec{v}$ (see figure 4.35(a)), then the components of the velocity are:

$$v_x = \frac{dx}{dt}, \quad v_y = \frac{dy}{dt}, \quad v_z = \frac{dz}{dt} \tag{4.130}$$

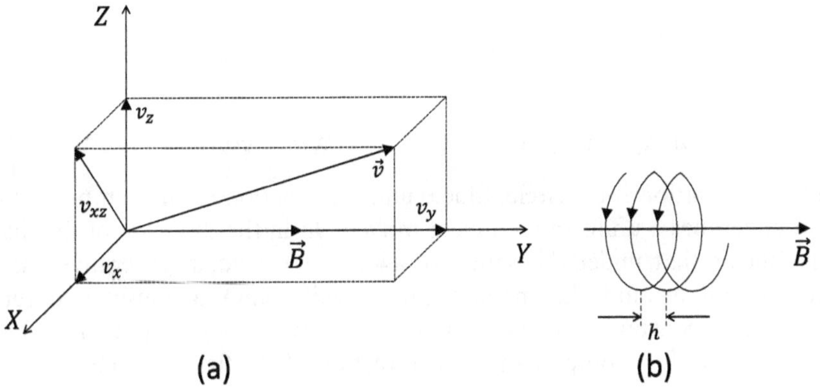

Figure 4.35. (a) A particle that penetrates a magnetic field of induction $\vec{B}$ at a velocity of $\vec{v}$. (b) Trajectory of the particle.

The v_y component of the velocity, parallel to the magnetic flux density, does not determine the action of the Lorentz force. As a result, the electron moves in the direction of the magnetic field with $v_y = \text{const}$. For the other two directions, i.e. those perpendicular to $\vec{B}$, the result is:

$$\begin{cases} m\dfrac{d^2x}{dt^2} = +e\dfrac{dz}{dt}B \\ m\dfrac{d^2z}{dt^2} = -e\dfrac{dx}{dt}B \end{cases} \tag{4.131}$$

Multiplying these terms diagonally, we obtain:

$$-emB\frac{d^2x}{dt^2}\cdot\frac{dx}{dt} = emB\frac{d^2z}{dt^2}\cdot\frac{dz}{dt} \Longrightarrow \frac{dx}{dt}\cdot\frac{d^2x}{dt^2} + \frac{dz}{dt}\cdot\frac{d^2z}{dt^2} = 0 \tag{4.132}$$

Integrating once in relation to time leads to:

$$\left(\frac{dx}{dt}\right)^2 + \left(\frac{dz}{dt}\right)^2 = v_{xz}^2 = \text{const} \tag{4.133}$$

which illustrates that under the action of the magnetic field, the velocity of the electron is kept constant; therefore, a magnetic field does not produce any variation in the kinetic energy of the electron, unlike an electric field. Using the equations of motion, (4.131), one can deduce the equation for the electron's trajectory, as follows:

$$\begin{cases} m\dfrac{dx}{dt} = eBz + A \\ m\dfrac{dz}{dt} = -eBx + C \end{cases} \tag{4.134}$$

Then:

$$m^2\left[\left(\frac{dx}{dt}\right)^2 + \left(\frac{dz}{dt}\right)^2\right] = (A + eBz)^2 + (C - eBx)^2 \tag{4.135}$$

or:

$$m^2 v_{xz}^2 = (A + eBz)^2 + (C - eBx)^2 = \text{const} \tag{4.136}$$

which is the equation for a circle. Since, under the action of the component v_y, the electron undergoes a uniform rectilinear movement in the direction of the magnetic field, it follows that under the action of a magnetic field, it generally describes a circular motion around the magnetic field while simultaneously undergoing a translation at a constant velocity v_y along the magnetic field (see figure 4.35(b)). If its velocity has only a component in the direction of the field, v_y, and $v_{xz} = 0$, then the electron moves parallel to $\vec{B}$ with velocity v_y, while if $v_y = 0$ and $v_{xz} \neq 0$, then the electron describes a circle in a plane perpendicular to $\vec{B}$.

When $\vec{v_0} \perp \vec{B} \Rightarrow F_L = qv_0B = \dfrac{mv_0^2}{r} \Rightarrow r = \dfrac{mv_0}{qB}$, the radius of the trajectory, i.e. $\omega = \dfrac{v_0}{r} = \dfrac{qB}{m} = \dfrac{2\pi}{T_0} \Rightarrow T_0 = \dfrac{2\pi m}{qB}$; the period of the trajectory is therefore independent of speed. The operation of particle accelerators is based on this phenomenon; here, we present the cyclotron (discovered in 1932). It is formed from a metal cylinder separated into two halves, called semi-cylinders (see figure 4.36). An alternating voltage is applied between them that has an amplitude of U and a period of:

$$T = T_0 = \frac{2\pi m}{qB} \tag{4.137}$$

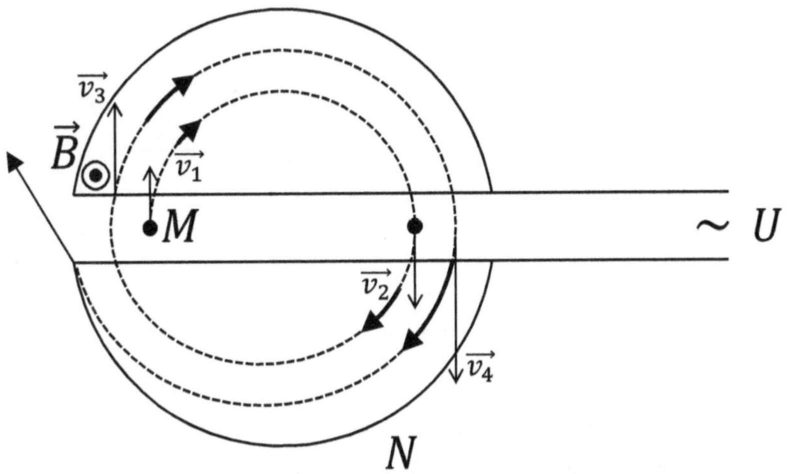

Figure 4.36. Schematic representation of the cyclotron.

A positive particle enters the cyclotron at point M when the voltage U has a positive polarity on the lower semi-cylinder and a negative polarity on the upper one), and is accelerated by a voltage of the order of 10^4 V. After half a period of time, the particle reaches point N; at this time, the polarity is changed (so that negative polarity is applied to the lower semi-cylinder and positive polarity to the upper one), and the particle acquires a new impulse proportional to the voltage U. The maximum energy at which it can be accelerated is:

$$E = \frac{mv_{\max}^2}{2} = m\frac{4\pi^2 R^2}{2T^2} = \frac{q^2 B^2 R^2}{2m} \tag{4.138}$$

which depends on B and R. The increased velocity of the particle after several cycles of acceleration increases the radius of the trajectory and also increases the mass m of the particle according to the relativistic relationship of mass:

$$m = \frac{m_0}{\sqrt{1 - \frac{v^2}{c^2}}} \tag{4.139}$$

Therefore, the duration of the cycle (T_0) also automatically increases, which limits the performance of the cyclotron in the sense that the cyclotron is desynchronized—the rotational period of the particle diverges from the period of the alternating voltage (f). Another particle accelerator is the betatron (invented in 1941), whose principle of operation is based on the permanent acceleration of charge carriers in an induced electric field.

4.3.2 The Hall effect

One of the most interesting applications of the Lorentz force is the Hall effect, which consists in the appearance of a potential difference U_H between the faces of a metal

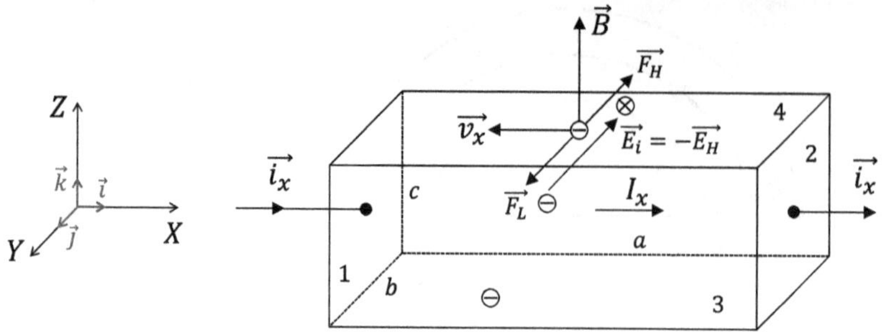

Figure 4.37. Explanation of the Hall effect.

or semiconductor plate, placed in a magnetic field, carrying a current I. The voltage U_H appears in a direction perpendicular to both the direction of the density of the electric current and that of the magnetic field induction $\vec{B}$, as seen in figure 4.37. When the conduction current with a current density of $\vec{j_x}$ flows in the direction Ox, an induced electric field $\vec{E_i} = \frac{\vec{F_L}}{q}$ appears; it is proportional to the conduction current density $\vec{j_x}$ and the magnetic induction $\vec{B_z}$:

$$\vec{E_i} = \frac{\vec{F_L}}{q} = \frac{q(\vec{v} \times \vec{B})}{q} \Rightarrow \vec{E_i} = \vec{v} \times \vec{B} \Rightarrow E_i = -\frac{j_x}{ne} B_z = R j_x B_z \tag{4.140}$$

where $q = -e$ for electrons and R represents the Hall constant. The appearance of the Hall field is due to the fact that in the magnetic field, movable charges (here, the quasi-free electrons) are deflected by the action of the Lorentz force in a direction perpendicular to their speed and the magnetic field $\vec{B}$, respectively. They accumulate on one side of the sample, charging it with electric charge of one sign; at the other side, an equal charge appears that has the opposite sign. The process of charge separation continues until the field created by the accumulated charges on the faces gives rise to a force $\vec{F_H}$ large enough to equal the Lorentz force acting in the opposite direction:

$$\vec{F_H} = -\vec{F_L} \tag{4.141}$$

In the stationary state (without a current in the Oy direction), the following relation must be satisfied:

$$q\vec{E_H} = -q(\vec{v} \times \vec{B}) \tag{4.142}$$

In this case, the Hall field is:

$$E_H = -E_i = -v_x B_z \tag{4.143}$$

However, the speed of the electric charge v_x can be determined from the expression for the conduction current density $i_x = nqv_x$, as follows:

$$v_x = \frac{i_x}{nq} \tag{4.144}$$

Thus, the Hall field is:

$$E_H = -\frac{1}{nq} i_x B_z \tag{4.145}$$

It is assumed that in the material there is only one type of charge carrier q that has a given polarity. From the expression for the Hall field (4.145), we can obtain a first expression for the Hall constant:

$$R = \frac{1}{nq} \tag{4.146}$$

It can be seen that the sign of the Hall constant is determined by the sign of the charge carrier:

- $R > 0$ if $q > 0$ (for holes in p-type semiconductors)
- $R < 0$ if $q < 0$ (for electrons in metals and n-type semiconductors)

So, from the determination of the sign of the Hall constant, one can determine the sign of the majority charge carrier in extrinsic semiconductors. In intrinsic semiconductors (with mixed conduction of electrons and holes), the Hall constant is modified so that it also accounts for the mobilities of the charge carriers, μ_n (electrons) and μ_p (holes), and their concentrations n and p, respectively:

$$R = \frac{1}{q} \frac{p\mu_p^2 - n\mu_n^2}{(n\mu_n + p\mu_p)^2} \tag{4.147}$$

This expression is reduced to expression (4.146) (with q negative) for n-type semiconductors in which $n \gg p$:

$$R = \frac{1}{q} \cdot \frac{-n\mu_n^2}{n^2\mu_n^2} = -\frac{1}{nq} \tag{4.148}$$

For mixed conduction, the sign of Hall constant, as determined by (4.147), depends on the difference $p\mu_p^2 - n\mu_n^2$. It can thus be seen that the sign of the Hall constant does not always correspond to the majority electric charge carrier, especially if the mobilities of the charge carriers are very different. In intrinsic semiconductors ($n = p$), even though the electrons and holes are deflected towards the same face of the semiconducting sample, the Hall field is not canceled. The reason for this is that, according to (4.147), for $n = p = n_i$, the Hall constant depends on the difference in mobilities:

$$R_H = \frac{1}{qn_i} \cdot \frac{\mu_p - \mu_n}{\mu_p + \mu_n} \tag{4.149}$$

As a rule, $\mu_n \gg \mu_p$, so that for intrinsic semiconductors, the sign of the Hall constant is negative. If $\mu_n \gg \mu_p$, which is the case in many semiconductor compounds, then

$$R_H = -\frac{1}{qn_i} \tag{4.150}$$

In other words, at equal concentrations, the electrons predominate due to their superior mobility compared to that of holes. Because the carrier concentration is much lower in semiconductors than in metals ($10^{17} - 10^{18}$ cm^{-3} in semiconductors compared to $10^{22} - 10^{23}$ cm^{-3} in metals), it follows that R_H and U_H are higher in semiconductors than in metals by several orders of magnitude.

The above expressions for the Hall constant are not rigorous, being obtained on the assumption that all carriers have the same speed, without taking into account their distributions resulting from their velocities or energies. A more correct calculation, based on solving the Boltzmann transport equation and using a relaxation time dependent on energy, i.e. on the form $\tau = A(T)\varepsilon^{\gamma}$, gives the following expression for the Hall constant in the case of small magnetic fields B:

$$R_H = \frac{\langle \tau^2 \rangle}{\langle \tau \rangle^2} \cdot \frac{1}{nq} \tag{4.151}$$

where the ratio $\frac{\langle \tau^2 \rangle}{\langle \tau \rangle^2}$ has different values for different scattering mechanisms.

The determination of the Hall constant and its dependence on temperature is particularly important in the study of semiconductors, allowing the determination of the concentration of charge carriers, their sign, the ionization energy of impurities, and the value of the band gap. By combining measurements of the Hall constant with electrical conductivity measurements, i.e. $\sigma = ne\mu$, the Hall mobility values of the carriers can be determined:

$$\mu_H = R_H \sigma \tag{4.152}$$

In the case of the sample in figure 4.37, multiplying the Hall field by the width of the sample b gives the Hall voltage:

$$U_H = bE_H = bR_H i_x B_z = bR_H \frac{I_x}{bc} B_z = \frac{R_H}{c} I_x B_z \tag{4.153}$$

The slope of the straight line, $\tan\alpha = \frac{dU_H}{dI_x}$, obtained by graphically plotting $U_H = f(I_x)$ at B_z = constant, allows the determination of the Hall constant:

$$\frac{dU_H}{dI_x} = \frac{R_H}{c} B_z \Longrightarrow R_H = \frac{c \tan\alpha}{B_z} \tag{4.154}$$

4.4 Force of interaction between two circuits carrying electric currents

If, in section 4.1, the electrodynamic force was presented as introduced experimentally by Ampére, now is the time to present it by a logical deduction in accordance with the formalism of the magnetic field associated with the stationary electrokinetic state. The magnetic field created by the current that flows through the first conductor (see figure 4.38, where it is represented by the current element $I\overrightarrow{dl}$) acts on a current

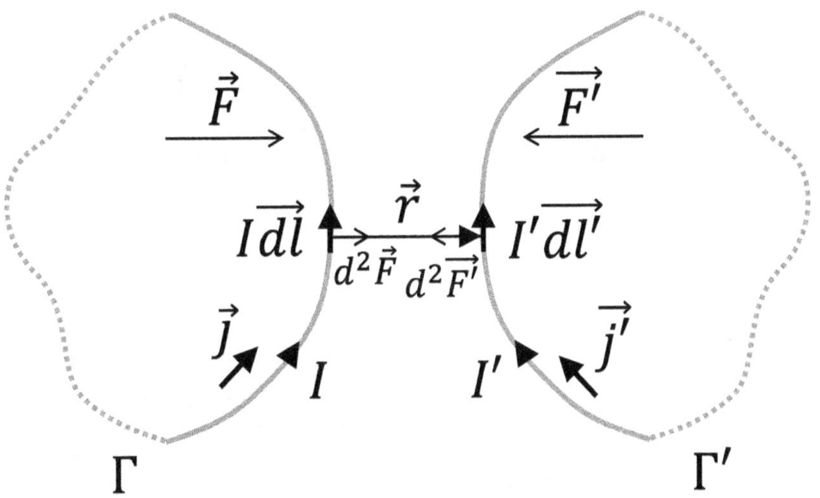

Figure 4.38. Electrodynamic force between two circuits carrying electric currents.

element $I'\vec{dl'}$ from the second circuit with a force $d^2\vec{F'}$. Taking into account the expression for electromagnetic force, it is given by:

$$d^2\vec{F'} = I'\vec{dl'} \times \frac{\mu_0}{4\pi}\frac{I\vec{dl} \times \vec{r}}{r^3} = \frac{\mu_0}{4\pi}II'\frac{\vec{dl'} \times (\vec{dl} \times \vec{r})}{r^3} \tag{4.155}$$

In turn, the current in the second conductor ($I'\vec{dl'}$) acts upon the element $I\vec{dl}$ with a force $d^2\vec{F} = -d^2\vec{F'}$. In the case of two circuits, the force $\vec{F'}$ with which the current I from the first circuit acts on the current I' flowing through second circuit is:

$$\vec{F'} = \frac{\mu_0}{4\pi}I'I\int_{\Gamma}\int_{\Gamma'}\frac{\vec{dl'} \times (\vec{dl} \times \vec{r})}{r^3} \tag{4.156}$$

or, taking into account the relationship between current and current density, the following relation is obtained for the force of interaction:

$$\vec{F'} = \frac{\mu_0}{4\pi}\int_{V'}\int_{V}\frac{\vec{j'} \times (\vec{j} \times \vec{r})}{r^3}dv\, dv' \tag{4.157}$$

Relation (4.157) for the electrodynamic force $\vec{F'}$ underlies the definition of the ampere (A), one of the fundamental units of measurement in the SI, a definition that was given in the first paragraph of this chapter.

4.5 Magnetic dipole moment

Using the Biot–Savart–Laplace law to compute the magnetic induction produced by a circular circuit carrying a current, we observed that the result for B can be also be expressed by a vector physical quantity associated with the circuit; this is called the magnetic moment, $\vec{m}$. This vector physical quantity, whose magnitude is $|\vec{m}| = I \cdot A$, is called the moment of the magnetic dipole. Here, we analyze a magnetic dipole and the moment of the associated magnetic dipole in detail using a closed electrical circuit of rectangular, flat shape; then, the notion of the moment of the magnetic dipole is extended to a specific electrical circuit.

The intensity of the electric current in the circuit is I, the sides of the circuit being a and b, respectively; see figure 4.39.

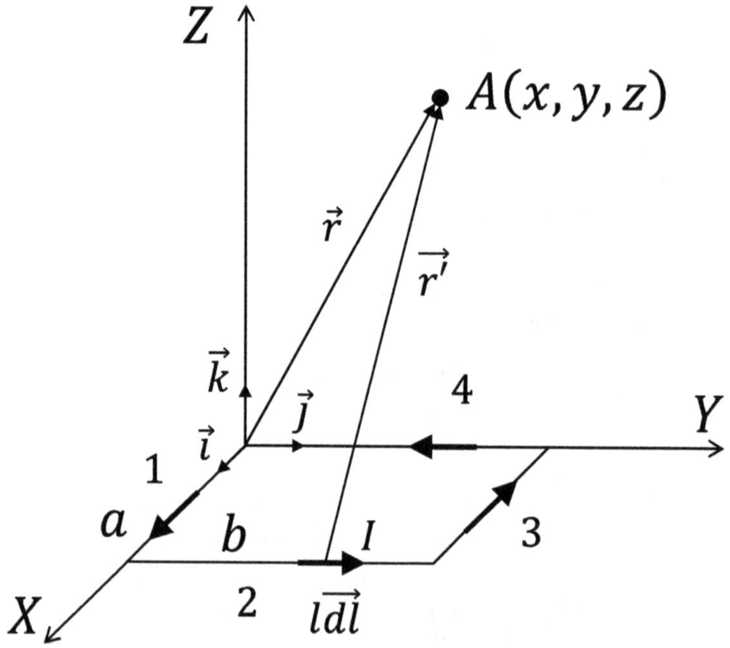

Figure 4.39. Rectangular circuit whose sides a and b are carrying electric currents of intensity I.

When the following conditions are met:

$$\begin{cases} \dfrac{a}{r} \ll 1, \ \dfrac{b}{r} \ll 1 \\ \dfrac{a}{r'} \ll 1, \ \dfrac{b}{r'} \ll 1 \end{cases} \tag{4.158}$$

the vector potential $\vec{A}(x, y, z)$ at a point placed at a distance $\vec{r'}$ from the current element $I\vec{dl}(x', y', z')$ is:

$$\vec{A} = \frac{\mu_0}{4\pi} I \oint_C \frac{\vec{dl}(x', y', z')}{r} \tag{4.159}$$

In the case of a straight-angled circuit (where we consider a positive sense of contour starting from the origin), we can write:

$$\vec{A} = \frac{\mu_0}{4\pi} I \left[\vec{i} \int_0^a \left(\frac{dx'}{r_1'} - \frac{dx'}{r_3'} \right) + \vec{j} \int_0^b \left(\frac{dy'}{r_2'} - \frac{dy'}{r_4'} \right) \right] \tag{4.160}$$

where $\vec{i}$ and $\vec{j}$ are the unit vectors of the Ox and Oy axes. Taking into account relationship (4.158) and developing $\frac{1}{r'}$ as a Taylor series (neglecting the terms of higher order), we can write:

$$\frac{1}{r'} = \frac{1}{r} + \frac{\partial}{\partial x'}\left(\frac{1}{r'}\right)\Bigg|_0^{x'} x' + \frac{\partial}{\partial y'}\left(\frac{1}{r'}\right)\Bigg|_0^{y'} y' + \frac{\partial}{\partial z'}\left(\frac{1}{r'}\right)\Bigg|_0^{z'} z' + \ldots. \tag{4.161}$$

Taking into account:

$$\begin{cases} r' = [(x - x')^2 + (y - y')^2 + (z - z')^2]^{1/2} \quad \Rightarrow \\ \frac{1}{r'} = (r')^{-1} = [(x - x')^2 + (y - y')^2 + (z - z')^2]^{-1/2} \end{cases}$$

and by taking only the partial derivative in relation to x', we obtain:

$$\begin{cases} \frac{\partial}{\partial x'}\left(\frac{1}{r'}\right) = \frac{-\frac{1}{2} \cdot 2(x - x')(-1)}{[(x - x')^2 + (y - y')^2 + (z - z')^2]^{3/2}} = \frac{x - x'}{r'^3} \\ \frac{\partial}{\partial x'}\left(\frac{1}{r'}\right)\Bigg|_{x',\, y',\, z'=0} = \frac{x}{(x^2 + y^2 + z^2)^{3/2}} = \frac{x}{r^3} \end{cases} \tag{4.162}$$

Given the above, (4.161) becomes:

$$\frac{1}{r'} = \frac{1}{r} + \frac{x}{r^3}x' + \frac{y}{r^3}y' + \frac{z}{r^3}z' \tag{4.163}$$

Using this expression for each of the quantities $\frac{1}{r_1'}, \frac{1}{r_2'}, \frac{1}{r_3'}, \frac{1}{r_4'}$ allows us to obtain the following for the vector potential:

$$\begin{aligned} \vec{A} &= \frac{\mu_0}{4\pi} I \left\{ \vec{i} \int_0^a \left(\frac{1}{r} + \frac{x}{r^3}x' - \frac{1}{r} - \frac{x}{r^3}x' - \frac{yb}{r^3} \right) dx' \right. \\ &\quad \left. + \vec{j} \int_0^b \left(\frac{1}{r} + \frac{xa}{r^3} + \frac{y}{r^3}y' - \frac{1}{r} - \frac{y}{r^3}y' \right) dy' \right\} \\ &= \frac{\mu_0 I}{4\pi} \left[\vec{i}(-) \int_0^a \frac{yb}{r^3} dx' + \vec{j} \int_0^b \frac{xa}{r^3} dy' \right] \\ &= \frac{\mu_0 I}{4\pi} \left(-\vec{i}\frac{aby}{r^3} + \vec{j}\frac{abx}{r^3} \right) = \frac{\mu_0 I ab}{4\pi r^3} (-y\vec{i} + x\vec{j}) \end{aligned} \tag{4.164}$$

Relation (4.164) allows us to regard the considered electrical circuit as equivalent to a magnetic dipole that has a moment of:

$$\vec{m} = Iab\vec{k} \tag{4.165}$$

where ab is the area limited by the flat circuit carrying the current I. We can see that relation (4.164) for the vector potential is a particular case of the relation:

$$\vec{A}(x, y, z) = \frac{\mu_0}{4\pi}\frac{\vec{m} \times \vec{r}}{r^3} \tag{4.166}$$

where $\vec{m}$ is given by equation (4.165). Indeed:

$$\vec{A}(x, y, z) = \frac{\mu_0}{4\pi r^3}\begin{vmatrix} \vec{i} & \vec{j} & \vec{k} \\ 0 & 0 & Iab \\ x & y & z \end{vmatrix} = \frac{\mu_0}{4\pi r^3} Iab(-y\vec{i} + x\vec{j}\,) \tag{4.167}$$

Thus, the moment of the magnetic dipole is a vector perpendicular to the electrical circuit, whose positive sense is given by the direction of travel of a right-handed screw placed normal to the plane of the circuit, when it is rotated in the direction of the current through the circuit. Moving now to a circuit of an arbitrary shape, it should be noted that even in this case, the notion of the moment of the magnetic dipole is used only when the vector potential is calculated at a point located at a great distance compared to the dimensions of the circuit. The circuit can be replaced by a network with many rectangular meshes such that every rectangular loop of the network passes a current of the same intensity in the same direction (figure 4.40). At the linear sides of the rectangular meshes of the network, carrying a current in the trigonometric direction, the currents passing through neighboring branched sides cancel reciprocally, except for the sides belonging to the circuit; this scenario can be used to approximate the electrical circuit quite well if the sides of the meshes are assumed to be sufficiently small. For any loop of the network, the moment of the magnetic dipole is:

$$\overrightarrow{dm} = I\vec{n}\,ds \tag{4.168}$$

where I is the intensity of the electric current in the network loop and $\vec{n}$ is the normal vector at the surface element ds. At a point far from the considered rectangular loop, the magnetic vector potential it creates is:

$$\overrightarrow{dA} = \mu_0 \frac{\overrightarrow{dm} \times \vec{r}}{4\pi r^3} \tag{4.169}$$

The moment of the magnetic dipole associated with the finite circuit can be obtained by summing the elementary moments associated with the rectangular meshes of the equivalent circuit:

$$\vec{m} = \int_S \overrightarrow{dm} = I\int_S \vec{n}\,ds \tag{4.170}$$

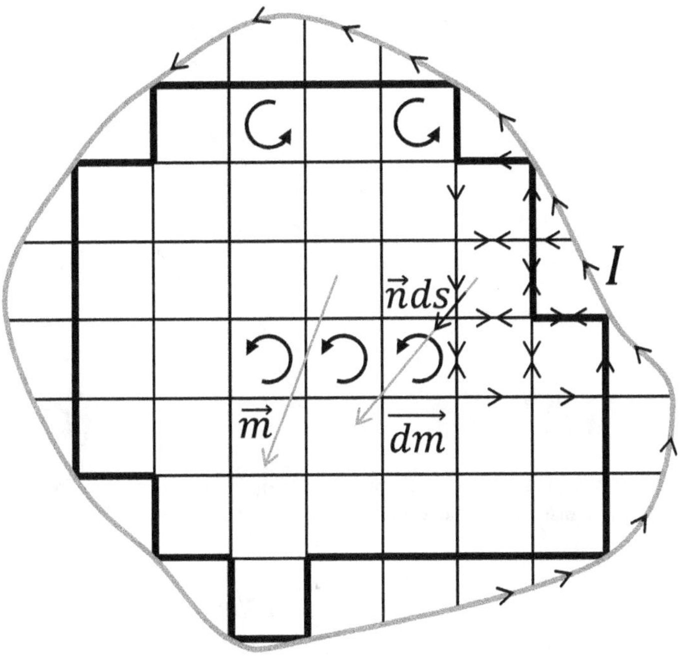

Figure 4.40. Flat network consisting of several rectangular meshes of network associated with a current circuit of arbitrary form.

where S is the area bounded by the edges of the rectangular grid (which can be taken to be equal to the area of the actual circuit). For the final circuit, the magnetic vector potential can be written as:

$$\vec{A} = \frac{\mu_0}{4\pi}\int_S \frac{\overrightarrow{dm} \times \vec{r}}{r^3} = \frac{\mu_0}{4\pi}\int_S \overrightarrow{dm} \times \frac{\vec{r}}{r^3} = \frac{\mu_0}{4\pi}\frac{\vec{m} \times \vec{r}}{r^3} \tag{4.171}$$

which is the same expression as in the case of the rectangular circuit.

4.6 Force and torque exerted by a magnetic field on an electric circuit

Let us consider a closed electrical circuit of arbitrary form in which a current of intensity I has been established. The circuit is placed in an external magnetic field that has the induction $\vec{B}$; see figure 4.41(a). The field subjects the circuit to different forces that cause it to move, rotate, and deform. Under the action of the force exerted by the external magnetic field, a portion $\overrightarrow{dl}$ move over a distance $\overrightarrow{dr}$ to the position where the portion is denoted by $\overrightarrow{dl'}$; see figure 4.41(b). To produce this movement, the magnetic field $\vec{B}$ performs mechanical work expression by:

$$d^2w = \overrightarrow{dF} \cdot \overrightarrow{dr} \tag{4.172}$$

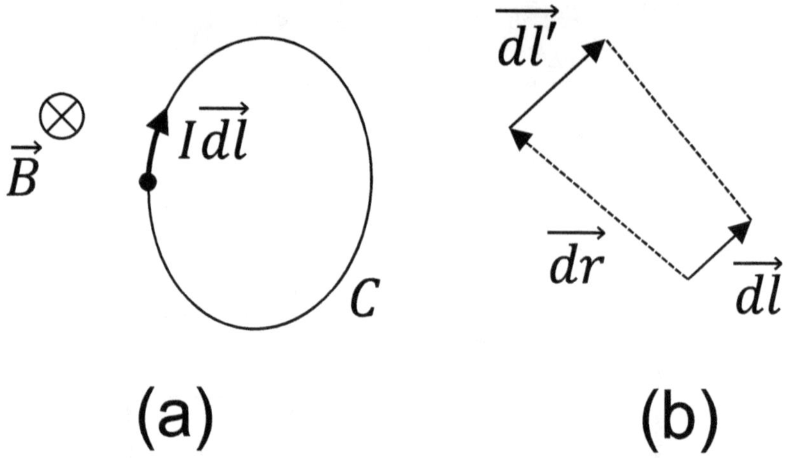

Figure 4.41. Actions of a magnetic field on a circuit carrying an electric current (a). The length element $\vec{dl}$ of the circuit is moved over a distance $\vec{dr}$ due to the Laplace force (b).

However, the force $\vec{dF}$ with which the magnetic field acts on the circuit element $I\vec{dl}$ is determined by the Laplace force:

$$\vec{dF} = I\vec{dl} \times \vec{B} \tag{4.173}$$

Substituting (4.173) into (4.172) give the result:

$$d^2w = \vec{dF} \cdot \vec{dr} = I(\vec{dl} \times \vec{B})\vec{dr} \tag{4.174}$$

or:

$$d^2w = I\vec{B}(\vec{dr} \times \vec{dl}) \tag{4.175}$$

The vector product in parentheses $(\vec{dr} \times \vec{dl})$ represents the surface swept by $\vec{dl}$ during its displacement. The scalar product of this surface and the magnetic field induction gives the elementary flux of the magnetic field through the surface generated by $\vec{dl}$ in its elementary displacement $\vec{dr}$. To obtain the value of the flux through the entire circuit, the integral must be calculated:

$$\int_C \vec{B}(\vec{dr} \times \vec{dl}) = d\Phi \tag{4.176}$$

In the second member, we still have the first-order differential of the flux, because on the left-hand side of equation, the integral is taken only after $\vec{dl}$, leaving the differential element $\vec{dr}$. Taking into account (4.176), the work done by the magnetic field in moving the entire circuit over the still elementary distance $\vec{dr}$ is:

$$dw = Id\Phi \tag{4.177}$$

Relation (4.177) takes different forms, depending on the type of movement of the circuit. If the circuit performs a translation and the components of the forces in the three directions are F_x, F_y, and F_z, the mechanical work done by these forces is:

$$dw = F_x dx + F_y dy + F_z dz \tag{4.178}$$

On the other hand:

$$Id\Phi = I\frac{\partial\Phi}{\partial x}dx + I\frac{\partial\Phi}{\partial y}dy + I\frac{\partial\Phi}{\partial z}dz \tag{4.179}$$

Now, from the identification of relations (4.178) and (4.179), we obtain the results:

$$F_x = I\frac{\partial\Phi}{\partial x}, \quad F_y = I\frac{\partial\Phi}{\partial y}, \quad F_z = I\frac{\partial\Phi}{\partial z} \tag{4.180}$$

When the circuit as a whole performs a rotation around its axis under the action of the forces of the field, then the elementary variation of the flux can be expressed by means of the angular coordinate θ, as follows:

$$d\Phi = \frac{\partial\Phi}{\partial\theta}d\theta \tag{4.181}$$

so that the mechanical work done by the magnetic field is:

$$dw = I\frac{\partial\Phi}{\partial\theta}d\theta = M_\theta d\theta \tag{4.182}$$

Thus, the moment of torque exerted by the magnetic field on the circuit is:

$$M_\theta = I\frac{\partial\Phi}{\partial\theta} \tag{4.183}$$

In both situations analyzed above, the deformation of the circuit was neglected.

The expression $dw = Id\Phi$ is also useful for expressing the force exerted by the magnetic field on a circuit when the circuit is equivalent to a magnetic dipole moment $\vec{m}$. If a current of constant intensity is established in the circuit, the work performed by the field can be written as:

$$dw = d(I\Phi) \tag{4.184}$$

However, $\Phi = \int_S \vec{B}\vec{ds}$. Therefore:

$$I\Phi = I\int_S \vec{B}\vec{n}ds \tag{4.185}$$

If the magnetic field is constant, along with the magnitude and orientation on the surface S of the circuit, then (4.185) becomes:

$$I\Phi = \vec{B}\int_S I\vec{n}ds = \vec{m}\vec{B} \tag{4.186}$$

In this situation, the work done by the magnetic field on a magnetic dipole moment $\vec{m}$, placed in the field, is:

$$dw = d(\vec{m}\vec{B}) = \vec{m}\overrightarrow{dB} + \vec{B}\overrightarrow{dm} \tag{4.187}$$

In the case of a constant magnetic field combined with constant magnitude and orientation, the work done becomes $dw = \vec{B}\overrightarrow{dm}$, suggesting the variation of the magnetic moment, $\overrightarrow{dm}$, associated with the circuit. If the moment of the magnetic dipole is constant in magnitude and orientation, then in a nonuniform magnetic field the circuit undergoes a translation, and the components of the forces are found to be as follows:

$$\begin{aligned} dw &= d(\vec{m}\vec{B}) = d(m_x B_x + m_y B_y + m_z B_z) \\ &= \left(m_x \frac{\partial B_x}{\partial x} + m_y \frac{\partial B_y}{\partial x} + m_z \frac{\partial B_z}{\partial x}\right) dx \\ &+ \left(m_x \frac{\partial B_x}{\partial y} + m_y \frac{\partial B_y}{\partial y} + m_z \frac{\partial B_z}{\partial y}\right) dy \\ &+ \left(m_x \frac{\partial B_x}{\partial z} + m_y \frac{\partial B_y}{\partial z} + m_z \frac{\partial B_z}{\partial z}\right) dz \\ &= F_x dx + F_y dy + F_z dz \end{aligned} \tag{4.188}$$

from which the components of the forces (F_x, F_y, and F_z) are extracted. The resultant force F is oriented in the direction of increasing field strength. If the magnetic field $\vec{B}$ is uniform (constant in magnitude and orientation throughout the space), no force ($\vec{F} = 0$) is exerted on the corresponding magnetic dipole $\vec{m}$. When the dipole keeps the magnitude of the moment constant but the orientation is inconstant, then a torque of forces is exerted on the dipole that rotates it. The expression for the torque moment can be established by starting from the relation $dw = d(\vec{m} \cdot \vec{B})$ and using the angular variable $d\theta$:

$$dw = M_c d\theta = d(\vec{m} \cdot \vec{B}) = \vec{B}\overrightarrow{dm}$$

However, from figure 4.42, it follows that:

$$\overrightarrow{dm} = -\vec{m} \times \overrightarrow{d\theta}$$

Hence:

$$dw = M_c d\theta = -\vec{B}(\vec{m} \times \overrightarrow{d\theta}) = (\vec{m} \times \vec{B})d\theta$$

resulting in:

$$\overrightarrow{M_c} = \vec{m} \times \vec{B} \tag{4.189}$$

The resultant moment $\overrightarrow{M_c}$ is normal to the vectors $\vec{m}$ and $\vec{B}$ and it constantly tends to rotate $\vec{m}$ over $\vec{B}$. When $\theta = \frac{\pi}{2}$, the torque M_c is maximized and the current loop on the moment $\vec{m}$ is in unstable equilibrium; conversely, when $\theta = 0$, $M_c = 0$ and the current loop is in stable equilibrium.

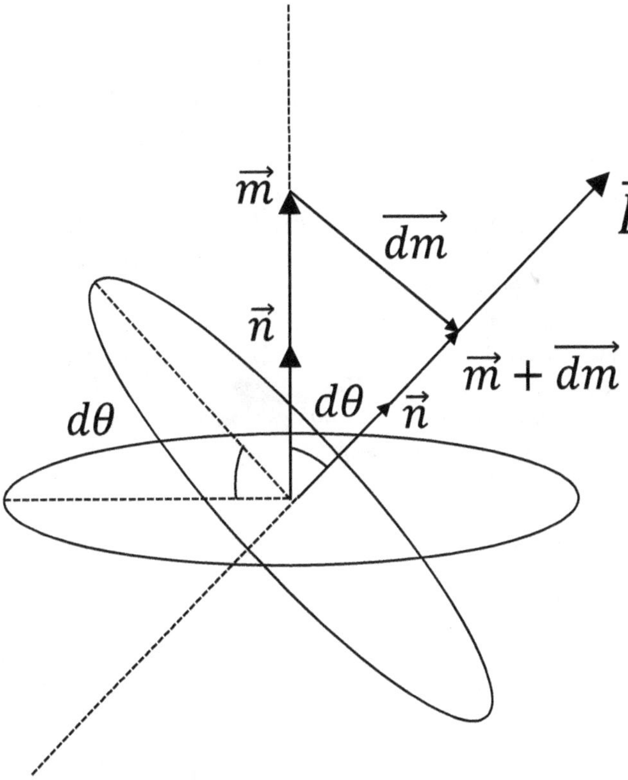

Figure 4.42. Action of the magnetic field on a magnetic moment.

4.7 Energy of a magnetic moment placed in a magnetic field

The expression for the potential energy W_m of a magnetic dipole moment $\vec{m}$, located in a magnetic induction field $\vec{B}$, can be obtained from the general formula for the mechanical work performed by a magnetic field on a circuit carrying a current of intensity I, i.e. $dw = Id\Phi = d(\vec{m} \cdot \vec{B})$. When a circuit carrying the current I is brought from infinity in a magnetic field, the external forces that produce the movement of the circuit perform mechanical work equal and of opposite sign to the work performed by the electromagnetic forces. This mechanical work done is the potential energy:

$$dw_m = -dw = -Id\Phi = -d(\vec{m}\vec{B}) = d(-\vec{m}\vec{B})$$

Thus, the expression for the magnetic energy of interaction between a circuit carrying a current I and the magnetic field in which it is located is:

$$W_m = -\vec{m} \cdot \vec{B} = -mB\cos\theta \tag{4.190}$$

where θ is the angle between the magnetic moment and $\vec{m}$ and the magnetic induction $\vec{B}$.

References and further reading

[1] Antohe S 2002 *Electricity and Magnetism* **vol II** (Bucharest: University of Bucharest Press)

[2] Antohe Ş and Antohe V-A 2023 *Electrostatics. Formalism of the Electrostatic Field in Vacuum and Matter* 1st edn (Bristol: IOP Publishing) 266 pages

[3] Purcell E M and Morin D J 2013 *Electricity And Magnetism* 3rd edn (Cambridge: Cambridge University Press) 853 pages

[4] Feynman R P, Leighton R B and Sands M 2011 *Feynman Lectures on Physics, Vol II: The New Millennium Edition: Mainly Electromagnetism and Matter* (New York: Basic Books) (Feynman Lectures on Physics) 589 pages

[5] Jackson J D 1998 *Classical Electrodynamics* 3rd edn (New York: Wiley) 832 pages

[6] Bleaney B I and Bleaney B 1965 *Electricity and Magnetism* 3rd edn vol 1 (Oxford: Oxford University Press)

[7] Blanpied W A 1971 *Modern Physics: An Introduction to Its Mathematical Language* (New York: Holt, Rinehart and Winston) 724 pages

[8] Griffiths D J 2017 *Introduction to Electrodynamics* 4th edn (Cambridge: Cambridge University Press)

[9] Clayton R, Keith W and Nasar S 1997 *Introduction to Electromagnetic Fields* (New York: McGraw-Hill)

[10] Fernow R C 2016 *Principles of Magnetostatics* 1st edn (Cambridge: Cambridge University Press)

IOP Publishing

Electromagnetism and Special Methods for Electric Circuits Analysis

Ştefan Antohe and Vlad-Andrei Antohe

Chapter 5

Electromagnetic induction and self-induction

Electromagnetic induction is one of the most important discoveries of the 19th century. This phenomenon is related to the generation of an electromotive force (EMF) and a current in a closed circuit crossed by a variable magnetic flux. It was discovered in 1831 by Michael Faraday[1] based on several experiments, of which only a few are described in this chapter. This chapter also studies the particular case of the electromagnetic induction phenomenon known as self-induction, where an induced electric field and an induced current appear in a circuit as a result of the temporal variation of its own magnetic flux passing through the circuit.

5.1 Electromagnetic induction

5.1.1 Faraday's experiments

The phenomenon of electromagnetic induction was evidenced by several experiments conducted by Faraday.

1. If an electric current I is established through a closed circuit C_1, a short-term current (transitory current) appears in a neighboring closed circuit C_2. This current can be measured by a galvanometer (see figure 5.1).

2. If a single closed conducting wire or coil consisting of several wires is in close proximity to a mobile source of a magnetic field (which can be a permanent magnet, as in figure 5.2, or a solenoid carrying an electric current), a current appears in it that flows as long as the source continues to move.

[1] Michael Faraday (1791–1867) was an English scientist who significantly contributed to the study of electromagnetism and electrochemistry.

doi:10.1088/978-0-7503-5854-5ch5

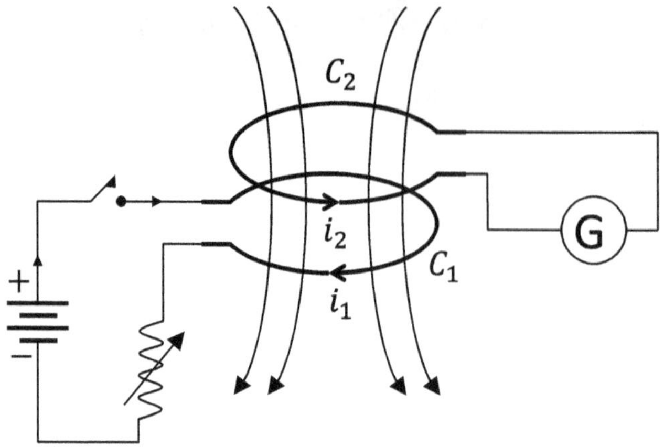

Figure 5.1. Two neighboring circuits used to demonstrate the phenomenon of electromagnetic induction.

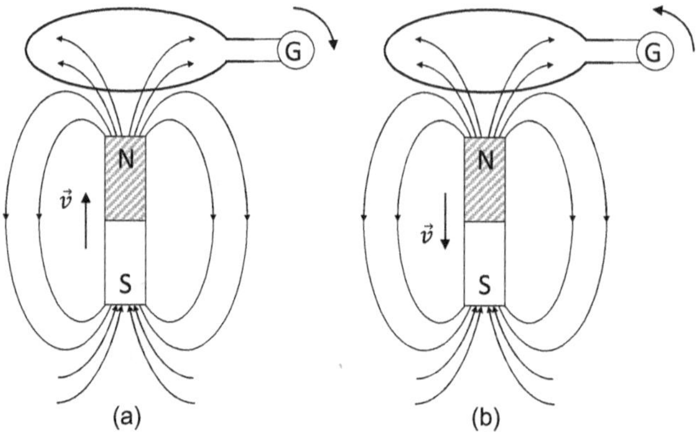

Figure 5.2. Permanent magnets moving with respect to a closed circuit induce an electric current in it, as indicated by the galvanometer.

In both experiments, the current indicated by the galvanometer was small and present only for a short time, which is why contemporary scientists minimized Faraday's discovery. Trying to counter this, Faraday designed an experiment in which he managed to continuously produce electric current in a circuit. The disk bearing his name (see figure 5.3) was the first DC generator to be based on a principle other than galvanic elements. The basis of this experiment was the fact that a portion of the electrical circuit, in this case, the radius $\vec{r}_0$, cuts the magnetic field lines $\vec{a}$, inducing an EMF. Since the currents induced under the same conditions are proportional to the circuit's conductance, it is more appropriate to speak of an induced EMF than an induced current.

With these experimental findings, Faraday's electromagnetic induction phenomenon can be summarized thus: *an electromotive force is induced: (i) in a stationary rigid circuit that is in a time-varying magnetic flux; (ii) in a rigid circuit moving in a stationary magnetic field, or in a magnetic field whose sources are in motion, so that*

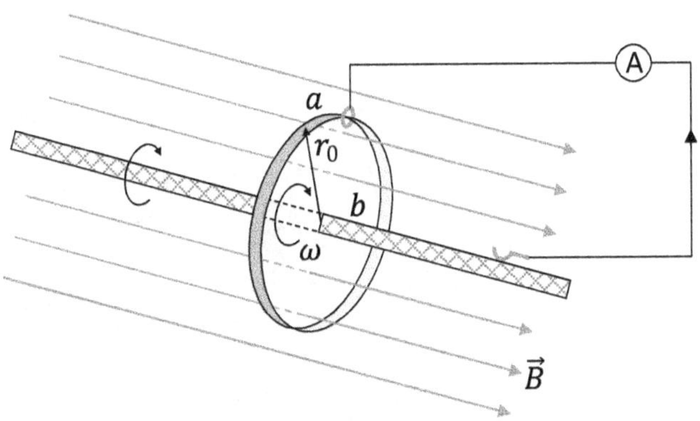

Figure 5.3. Conductive disc rotating in a magnetic field perpendicular to its surface originating an electromotive voltage between its center and its periphery.

the flux passing through the circuit changes; and (iii) in a portion of a circuit that cuts any line of the magnetic field.

All these experimental findings found their physical explanation much later. Benefiting from the discoveries of Einstein and Lorentz, it will be shown that Faraday's experimental observations of the induced electromotive force and the induced current, respectively, are natural consequences of the action of the Lorentz force.

5.1.2 The law of electromagnetic induction (Faraday's Law)

Faraday's law of electromagnetic induction can be formulated as follows: *the induced electromotive voltage that appears in a circuit due to the variation over time of the magnetic flux enclosed by the circuit is proportional to the rate of variation of the magnetic flux.*

According to figure 5.4, the displacement of circuit C from position 1 to position 2, in a magnetic induction field $\vec{B}$ within the time dt, determines the variation of the magnetic flux enclosed by the circuit at the rate:

$$\frac{\Phi_2 - \Phi_1}{\Delta t} = \frac{d\Phi}{dt} \tag{5.1}$$

The mathematical expression of Faraday's law is:

$$E_i = k\frac{d\Phi}{dt} \tag{5.2}$$

where E_i is the induced electromotive force in the circuit and k is a proportionality factor whose significance will be described later. *The induced EMF is independent of the nature of the material from which the circuit is built and the manner in which the magnetic flux varies over time. The value of the induced EMF depends only on the rate of variation of the magnetic flux enclosed by the closed circuit.*

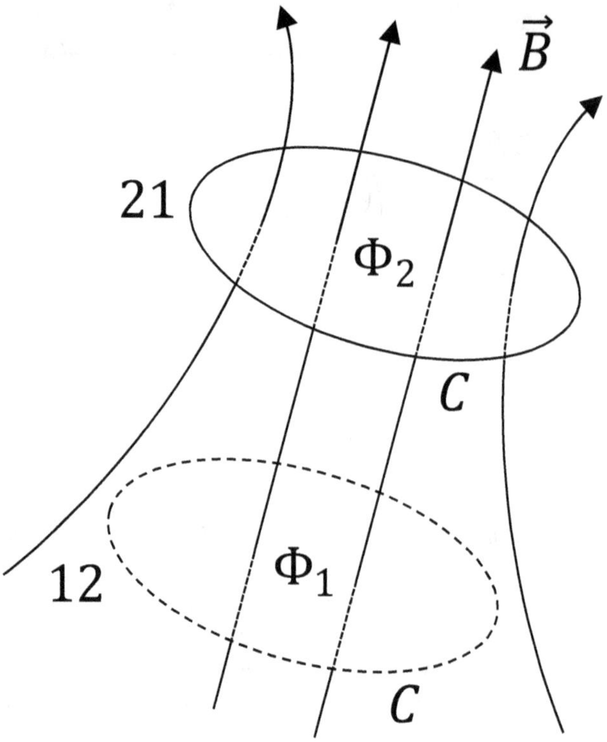

Figure 5.4. Circuit moving in a magnetic induction field $\vec{B}$.

5.1.3 Lenz's law

Lenz's law establishes that *the sense of the EMF is such that the induced electric current produced in a closed circuit gives rise to a magnetic flux that opposes the variation of the inducing magnetic flux, thus opposing the cause that determines the occurrence of the EMF itself.*

Both Faraday's law and Lenz's law have been established experimentally, so that on the basis of experimental determinations, the value of k is found to be:

$$k = -1 \tag{5.3}$$

In figure 5.5, experimental illustrations of Lenz's law are provided. A magnetic field $\vec{B}$ oriented perpendicular to the plane of the figure and entering it interacts with a closed conducting circuit. When this circuit either enters the magnetic field (as shown in figure 5.5(a)) or leaves the field (as illustrated in figure 5.5(b)), a current is induced in it. This current flows in such a direction that the magnetic field it generates opposes the variation of the primary magnetic flux. If the closed circuit penetrates the magnetic field $\vec{B}$, as shown in figure 5.5(a) (i.e. the magnetic flux passing through it increases), the induced current generates a magnetic field $\vec{B_i}$ in the opposite direction to $\vec{B}$, thus leading to a decrease in the variation of the magnetic flux. If the

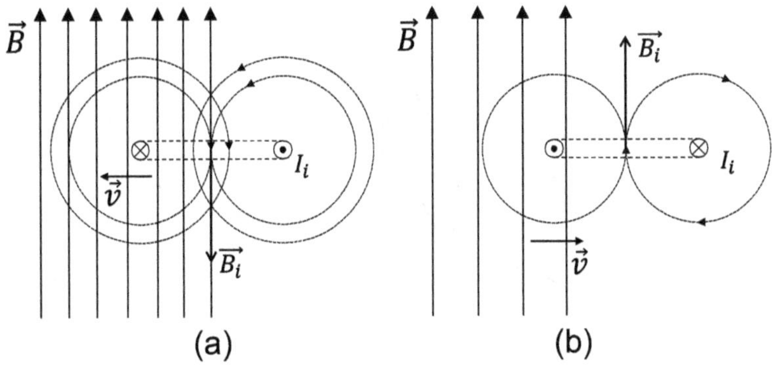

Figure 5.5. Illustration of Lenz's rule.

closed circuit leaves the magnetic field $\vec{B}$, as shown in figure 5.5(b) (i.e. the magnetic flux through it decreases), the induced current generates a magnetic field $\vec{B_i}$ that has the same direction and sense as $\vec{B}$, thus leading to an increase in the variation of the magnetic flux.

5.1.4 The developed integral form of the electromagnetic induction law (Faraday's law)

The combination of the two laws, i.e. Faraday's law and Lenz's law, leads to the law of electromagnetic induction, whose mathematical formulation is:

$$E_i = -\frac{d\Phi}{dt} \tag{5.4}$$

The law of electromagnetic induction states that *the induced electromotive force along a closed curve* Γ *is proportional to the rate of decrease of the magnetic flux through any of the surfaces bounded by this curve* Γ.

It should be noted that the rate of flux variation and the induced EMF may take different forms depending on how the flux variation is obtained. However, the EMF E_i can be expressed as the line integral of the electric field $\vec{E}$ induced along the closed curve Γ, while the magnetic flux passing through any surface that rests on the curve Γ is expressed by the surface integral of $\vec{B}$ through this surface (figure 5.6), so that the law of electromagnetic induction can be explicitly written:

$$E_i = \int_{\Gamma} \vec{E}\vec{dl} = -\frac{d}{dt}\int_{S_\Gamma} \vec{B}\vec{ds} \tag{5.5}$$

When applying the law of electromagnetic induction, the following aspects must be taken into account:

(a) The closed curve Γ is often taken along a filiform electrical conductor, but generally, this is not necessary, and a curve Γ, of any shape, can also be drawn through insulating media or vacuum.
(b) If the considered medium is in motion, the curve Γ is attached to the bodies in their movement (as in the case of the law of magnetic circuits, i.e. Ampère's law).

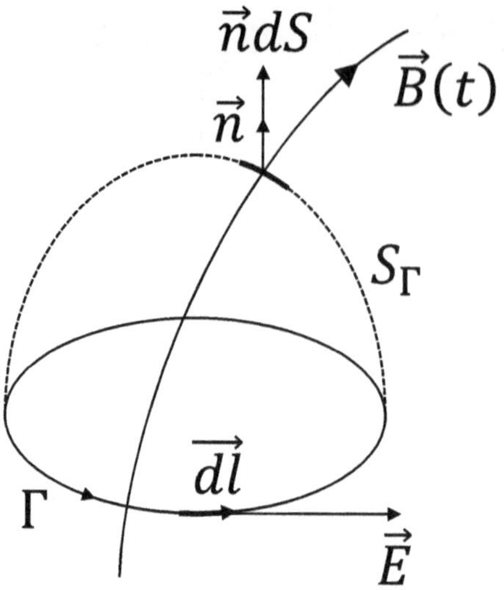

Figure 5.6. The surface S_Γ bounded by the closed curve Γ.

(c) The direction of integration on the curve Γ (the sense of $\overrightarrow{dl}$) and the sense of the normal vector $\vec{n}$ on the surface S_Γ, in relation to which the flux is calculated (i.e. the sense of $\overrightarrow{ds} = \vec{n}\,ds$), are associated according to the right screw rule (RSR); that is, the normal vector $\vec{n}$ is oriented in the sense in which a screw advances when rotated in the direction of $\overrightarrow{dl}$ on curve Γ.

(d) In the case where the contour Γ is taken along the conductor of a coil with N overlapping turns, the magnetic flux that intervenes in the calculation of the induced EMF is the flux through a surface that rests on the entire contour, i.e. the flux through all the wire loops $\Phi_t = N\Phi$ and therefore: $E = -N\frac{d\Phi}{dt}$.

We stated that the induced EMF may take different forms, depending on how the flux variation takes place, i.e. $\Phi = \int_{S_\Gamma} \vec{B}\,\overrightarrow{ds}$, and this can be explicitly viewed if the temporal derivative of the flux, appearing on the right-hand side of equation (5.5), is computed. The derivative of the magnetic flux in relation to time is a *substantial derivative*, and it includes both the variation of the flux produced by a time-dependent magnetic field $\vec{B}(t)$ through a fixed surface, as well as the variation of the flux of a constant (time-independent) magnetic field $\vec{B}$ through a movable surface (time-dependent surface). Such a substantial derivative was also calculated in the case of the displacement vector $\vec{D}$ used to write the developed integral form of the magnetic circuit law:

$$\int_\Gamma \vec{B}\,\overrightarrow{dl} = \mu_0\left(\int_{S_\Gamma} \vec{j}\,\overrightarrow{ds} + \frac{d}{dt}\int_{S_\Gamma} \vec{D}\,\overrightarrow{ds}\right)$$

in chapter 4. It is considered that in a time-varying magnetic field $\vec{B}(t)$, there is a closed circuit that has the contour Γ (figure 5.7), on which the surface S_Γ is

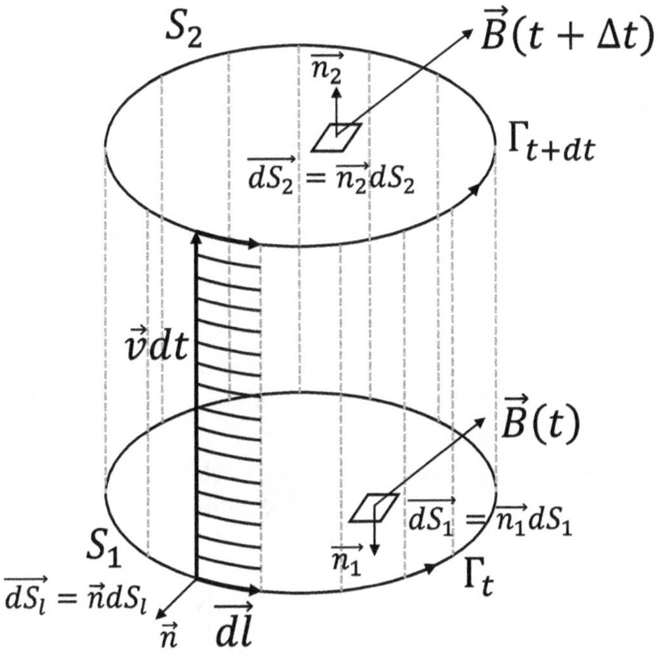

Figure 5.7. A closed rigid circuit Γ moving in a time-dependent magnetic field $\vec{B}(t)$.

supported, which is moving at speed $\vec{v}$ together with the moving contour Γ. At time t, the flux through the surface S_Γ bounded by the curve Γ_t is:

$$\Phi_t = \int_{S_1} \vec{B}(t)\overrightarrow{ds_1} \tag{5.6}$$

and at time $t + dt$, the flux is:

$$\Phi_{t+dt} = \int_{S_2} \vec{B}(t+dt)\overrightarrow{ds_2} \tag{5.7}$$

Figure 5.7 shows a closed conductive wire moving in a time-dependent magnetic field. By definition, the time derivative of flux $\frac{d\Phi}{dt}$ is:

$$\frac{d\Phi}{dt} = \lim_{\Delta t \to 0} \frac{\Delta\left(\int \vec{B}\overrightarrow{ds}\right)}{\Delta t} \tag{5.8}$$

where:

$$\frac{\Delta\left(\int \vec{B}\overrightarrow{ds}\right)}{\Delta t} = \frac{\int_{S_2} \vec{B}(t+\Delta t)\overrightarrow{ds_2} - \int_{S_1} \vec{B}(t)\overrightarrow{ds_1}}{\Delta t} \tag{5.9}$$

To calculate the difference appearing in the numerator of the fraction in equation (5.9), Gauss's theorem will be used, which will be applied to the cylindrical volume generated by the moving surface S_Γ bounded by the closed contour Γ.

$$\int_{S_{tot}} \vec{B}\vec{ds} = \int_{V_{S_{tot}}} \nabla\vec{B}\,dv$$

$$= \int_{S_2} \vec{B}(t+\Delta t)\vec{ds_2} - \int_{S_1} \vec{B}(t)\vec{ds_1} - \int_{\Gamma} \vec{B}(t)(\vec{v}dt \times \vec{dl}) \tag{5.10}$$

Here, we have consistently assumed that the vector normal to the surface is positive only when it is oriented from the inside to the outside of a closed surface, and the volume element in this cylinder is $dv = \vec{ds} \cdot \vec{v}dt$.

Note: Gauss's theorem applies to instantaneous field values, $\vec{B}(t)$. By developing this as a Taylor series, we can obtain the induction of the magnetic field at the moment $t + \Delta t$, as follows:

$$\vec{B}(t+\Delta t) = \vec{B}(t) + \left(\frac{\partial \vec{B}}{\partial t}\right)\Delta t \tag{5.11}$$

With this, equation (5.9) becomes:

$$\frac{\Delta\Phi}{\Delta t} = \frac{\int_{S_2} \vec{B}(t)\vec{ds_2} + \int_{S_2} \frac{\partial \vec{B}}{\partial t}\Delta t\vec{ds_2} - \int_{S_1} \vec{B}(t)\vec{ds_1}}{\Delta t}$$

$$= \frac{\int_{S_2} \vec{B}(t)\vec{ds_2} - \int_{S_1} \vec{B}(t)\vec{ds_1}}{\Delta t} + \int_{S_2} \frac{\partial \vec{B}}{\partial t}\vec{ds_2} \tag{5.12}$$

However, from equation (5.10), it follows that:

$$\int_{S_2} \vec{B}(t)\vec{ds_2} - \int_{S_1} \vec{B}(t)\vec{ds_1} = \int_{V} \nabla\vec{B}\,dv + \int_{\Gamma} \vec{B}(\vec{v}dt \times \vec{dl})$$

The result is that:

$$\frac{\Delta\Phi}{\Delta t} = \frac{1}{\Delta t}\left[\int_{S} \nabla\vec{B}\vec{ds}\vec{v}\Delta t + \int_{\Gamma} \vec{B}(\vec{v}\Delta t \times \vec{dl}) + \int_{S} \frac{\partial \vec{B}}{\partial t}\Delta t\vec{ds}\right]$$

$$= \int_{S} \frac{\partial \vec{B}}{dt}\vec{ds} + \int_{S} \vec{v}\,\nabla\,\vec{B}\vec{ds} - \int_{\Gamma} (\vec{v} \times \vec{B})\vec{dl} \tag{5.13}$$

By transforming the line integral from equation (5.13) into a surface integral with the help of Stokes's theorem, we obtain:

$$\frac{\Delta \Phi}{\Delta t} = \int_S \left[\frac{\partial \vec{B}}{\partial t} + \vec{v} \nabla \vec{B} - \nabla \times (\vec{v} \times \vec{B}) \right] \vec{ds} \tag{5.14}$$

In general, it can be seen that the rate of magnetic flux variation can be given: (i) by the variation in time of $\vec{B}(t)$ through a surface independent of time, as well as (ii) by the movement of the surface in a magnetic field independent of time, i.e. the last term of equation (5.14). If the magnetic field is also nonuniform, then the term $(\vec{v} \nabla \vec{B})$ can participate in the flux variation (in regions far from magnetic field sources, where it is supposed that there are open lines for the magnetic field). But generally, in the case of a magnetic field $\nabla \vec{B} = 0$, so that the rate of variation of the magnetic flux is reduced to:

$$\frac{d\Phi}{dt} = \frac{d}{dt} \int_S \vec{B}\vec{ds} = \int_S \left[\frac{\partial B}{\partial t} - \nabla \times (\vec{v} \times \vec{B}) \right] \vec{ds} \tag{5.15}$$

Thus, the law of electromagnetic induction in developed integral form becomes:

$$E_\Gamma = \int_\Gamma \vec{E}\vec{dl} = -\int_{S_\Gamma} \frac{\partial \vec{B}}{\partial t} \vec{ds} + \int_{S_\Gamma} \nabla \times (\vec{v} \times \vec{B}) \vec{ds} \tag{5.16}$$

in which the first term represents the *EMF induced by transformation* (or pulsation) due to the variation over time of $\vec{B}$:

$$E_{\text{transformational}} = -\int_S \frac{\partial \vec{B}}{\partial t} \vec{ds} \tag{5.17}$$

and the second term represents the *EMF induced by movement* (translation, rotation, or deformation in a time-independent magnetic field):

$$E_{\text{motional}} = \int_S \nabla \times (\vec{v} \times \vec{B}) \vec{ds} \tag{5.18}$$

The motional component of the electromotive voltage, i.e. E_{motional} can be rewritten, based on the Stokes relationship, as the integral along the contour Γ which moves with the speed $\vec{v}$, as follows:

$$E_{\text{motional}} = \int_\Gamma (\vec{v} \times \vec{B}) \vec{dl} \tag{5.19}$$

Therefore, in the particular case of the movement of a conductor of length $\vec{l}$, which transversely cuts magnetic field lines $\vec{B}$ at the speed $\vec{v}$ (see figure 5.8), the angle between $(\vec{v} \times \vec{B})$ and $\vec{dl}$ is zero. As a result:

$$E_{\text{motional}} = vBl \tag{5.20}$$

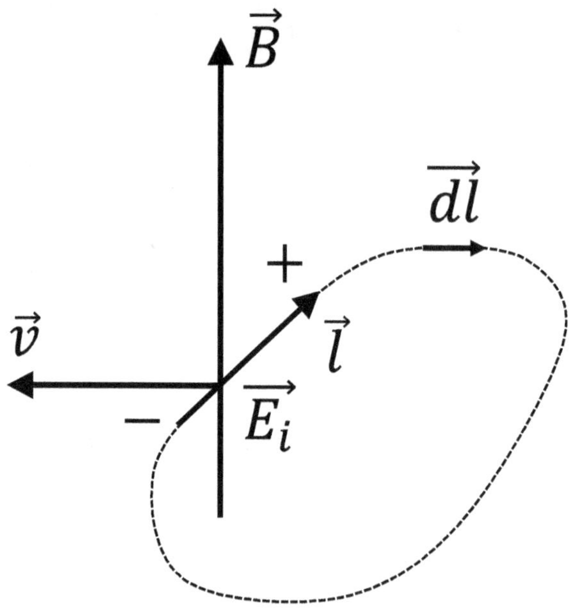

Figure 5.8. Linear conductor of length l that cuts the lines of a magnetic field.

with the sense of the vector product $(\vec{v} \times \vec{B})$ shown on the circuit. If two of the vectors $\vec{v}$, $\vec{B}$, and $\vec{dl}$ are parallel or antiparallel, the product $(\vec{v} \times \vec{B})\vec{dl} = 0$; as a result, the motional component of the electromotive voltage is zero. The following rule can therefore be formulated: *the motional voltage is induced only if the conductor cuts the magnetic field lines in its movement.*

5.1.5 Developed local form of the electromagnetic induction law (Faraday's law)

In the case of continuity domains, applying Stokes' theorem to the left-hand side of equation (5.16) yields the relation:

$$\int_{S_\Gamma} (\nabla \times \vec{E})\vec{ds} = -\int_{S_\Gamma} \frac{\partial \vec{B}}{\partial t}\vec{ds} + \int_{S_\Gamma} \nabla \times (\vec{v} \times \vec{B})\vec{ds} \tag{5.21}$$

However, the surface S_Γ is arbitrary; therefore, the developed local form of the law of electromagnetic induction is:

$$\nabla \times \vec{E} = -\frac{\partial \vec{B}}{\partial t} + \nabla \times (\vec{v} \times \vec{B}) \tag{5.22}$$

For static bodies (i.e. $\vec{v} = 0$), the local form of the law of electromagnetic induction becomes:

$$\nabla \times \vec{E} = -\frac{\partial \vec{B}}{\partial t} \tag{5.23}$$

The equation thus obtained can be recognized as one of Maxwell's equations. This equation has special significance, showing that in a space where there is a magnetic field that varies over time, an electric field whose field lines are closed around the magnetic field also arises.

Note: in an electrostatic field, the field lines are open (circulation on a closed path is zero in an irrotational field, since $\nabla \times \vec{E} = 0$). In the case of a time-varying electromagnetic field, for any variation of the magnetic field, lines of electric field closed around the magnetic field may occur. This produces a rotational field (a field with a nonzero curl), conditioned, however, by the temporal variation of the magnetic field. In the inducing electric field, the electric voltage between two points depends on the path and can no longer be stated in the form of the potential difference between two points; therefore, in this type of field, we cannot define an electrostatic potential.

5.1.6 Energetic interpretation of the electromagnetic induction phenomenon

One can find the expression for EMF given by the law of electromagnetic induction on the basis of energetic reasoning. Let us consider the scheme in figure 5.9, in which the conductor C moves frictionlessly on two linear conductors 1 and 2. Conductor C, in sliding on the other two conductors, cuts the lines of a magnetic field $\vec{B}$, which is perpendicular to the plane of the figure. Conductor C is connected to a galvanic source of EMF E and, as a result, it passes a current I that depends on the resistance R of conductor C (ignoring the internal resistance of the source and the resistance of the connecting conductors). While conductor C carries the current I, a magnetic field acts on it with a force $\vec{F} = I(\vec{l} \times \vec{B})$, oriented to the left. This force determines the movement of the conductor, which takes place at the speed of $\vec{v}$. In a very short period of time dt, the source supplies the circuit with the energy:

$$dw = EIdt \tag{5.24}$$

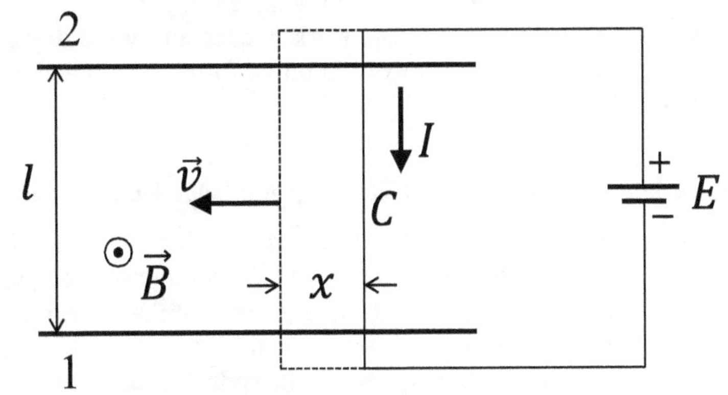

Figure 5.9. Conductor C sliding frictionlessly on two conductors 1 and 2 cuts the lines of a magnetic field $\vec{B}$ perpendicular to the plane of the figure.

On the one hand, this energy is found to be the energy dissipated in the resistance R of conductor C by the electrocaloric effect:

$$dw_J = RI^2dt \tag{5.25}$$

while, on the other hand, it is the energy of the mechanical work performed in the movement of the conductor:

$$dw_L = Fdx = IlBdx = IlBvdt \tag{5.26}$$

Thus, the equation for the energy balance is:

$$dw = dw_J + dw_L \text{ or } EIdt = RI^2dt + IlBvdt \tag{5.27}$$

Dividing both members of the equation (5.27) by Idt gives:

$$E = RI + lvB \tag{5.28}$$

It is found that the term lvB has the dimensions of an electromotive voltage. Passing the last term on the right-hand side to the term on the left, we obtain:

$$E - lvB = RI \tag{5.29}$$

This relationship can also be written in the form:

$$E + E_i = RI \tag{5.30}$$

where:

$$E_i = -lvB \tag{5.31}$$

However, we may observe that $v = \frac{dx}{dt}$. As a result, equation (5.31) becomes:

$$E_i = =-Bl\frac{dx}{dt} = -B\frac{d}{dt}(lx) = -\frac{d}{dt}(Blx) = -\frac{d\Phi}{dt} \tag{5.32}$$

The expression for the induced EMF, i.e. E_i, in the conductor C was found as a result of its displacement in the magnetic field while cutting its field lines. From the reasoning given above, it follows that the work consumed by the displacement of conductor C in the magnetic field is converted into a form of energy directly related to the induction of the EMF E_i.

5.1.7 Electronic interpretation of the electromagnetic induction phenomenon

Let us consider a metallic conductor of length $\vec{l}$ moving in a magnetic field with a speed $\vec{v}$ perpendicular to the field (see figure 5.10). The conductor is perpendicular to the plane ($\vec{v} \times \vec{B}$). Taking into account the orientations of the vectors $\vec{v}$ and $\vec{B}$ and also the fact that the free charge carriers in the metal are electrons, it is possible to establish the signs of the charges that appear at the ends of the conductor due to the action of the magnetic field. The quasi-free electrons in the conductor, when it moves

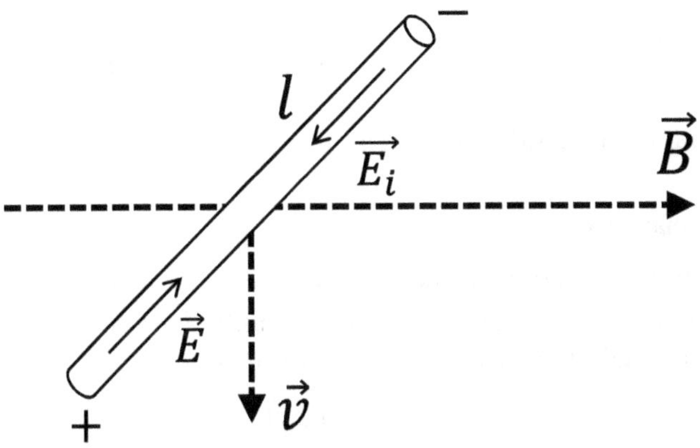

Figure 5.10. Conductor that cuts the lines of a magnetic field $\vec{B}$ at a speed $\vec{v}$ perpendicular to $\vec{B}$.

at speed $\vec{v}$ perpendicular to the induction of the magnetic field, are each individually subjected to the action of the Lorentz force:

$$\vec{F_i} = -q(\vec{v} \times \vec{B}) \tag{5.33}$$

Due to this force, the electrons move toward one end of the conductor, thus charging it negatively; the other end is charged with a positive charge consisting of the uncompensated charges of the nodes of the crystal network of the metal. Due to this polarization of the conductor's ends, an electric field appears that acts on the electron in the opposite direction to that of the Lorentz force, with the force:

$$\vec{F_E} = -q\vec{E} \tag{5.34}$$

Due to this force $\vec{F_E}$, the stationary state of the system is achieved when:

$$\vec{F_E} + \vec{F_L} = 0, \quad -q\vec{E} - q(\vec{v} \times \vec{B}) = 0 \tag{5.35}$$

This results in:

$$-q\vec{E} = +q(\vec{v} \times \vec{B}) \tag{5.36}$$

Thus, an induced electric field appears that is associated with the movement of the conductor with the velocity $\vec{v}$ in the magnetic field $\vec{B}$. The induced field, as defined in the previous chapters, is:

$$\vec{E_i} = -(\vec{E})_{\text{electrostatic equilibrium}} \Longrightarrow \vec{E_i} = (\vec{v} \times \vec{B}) \tag{5.37}$$

In general, the induced field is oriented from (−) to (+) inside a conductor, while the same is true for the electric field inside a source. Multiplying both sides of the equation (5.37) by $\vec{l}$, we obtain:

$$\vec{E}\vec{l} = -\vec{l}(\vec{v} \times \vec{B}) \tag{5.38}$$

On the left-hand side, we recognize a quantity that has the dimensions of electrical voltage, while on the right, taking into account the geometry involved, we have the variation of the magnetic flux, or, in other words, the number of magnetic field lines cut per unit time by the conductor moving with speed $\vec{v}$:

$$E_i = -\vec{l}\left(\frac{d\vec{x}}{dt} \times \vec{B}\right) = = -\vec{B}\frac{d}{dt}(\vec{l} \times \vec{x}) = -\frac{d}{dt}(Bxl) = -\frac{d\Phi}{dt} \tag{5.39}$$

hence recovering the integral form of Faraday's law of electromagnetic induction.

5.1.8 Foucault currents

A metal disc rotates in a magnetic induction field $\vec{B}$ perpendicular to the plane of the disk, as presented in figure 5.11. A portion of length $\vec{dr}$ from the radius of the disk cuts the magnetic field lines with a tangential velocity of:

$$\vec{v} = \vec{\omega} \times \vec{r} \tag{5.40}$$

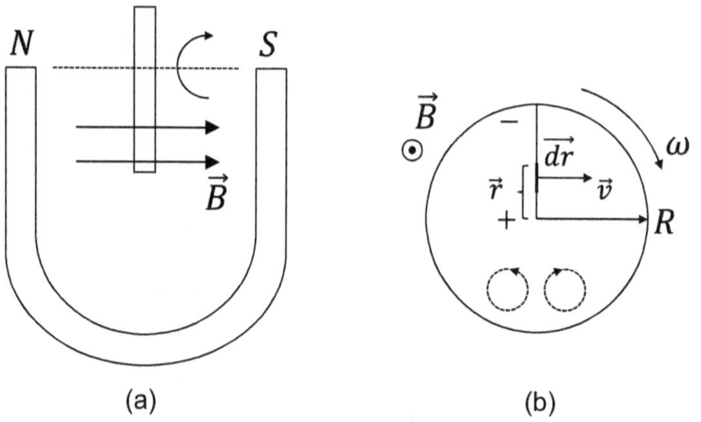

Figure 5.11. (a) Metal disc with a radius R rotating at an angular velocity ω in a magnetic induction field $\vec{B}$ created by a permanent magnet. (b) Element of length $\vec{dr}$, a portion of the radius that cuts the magnetic field lines at a speed of $\vec{v} = \vec{\omega} \times \vec{r}$.

Between the ends of the radial element of length $\vec{dr}$, an EMF is induced which, according to the law of electromagnetic induction, is:

$$dE_i = -drvB \tag{5.41}$$

Because $v = \omega r$, we obtain:

$$dE_i = -\omega rBdr \tag{5.42}$$

The induced voltage between the center and the periphery of the disc is:

$$E_i = -\omega B\int_0^R rdr = -\frac{B\omega R^2}{2} \tag{5.43}$$

assuming that $\vec{B}$ is constant along the radius of the disk. This voltage causes the occurrence of eddy currents as seen in figure 5.11(b), referred to as ***Foucault currents***. In a disk where the power lines close, Joule heat dissipates. These currents are unwanted in various electrical and electronic devices; therefore, technical solutions are sought that can decrease the density of the eddy currents and thus help to reduce the Joule losses.

Another case of eddy currents occurs in conductors carrying alternating currents. Let us consider a linear conductor with a circular cross-section of radius R_0, carrying an alternating current. Assuming that the current density is uniform inside the conductor and applying the law of magnetic circuits, the induction of the magnetic field $\vec{B}$ (i.e. the magnetic flux density) inside and outside the conductor can be calculated. Considering the electric field lines induced in a plane parallel to the axis of the conductor (see figure 5.12(a)), that is, perpendicular to $\vec{B}$ and encircling it, the induced current with density $\vec{j_i}$ creates an induced magnetic field of induction $\vec{B_i}$ acting in the opposite direction to the magnetic field $\vec{B}$, leading to a decrease in the current $\vec{i}$, the resultant magnetic field inside the conductor, and their growth outside it.

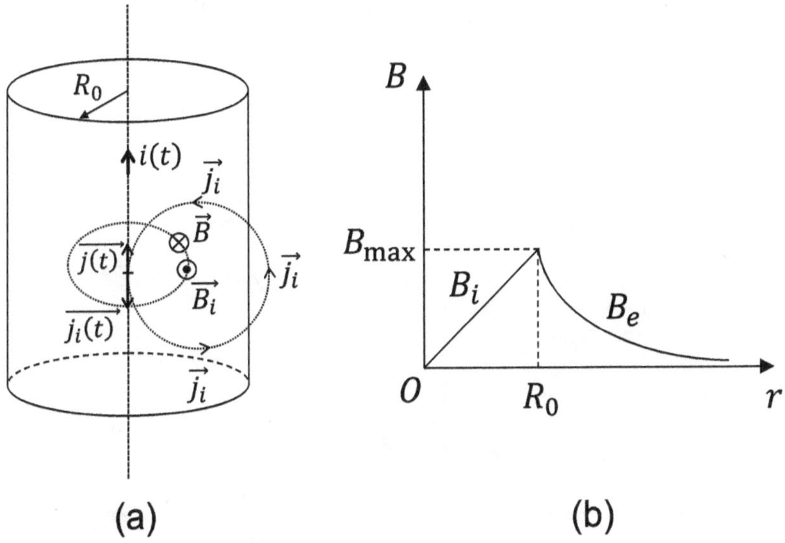

Figure 5.12. Dependence of magnetic field induction on the distance from the axis of a thick conductor carrying an electric current.

Inside:

$$\int_{\Gamma_i} \vec{B_i}\vec{dl} = \mu_0 i' = \mu_0 \frac{i}{\pi R_0^2}\pi r^2 \quad \Rightarrow B_i 2\pi r = \mu_0 \frac{i}{\pi R_0^2}\pi r^2 \Rightarrow B_i = \frac{\mu_0 i r}{2\pi R_0^2}$$

Outside:

$$\int_{\Gamma_e} \vec{B_e}\vec{dl} = \mu_0 i \quad \Rightarrow \quad B_e 2\pi r = \mu_0 i \quad \Rightarrow \quad B_e = \frac{\mu_0 i}{2\pi r}$$

A graphic representation of $B(r)$ is shown in figure 5.12(b). However, due to the variation of $i(t)$ over time, $\vec{B}$ itself also varies over time; hence, an induced voltage appears inside and outside the conductor associated with an induced circular electric field that generates induction currents $\vec{j_i}$ in the opposite direction to $i(t)$. These induction currents $\vec{j_i}$, according to Lenz's rule, oppose the variation $\frac{d\Phi}{dt}$ of the magnetic flux and therefore cause it to decrease by $\frac{d\Phi}{dt}$. This occurs only when Φ inside the conductor is reduced. This reduction is possible only if the current concentrates on the surface of the conductor, which leads to a new dependence of the magnetic field induction $\vec{B}$ as a function of the distance from the axis of the conductor. Inside the conductor, the magnetic field is zero without affecting the magnetic field outside the conductor. Thus, eddy currents tend to limit the distribution of alternating currents to a thin film at the surface of the conductor, automatically leading to increased resistance of the conductor (see figure 5.13). The effect becomes even greater as the frequency of the alternating current increases. This effect of eddy currents is called the pellicular effect or *skin effect*.

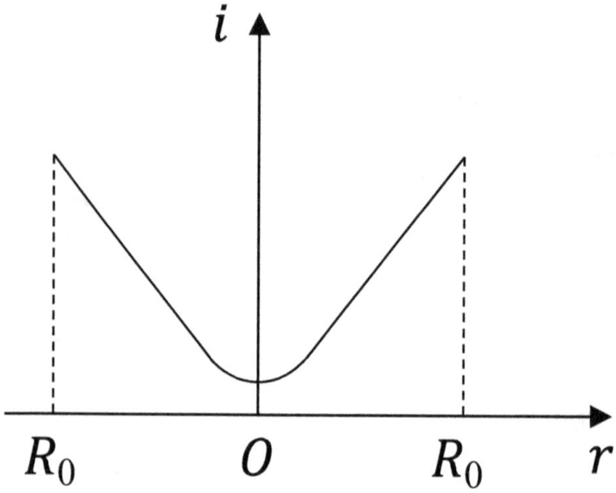

Figure 5.13. Current distribution inside a massive conductor carrying an alternating current of high frequency, as a result of the '*skin effect*'.

5.2 Self-induction

5.2.1 Self-induction coefficient (self-inductance)

In the absence of a magnetic material, the value of the magnetic field $\vec{B}$ created by a circuit carrying the time-dependent current $i(t)$ (see figure 5.14) follows the Biot–Savart–Laplace law and is proportional to the current intensity. As a result, the

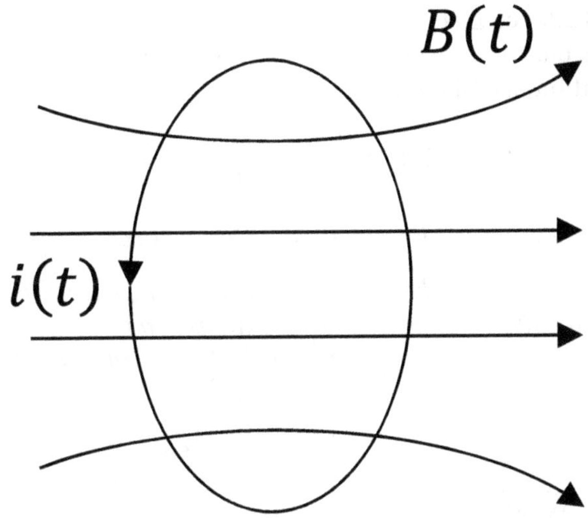

Figure 5.14. Magnetic field lines created by a circuit carrying a current of intensity $i(t)$.

magnetic flux created by an electric current passing through a surface bounded by the circuit contour is proportional to the current intensity carried:

$$\Phi(t) = Li(t) \tag{5.44}$$

Taking into account the convention for the sign of the flux passing through a given surface (the sign of the flux through a surface is given by the sign of the scalar product $\vec{B} \cdot \vec{S}$, where the vector $\vec{S}$ is perpendicular to the surface), we find that if the circuit is traveled in the sense of the current, then the self-flux (or 'auto flux') of the circuit is positive. Therefore, L is a positive constant that depends only on the geometric parameters of the circuit and is called the circuit's self-inductance. Since we assumed that the circuit carries a time-varying current, it follows that the self-flux that crosses it also varies over time. As a result, according to the law of electromagnetic induction, a self-induced voltage is induced in the circuit:

$$E = -\frac{d\Phi}{dt} = -L\frac{di}{dt} \tag{5.45}$$

Since the inductance L is always positive, relation (5.45) shows that the EMF induced by the variation of a current passing through a coil, or through an inductor, has a sense that determines an induced current that opposes changes in the primary current through the circuit. Thus, according to Lenz's rule, the EMF self-induced in an inductor causes the occurrence of a flux that tends to compensate for variations in the original magnetic flux. Therefore, relation (5.45) can be written:

$$L = -\frac{E}{\frac{di}{dt}} = -\frac{-\frac{d\Phi}{dt}}{\frac{di}{dt}} = \frac{d\Phi}{di} \tag{5.46}$$

Thus, the self-inductance L is given by the rate of change of the flux in relation to its own current. The unit of measurement for self-inductance is the '*Henry*' (H) and is defined by derivation, as follows:

$$1H = \frac{W_b}{A} = \frac{Tm^2}{A} \tag{5.47}$$

Example 1. *Calculate the self-inductance of a toroidal coil.*

Consider a toroidal coil that has n copper wires wound on a circular iron core with a square cross-section that has internal and external radii of r_i and r_e, respectively (see figure 5.15). The thickness of the core is h. Let us calculate the self-inductance, i.e. $L = \frac{d\Phi}{di}$.

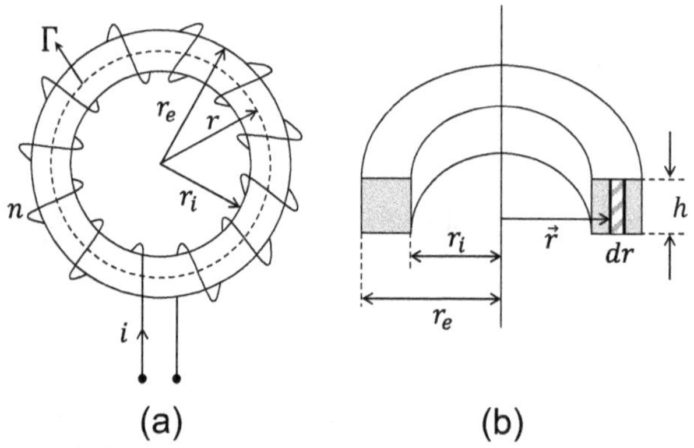

Figure 5.15. *(a) Toroidal coil with n copper turns wound on a circular core with a square cross-section; (b) cross-section through the iron core.*

Applying the law of magnetic circuits allows us to determine the induced magnetic field in the core. Since the field lines are concentric circles:

$$\int_{\Gamma} \vec{B}\vec{dl} = \mu ni \text{ or } B2\pi r = \mu_0\mu_r ni \Longrightarrow B_i = \frac{\mu_0\mu_r ni}{2\pi r} \tag{5.48}$$

The flux passing through the surface element $ds = hdr$ is (figure 5.15(b)): $d\Phi = Bds = \frac{\mu_0\mu_r ni}{2\pi r}hdr$. Therefore, the total flux passing through n copper turns is:

$$\Phi = n\frac{\mu_0\mu_r}{2\pi}nih\int_{r_i}^{r_e}\frac{dr}{r} = \frac{\mu_0\mu_r n^2 ih}{2\pi}\ln\frac{r_e}{r_i}$$

with the result that:

$$L = \frac{\mu_0\mu_r n^2 h \ln\frac{r_e}{r_i}}{2\pi} \tag{5.49}$$

Example 2. *Calculate the self-inductance of a very long solenoid.*

The self-inductance of a solenoid of length l, cross-section S, and n turns is:

$$\Phi = BnS = \frac{\mu_0 nI}{l} nS = \frac{\mu_0 n^2 S}{l} I \quad \Longrightarrow L = \frac{\mu_0 n^2 S}{l} \tag{5.50}$$

5.2.2 Mutual inductance

A time-dependent current carried by a circuit causes the magnetic flux to vary not only in the original circuit but also in a neighboring circuit. Let us consider two neighboring circuits near each other (see figure 5.16). At a certain point, currents i_1 and i_2 passing through the circuits are established, while magnetic fluxes $\Phi_1(i_1, i_2)$ and $\Phi_2(i_1, i_2)$ are established through the surfaces bounded by the two circuits. When the current in circuit C_1 varies by di_1, the flux passing through the surface bounded by C_2 varies by $d\Phi_2$. The mutual inductance coefficient of circuit C_2 versus that of circuit C_1 is given by:

$$M_{21} = \frac{d\Phi_2}{di_1} \tag{5.51}$$

When the current in circuit C_2 varies by di_2, the magnetic flux passing through the surface bounded by circuit C_1 varies by $d\Phi_1$. The mutual inductance of circuit C_1 versus that of circuit C_2 is given by the expression:

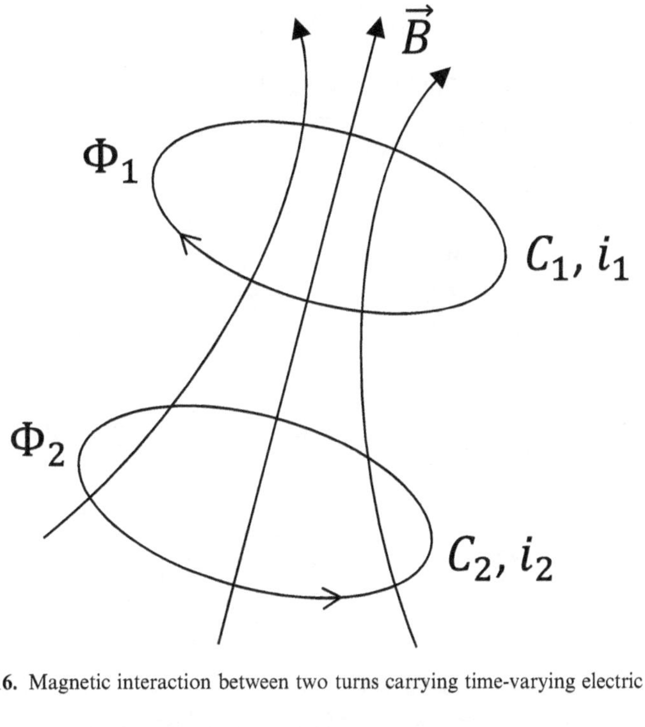

Figure 5.16. Magnetic interaction between two turns carrying time-varying electric currents.

$$M_{12} = \frac{d\Phi_1}{di_2} \tag{5.52}$$

The coefficients M_{21} and M_{12} are called coefficients of mutual induction (mutual inductances) and are constants that depend on the geometrical parameters of the two circuits and their relative position. Detailed calculations have been made of the two coefficients of mutual induction for a wide variety of coils of different shapes, in vacuum, in different positions relative to each other, and for non-ferromagnetic materials. These led to the conclusion that:

$$M_{12} = M_{21} = M \tag{5.53}$$

Next, we will show that relation (5.53) is generally valid, so that in the case of two coils of different shapes, located near each other, one can speak of a single mutual induction coefficient for the two circuits.

Let us consider two circuits, C_1 and C_2, of arbitrary shape placed close to each other, as shown in figure 5.17. In order to demonstrate relation (5.53), it is sufficient to show that the flux Φ_{12} that crosses circuit C_1 due to the variation of the current i in circuit C_2 is equal to the flux Φ_{21} that crosses circuit C_2 if the same variation of the current i takes place in circuit C_1. The value of the vector potential created by the current i, which flows through circuit C_1 at any point (x_2, y_2, z_2) belonging to circuit C_2 (see figure 5.17) is:

$$\overrightarrow{A_{21}}(x_2, y_2, z_2) = \frac{\mu_0 i}{4\pi} \int_{(C_1)} \frac{\overrightarrow{dl_1}}{r_{21}} \tag{5.54}$$

The circulation of the vector potential along circuit C_2 is $\int_{C_2} \overrightarrow{A_{21}}\overrightarrow{dl_2}$; upon applying Stokes' theorem, this becomes:

$$\int_{C_2} \overrightarrow{A_{21}}\overrightarrow{dl_2} = \int_{S_2} (\nabla \times \overrightarrow{A_{21}})\overrightarrow{ds_2} = \int_{S_2} \overrightarrow{B_1} \cdot \overrightarrow{ds_2} = \Phi_{21}$$

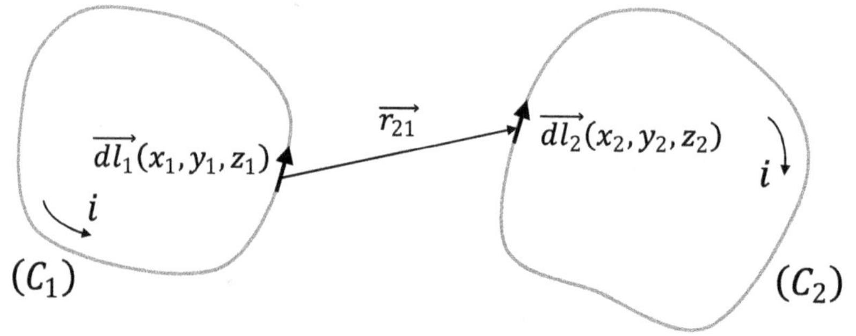

Figure 5.17. Two neighboring circuits of arbitrary shape used to demonstrate Neumann's formula.

The flux of the associated magnetic field ($\overrightarrow{B_1} = \nabla \times \overrightarrow{A_{21}}$) passing through surface S_2 bounded by circuit C_2 is:

$$\Phi_{21} = \int_{S_2} \overrightarrow{B_1} \cdot \overrightarrow{ds_2} = \int_{C_2} \overrightarrow{A_{21}dl_2} = \frac{\mu_0 i}{4\pi} \int_{C_2} \int_{C_1} \frac{\overrightarrow{dl_1 dl_2}}{r_{21}} \tag{5.55}$$

Similarly, the flux of the magnetic field $\overrightarrow{B_2}$ associated with the vector potential $\overrightarrow{A_{12}}$ created by the current i of circuit C_2 at one point of circuit C_1 is:

$$\Phi_{12} = \int_{S_1} \overrightarrow{B_2} \cdot \overrightarrow{ds_1} = \int_{C_1} \overrightarrow{A_{12}dl_1} = \frac{\mu_0 i}{4\pi} \int_{C_1} \int_{C_2} \frac{\overrightarrow{dl_2 dl_1}}{r_{21}} \tag{5.56}$$

Taking into account that $|\overrightarrow{r_{12}}| = |\overrightarrow{r_{21}}|$ and that the value of the integrals is not affected by the order in which the integration is performed along the two circuits, the two fluxes have the same value and therefore the validity of relationship (5.53) is demonstrated for the general case. The general relation between the mutual inductances of two circuits C_i and C_j is:

$$M_{ij} = M_{ji} = \frac{\mu_0}{4\pi} \int_{(C_j)} \int_{(C_i)} \frac{\overrightarrow{dl_j dl_i}}{r_{ij}} \tag{5.57}$$

Relation (5.57) is the mathematical expression of Neumann's theorem. Going back to figure 5.15, it can be observed that the variation of the flux passing through circuit C_1 is determined both by the variation of the current i_1 and by the variation of the current i_2 in circuit C_2, so that the rate of variation of the flux from C_1 over time can be expressed as follows:

$$\frac{d\Phi_1}{dt} = \frac{d\Phi_1}{di_1}\frac{\partial i_1}{\partial t} + \frac{\partial \Phi_1}{\partial i_2}\frac{di_2}{dt} \tag{5.58}$$

Similarly, the rate of variation of the flux passing through circuit C_2 due to the variations of i_2 and i_1 is:

$$\frac{d\Phi_2}{dt} = \frac{\partial \Phi_2}{\partial i_2}\frac{di_2}{dt} + \frac{\partial \Phi_2}{\partial i_1}\frac{di_1}{dt} \tag{5.59}$$

However, we notice that on the right-hand sides of relations (5.58) and (5.59), respectively, we have the self-inductances $L_1 = \frac{\partial \Phi_1}{\partial i_1}$ and $L_2 = \frac{\partial \Phi_2}{\partial i_2}$ of the two circuits; in addition, the coefficients of mutual induction (mutual inductances) $M_{12} = \frac{\partial \Phi_1}{\partial i_2}$ and $M_{21} = \frac{\partial \Phi_2}{\partial i_1}$ appear. Therefore, the rates of variation of the flux through the two circuits can be rewritten as follows:

$$\frac{d\Phi_1}{dt} = L_1\frac{di_1}{dt} + M_{12}\frac{di_2}{dt} \tag{5.60}$$

and:

$$\frac{d\Phi_2}{dt} = L_2\frac{di_2}{dt} + M_{21}\frac{di_1}{dt} \tag{5.61}$$

The rates of variation of the two fluxes determine the electromotive voltages induced in the two circuits:

$$E_1 = -L_1\frac{di_1}{dt} - M_{12}\frac{di_2}{dt} \tag{5.62}$$

and:

$$E_2 = -L_2\frac{di_2}{dt} - M_{21}\frac{di_1}{dt} \tag{5.63}$$

which express particular forms of the law of electromagnetic induction.

When electromagnetic influences occur between more than two circuits, or, in other words, when there are several inductively coupled circuits, the mutual inductance of any circuit i with respect to another circuit k is given by:

$$M_{ik} = \frac{\partial\Phi_i}{\partial i_k} \tag{5.64}$$

defined under the condition that all the currents in the circuits other than k are constant over time. We notice, however, that the flux Φ in circuit i depends on the currents in all the inductively coupled circuits. At the same time, circuit i is characterized by the self-inductance L_i, which is expressed as:

$$L_i = \frac{\partial\Phi_i}{\partial i_i} \tag{5.65}$$

When the circuits are in vacuum and are not made of ferromagnetic materials, when the quantities M_{ik} and L_i are constant over time, for a given geometry of the circuit system, and when the total flux passing through circuit C_i is equal to the sum of the fluxes induced in it by all other circuits, Maxwell's relation can be written as:

$$\Phi_i = \sum_k L_{ik} i_k \tag{5.66}$$

For $i = k$, L_{ii} represents the self-induction coefficient, while for $i \neq k$, $L_{ik} = L_{ki} = M_{ik}$ represents the mutual induction coefficients.

5.2.3 Inductances of some common electrical circuits

1. **Single-coil circuit**

   A portion of a circuit is called an ideal coil if its dominant characteristic is the self-inductance L. In addition to self-inductance, a real coil also has electrical resistance as well as electrical capacitance. Supposing that the coil has only a resistance R and an inductance L, when a voltage U is applied to the ends of the coil, its conducting wires carry a current of intensity i. If this current varies by di, a voltage of $E_i = -L\frac{di}{dt}$ is induced in the coil, so that Ohm's law for this circuit is written as follows:

$$U + E_i = Ri \tag{5.67}$$

or:

$$U = L\frac{di}{dt} + RI \tag{5.68}$$

2. **Circuit consisting of two inductively coupled coils**

   Applying Ohm's law to each coil according to (5.58), the voltage applied across their ends can be written as:

$$\begin{cases} U_1 = \dfrac{di}{dt}(L_1 \pm M) + R_1 i \\ U_2 = \dfrac{di}{dt}(L_2 \pm M) + R_2 i \end{cases} \tag{5.69}$$

   Adding up these voltages, we obtain:

$$U = U_1 + U_2 = (R_1 + R_2)i + (L_1 + L_2 \pm 2M)\frac{di}{dt} \tag{5.70}$$

   where the (+) sign is used when the fluxes passing through the two coils are summed together (additive coupling), while the (−) sign (differential coupling) is used when the fluxes passing through the two coils are subtracted. This circuit behaves as if it had the self-inductance:

$$L = L_1 + L_2 \pm 2M \tag{5.71}$$

   and the resistance:

$$R = R_1 + R_2 \tag{5.72}$$

   The coupling between the two coils can be modified by changing their relative positions. The maximum value of mutual inductance is reached with a specific coil geometry and a particular relative position between the coils.

3. **Two inductively coupled circuits, each containing one coil**

   A case of interest in the electricity laboratory is that of two different circuits, each containing just a single coil; however, these coils are inductively coupled. Suppose that a current i passing through a circuit is established by a voltage U applied between its ends; then, when this current varies by di, the voltage U at the ends of the circuit can be written as:

$$U = L\frac{di}{dt} + M\frac{di'}{dt} + Ri \tag{5.73}$$

   where i is the current intensity in the primary circuit (with the voltage source U) and I' is the current intensity in the secondary circuit (without a voltage source, $U' = 0$). The secondary circuit is a closed circuit (or short circuit), so:

$$U' = 0 = R'i' + L'\frac{di'}{dt} + M\frac{di}{dt} \tag{5.74}$$

From (5.74), the derivative $\frac{di'}{dt}$ can be extracted, as follows:

$$\frac{di'}{dt} = -\frac{1}{L'}\left(M\frac{di}{dt} + R'i'\right) \tag{5.75}$$

and substituting it into equation (5.73) yields:

$$U = L\frac{di}{dt} - \frac{M}{L'}\left(M\frac{di}{dt} + R'i'\right) + Ri \tag{5.76}$$

or:

$$U = \left(L - \frac{M^2}{L'}\right)\frac{di}{dt} + Ri - \frac{MR'}{L'}i' \tag{5.77}$$

When the last term from equation (5.77) is neglected, the following equation is obtained:

$$U = \left(L - \frac{M^2}{L'}\right)\frac{di}{dt} + Ri \tag{5.78}$$

So, the primary circuit in which the current has the intensity i behaves as if it has its own inductance:

$$L - \frac{M^2}{L'} \tag{5.79}$$

We notice that if $L - \frac{M^2}{L'} \geqslant 0$, then:

$$M \leqslant \sqrt{LL'} \tag{5.80}$$

If we define the coupling coefficient as:

$$K = \frac{M}{\sqrt{LL'}} \tag{5.81}$$

then, in the above case, $K \leqslant 1$. When $K \to 1$, the circuit behaves as if it has a self-inductance that tends to zero.

Further reading

[1] Antohe Ş and Antohe V-A 2023 *Electrostatics. Formalism of the Electrostatic Field in Vacuum and Matter* 1st edn (Bristol: IOP Publishing) 266 pages

[2] Purcell E M and Morin D J 2013 *Electricity and Magnetism* 3rd edn (Cambridge: Cambridge University Press) 853 pages

[3] Jackson J D 1998 *Classical Electrodynamics* 3rd edn (New York: Wiley) 832 pages

[4] Bleaney B I and Bleaney B 1965 *Electricity and Magnetism* 3rd edn vol 1 (Oxford: Oxford University Press)

[5] Blanpied W A 1971 *Modern Physics: An Introduction to its Mathematical Language* (New York: Holt, Rinehart and Winston) 724 pages

[6] Griffiths D J 2017 *Introduction to Electrodynamics* 4th edn (Cambridge: Cambridge University Press)

[7] Clayton R, Keith W and Nasar S 1997 *Introduction to Electromagnetic Fields* (New York: McGraw-Hill)

[8] Friesen D J 2023 *Electromagnetics for Practicing Engineers: Electrostatics and Magnetostatics* **vol 1** (Norwood, MA: Artech House) 234 pages

[9] Sadiku M and Nelatury S 2021 *Elements of Electromagnetics (Oxford Series Electric Computer Engineer)* (Oxford: Oxford University Press)

[10] Maxwell J C 2003 *A Treatise on Electricity and Magnetism* **vol 1** (Mineola, NY: Dover) 506 pages

IOP Publishing

Electromagnetism and Special Methods for Electric Circuits Analysis

Ştefan Antohe and Vlad-Andrei Antohe

Chapter 6

Electromagnetic energy produced by electric current

A complete analysis of a physical field also implies the study of the energy associated with that field. In the case of the electric field, it has been established that the energy of different charged or polarized bodies is stored in the whole volume where the electric field is present and can be computed as the volume integral of the density of the electrical energy. This density is expressed as a function of the physical quantities describing the electric field. As in the case of the electric field, the energy associated with the presence of a magnetic field is also stored in the magnetic field and can be computed by taking the volume integral of the density of the magnetic energy, which is related to physical quantities describing the magnetic field.

6.1 Energy stored in the magnetic fields of coils

6.1.1 The case of a single coil

In electrostatics, it has been shown that the energy stored in the electric field between the plates of a capacitor is:

$$W_e = \frac{1}{2}\frac{Q^2}{C} \tag{6.1}$$

This has also been expressed in terms of the intensity of the electric field created by the charged capacitor, establishing that the energy density in a homogeneous electric field of intensity $\vec{E}$ is:

$$w_e = \frac{1}{2}\varepsilon_0 E^2 \tag{6.2}$$

By analogy, we will establish an expression for the energy stored in a magnetic field inside a coil carrying a current of intensity i as a function of the magnetic flux density $\vec{B}$ in the coil.

When an electric current is established in a circuit containing an ideal coil that has an inductance L and a resistance R (see figure 6.1), part of the energy supplied by the source is converted into heat by the Joule effect, and another part is stored in the magnetic field that arises inside the coil. Multiplying the equation given by Ohm's law, $E = Ri + L\frac{di}{dt}$, (when the switch K is in position 1) by the current intensity i, we obtain:

$$Ei = Ri^2 + Li\frac{di}{dt} \tag{6.3}$$

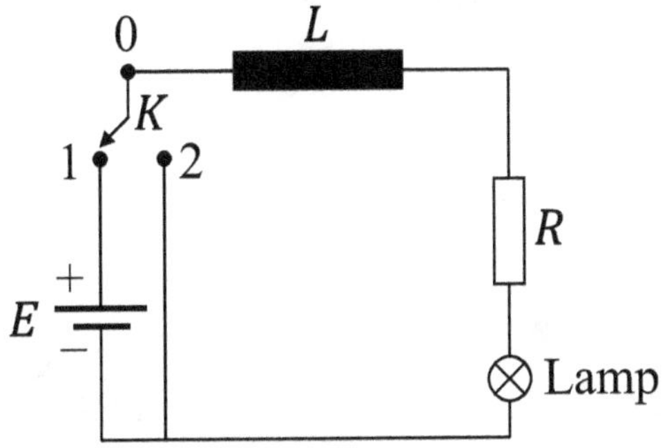

Figure 6.1. Series circuit with resistor and inductance.

The left-hand side of equation (6.3) represents the speed at which the source supplies energy to the circuit (source power). The first term on the right-hand side represents the speed at which energy is dissipated by the resistor in the form of heat, while the second term represents the speed at which energy is stored in the magnetic field of the coil. If W_m denotes the energy accumulated in the coil, the second term on the right-hand side of equation (6.3) represents the rate of variation of W_m over time:

$$\frac{dW_m}{dt} = Li\frac{di}{dt} \tag{6.4}$$

We wish to find the total energy stored in the magnetic field of the coil during the time from $t = 0$, when the current intensity in the circuit was $i = 0$, to the time t, when the current intensity is i. To do so, we integrate relationship (6.4) between the limits 0 and W_m, which gives:

$$\int_0^{W_m} dW_m = \int_0^i Lidi \quad \Rightarrow \quad W_m = \frac{Li^2}{2} \tag{6.5}$$

This energy is stored in the magnetic field produced by the electric current and is located throughout the space where this field manifests itself. This energy can return in various forms; generally, this takes the form of heat if the source is short-circuited while switch K is in position 2, or it can occur upon interruption of the electric current when the circuit is opened. This energy is therefore a potential energy that appears in the circuit when the electric current is established, i.e. during the period of current increase from 0 to the nominal value i, its total value being given by (6.5). It follows that in order to establish an electric current in a circuit consisting of a resistor connected in series with a coil, it is necessary for the source to transfer both the energy that is converted into heat by the Joule effect in resistance and a supplementary energy equal to $\frac{Li^2}{2}$. The latter energy is stored in the magnetic field produced by the current of intensity i and is fully returned upon suppression of the current. The cessation of the electric current in the circuit shown in figure 6.1 can be achieved either by short-circuiting the source, moving switch K to position 2, or by interrupting the circuit simply by opening switch K.

(1) In the first case, when the source is short-circuited, i.e. switch K is in position 2, the equation for the energy balance is:

$$0 = Ri^2 + Li\frac{di}{dt} \tag{6.6}$$

In other words:

$$-Lidi = Ri^2dt \tag{6.7}$$

Consequently, for the entire time interval t during which the current varies from i to 0, the total electromagnetic energy is:

$$\int_i^0 -Lidi = \int_0^t Ri^2dt \quad \Rightarrow \quad \frac{1}{2}Li^2 = Ri^2t \tag{6.8}$$

which shows that all the energy stored in the magnetic field of the coil is returned to the circuit and converted by its resistive elements into caloric energy, then dissipated into the environment (the lamp shown in figure 6.1 will stay on for a period).

(2) When the circuit is interrupted, the two ends of the circuit (for example, the contacts of the switch) become the plates of a capacitor that has air or another insulating medium as its dielectric. This capacitor charges at a potential difference equal to the sum of the electromotive force (EMF) E of the source and the induced EMF $\mathcal{E}$ at the ends of the coil, both of which have the same direction (indeed, the current decreases, so, according to Lenz's law, the sign of the induced EMF must be such as to participate in the increase and maintenance of i through the circuit). The difference in potential between the two ends of the switch may be large enough that it leads to the ionization of the insulating medium around it and the discharge of the capacitor, forming an electric arc through whose resistance the entire electromagnetic energy stored in the magnetic field is dissipated as heat. In this case, the energy balance can be written as follows:

$$Lidi + Ri^2dt + \frac{1}{C}qdq = 0 \tag{6.9}$$

and:

$$-\int_i^0 Lidi = \int_0^t Ri^2dt + \int_0^q \frac{1}{C}qdq \Longrightarrow \frac{1}{2}Li^2 = Ri^2t + \frac{1}{2}\frac{q^2}{C} \tag{6.10}$$

Therefore, this relationship shows that the electromagnetic energy that has been stored in the magnetic field of the coil is partially returned to the environment through the resistive elements of the circuit. Part of this energy is stored in the electric field of the capacitor, which, in turn, gives it to the circuit in the form of heat, and the other part is returned to the magnetic field of the coil. It follows, therefore, that there is a phenomenon of oscillation of electromagnetic energy between the electric field in the capacitor and the magnetic field in the coil, but the process is also accompanied by irreversible Joule energy dissipation, so that in the end this energy is transferred entirely to the surrounding environment, and the current is extinguished.

6.1.2 General expression for the potential energy of an electrical circuit

If the inductance of the circuit in figure 6.1 is not constant over time (constant inductances have been considered until now), then Ohm's law for the circuit, when K is in position 1, is written as:

$$E + E_i = Ri \Rightarrow E - \frac{d\Phi}{dt} = Ri \Longrightarrow E = Ri + \frac{d\Phi}{dt} \tag{6.11}$$

Multiplying equation (6.11) by idt, we get the equation for the energy balance, which in this case is:

$$Eidt = Ri^2dt + id\Phi \tag{6.12}$$

However, taking into account the relationship that defines the flux Φ through an inductance L carrying a current i, $\Phi = Li$, it follows that: $d\Phi = Ldi + idL$. The energy balance equation therefore becomes:

$$Eidt = Ri^2dt + Lidi + i^2dL \tag{6.13}$$

Upon integrating this from 0 to t, we obtain:

$$\int_0^t Eidt = \int_0^t Ri^2dt + \int_0^t Lidi + \int_0^t i^2dL \tag{6.14}$$

However:

$$\int_0^t Lidi = \left.\frac{Li^2}{2}\right|_0^t - \int_0^t \frac{i^2}{2}dL$$

thus:

$$\int_0^t Eidt = \int_0^t Ri^2dt + \frac{Li^2}{2} + \frac{1}{2}\int_0^t i^2dL \tag{6.15}$$

- The first term, $W = \int_0^t Eidt$, represents the energy supplied by the source during time t required for the establishment of the electric current;
- $W_R = \int_0^t Ri^2dt$ represents the energy dissipated by the Joule effect in the resistance R;
- $W_L = \frac{1}{2}Li^2$ is the intrinsic energy of the magnetic field, located in the entire space around the circuit (predominantly inside the coil);
- $W_m = \frac{1}{2}\int_0^t i^2dL$ is the electrical energy supplied by the source that produces mechanical work by the deformation of the inductance. Of course, in static circuits with $L =$ constant, this results in $W_m = 0$.

6.2 Electromagnetic energy of a system of electrical circuits

In the case of several circuits, each characterized by a resistance and an variable inductance carrying time-varying currents, an analysis of the system's energy balance makes it possible to establish an expression for the electromagnetic energy of the system.

We first consider only two inductively coupled circuits, characterized by the parameters $(R_1, L_1, i_1, M_{12} = M)$ and $(R_2, L_2, i_2, M_{21} = M)$, respectively. In this case, the energy balance can be expressed in terms of the power provided to each circuit, as follows:

$$\begin{cases} p_1 = e_1i_1 = R_1i_1^2 + i_1\dfrac{d\Phi_1}{dt} = R_1i_1^2 + i_1\dfrac{d(L_1i_1 + Mi_2)}{dt} \\ p_2 = e_2i_2 = R_2i_2^2 + i_2\dfrac{d\Phi_2}{dt} = R_2i_2^2 + i_2\dfrac{d(L_2i_2 + Mi_1)}{dt} \end{cases} \tag{6.16}$$

The total electrical power supplied to the system is equal to the sum of the powers transferred to each circuit separately, $P = p_1 + p_2$, while the total electricity supplied by the external generator to the system is equal to $W = \int_0^t Pdt$, such that:

$$W = \int_0^t \left(R_1i_1^2 + R_2i_2^2\right)dt + \int_0^t (L_1i_1di_1 + L_2i_2di_2 + Mi_2di_1 + Mi_1di_2)$$

$$+\int_0^t \left(i_1^2dL_1 + i_2^2dL_2 + 2i_1i_2dM\right) \tag{6.17}$$

However, following integration through partials, we obtain:

$$\int_0^t [L_1 i_1 di_1 + L_2 i_2 di_2 + Md(i_1 i_2)]$$

$$= \frac{1}{2}L_1 i_1^2 + \frac{1}{2}L_2 i_2^2 + Mi_1 i_2 \Big|_0^t - \int_0^t \left(\frac{i_1^2}{2} dL_1 + \frac{i_2^2}{2} dL_2 + i_1 i_2 dM \right)$$

Hence,

$$W = \int_0^t \left(R_1 i_1^2 + R_2 i_2^2\right)dt + \frac{1}{2}L_1 i_1^2 + \frac{1}{2}L_2 i_2^2 + Mi_1 i_2 \Big|_0^t$$

$$+ \int_0^t \left(\frac{i_1^2}{2} dL_1 + \frac{i_2^2}{2} dL_2 + i_1 i_2 dM \right) \tag{6.18}$$

Thus, the energy W supplied by the external generator to the system is divided into three parts, namely:

- $W_R = \int_0^t (R_1 i_1^2 + R_2 i_2^2)dt$—the energy converted into heat by the Joule effect;
- $W_i = \frac{1}{2}L_1 i_1^2 + \frac{1}{2}L_2 i_2^2 + Mi_1 i_2 \Big|_0^t$—the intrinsic energy of the system, which has a purely oscillating character and a zero average value and does not grow boundlessly over time;
- $W_m = \int_0^t \left(\frac{i_1^2}{2} dL_1 + \frac{i_2^2}{2} dL_2 + i_1 i_2 dM \right)$—the mechanical work performed in the deformation of circuits, where L and M are variable. This function is dependent only on the geometric elements of the circuits.

In the case of n circuits, the total electrical energy given to the system is also divided into three analogous parts:

$$W_R = \int_0^t \sum_{i=1}^n R_i i_i^2 dt$$

$$W_i = \sum_{i=1}^n \frac{1}{2} L_i i_i^2 + \sum_{i,k} M_{ik} i_i i_k \Big|_0^t \tag{6.19}$$

$$W_m = \int_0^t \left(\sum_{i=1}^n \frac{1}{2} i_i^2 dL_i + \sum_{i,k} i_i i_k dM_{ik} \right)$$

If one takes into account Maxwell's relationship expressing the total flux through a circuit C_i as a function of the currents in the neighboring circuits ($\Phi_i = \sum_{k=1}^n L_{ik} i_k$), the intrinsic magnetic energy becomes:

$$W_i = \frac{1}{2}\sum_i \sum_k L_{ik} i_i i_k = \frac{1}{2}\sum_i i_i \Phi_i \tag{6.20}$$

where, if $i = k$, then L_{ik} represents the self-inductance of one of the circuits of the system, whereas for $i \neq k$, it represents the mutual inductance $M_{ik} = M_{ki}$ between the two circuits.

6.3 Location of magnetic energy. Magnetic energy density

In the case of the magnetic field of an infinitely long solenoid, the magnetic energy stored in the magnetic field can easily be calculated:

$$W_i = \frac{1}{2}Li^2 = \frac{1}{2}\frac{\mu N^2 S i^2}{l}$$

However, as $B = \frac{\mu i N}{l} \Rightarrow i = \frac{Bl}{\mu N}$, it also follows that:

$$W_i = \frac{1}{2}\frac{B^2}{\mu} Sl = \frac{1}{2}\frac{B^2}{\mu} V \tag{6.21}$$

where V is the volume of the solenoid. Thus, the density of the magnetic energy stored in the magnetic field generated by the solenoid is:

$$w_m = \frac{1}{2}\frac{B^2}{\mu} = \frac{\mu H^2}{2} = \frac{\vec{B}\vec{H}}{2} \tag{6.22}$$

Magnetic energy is stored in the entire volume occupied by the magnetic field and can be characterized by the magnetic energy density, i.e. $w_m = \frac{\vec{B}\vec{H}}{2}$ (the energy per unit volume in the field). Having obtained the magnetic energy density, we can now calculate the energy of the magnetic field located in the volume V_Σ:

$$W_i = \int_{V_\Sigma} \frac{\vec{B}\vec{H}}{2} dv \tag{6.23}$$

Relation (6.22) for the magnetic energy density was obtained for a particular case; however, it is general, with its validity extending over all linear magnetic media and in nonuniform magnetic fields. To demonstrate this, we will start from the expression for the energy density in a magnetic field created by a circuit characterized by the self-inductance L, carrying a current that has a conduction current density of $\vec{j}$, as shown in figure 6.2. The energy located throughout the space is given by relationship (6.23). The nominator under the integral in equation (6.23) is written using the vector potential $\vec{A}$ as follows:

$$\vec{H}\vec{B} = \vec{H}(\nabla \times \vec{A}) = \vec{A}(\nabla \times \vec{H}) + \nabla(\vec{A} \times \vec{H})$$

In the above equation, the relationship from vector analysis was used:

$$\nabla(\vec{a} \times \vec{b}) = \vec{b}(\nabla \times \vec{a}) - \vec{a}(\nabla \times \vec{b})$$

Given these equations, we can obtain the energy stored in the magnetic field created by the circuit as follows:

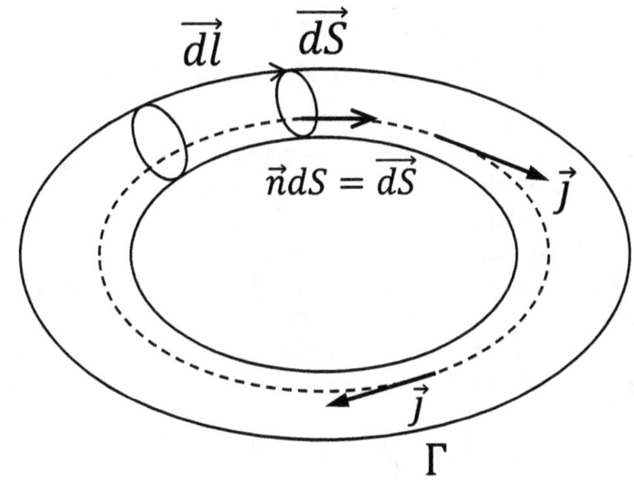

Figure 6.2. Circuit with a self-inductance L carrying a conduction current that has a density $\vec{j}$.

$$\begin{aligned} W_i &= \int_V \frac{1}{2}[\vec{A}(\nabla \times \vec{H}) + \nabla(\vec{A} \times \vec{H})]dv \\ &= \int_V \frac{1}{2}[\vec{A} \cdot \vec{j} + \nabla(\vec{A} \times \vec{H})]dv \\ &= \int_V \frac{1}{2}\vec{A}\vec{j}\, dv + \frac{1}{2}\int_{\Sigma_v} (\vec{A} \times \vec{H})\vec{ds} \end{aligned} \tag{6.24}$$

The last term in this relationship is zero because $\vec{A} \times \vec{H} = 0$. Indeed, the area delimiting the volume V is extended to infinity, and $\vec{A}$ and $\vec{H}$ are functions of $1/r^n$:

$$(\vec{A} \times \vec{H}) = \vec{A} \times \nabla \times \vec{A} = \nabla(\vec{A}\vec{A}) - (\vec{A}\vec{A})\nabla = 0$$

Under these conditions, the energy stored in the magnetic field created by the circuit is:

$$W_i = \int_\Gamma \frac{1}{2}\vec{A}\vec{j}\,\vec{ds}\vec{dl} = \int_\Gamma \frac{1}{2}\vec{A}\, i\vec{dl} = \frac{1}{2}i\int_\Gamma \vec{A}\,\vec{dl} = \frac{1}{2}i\int_{S_\Gamma} (\nabla \times \vec{A})\vec{ds}$$

$$= \frac{1}{2}i\int_{S_\Gamma} \vec{B}\vec{ds} = \frac{1}{2}i\Phi = \frac{1}{2}Li^2 \tag{6.25}$$

It has been shown that if we accept the magnetic energy density, expression (6.23), and the fact that the magnetic energy is extended throughout the whole space occupied by the magnetic field, the total magnetic energy has the general relation $\frac{1}{2}i\Phi = \frac{1}{2}Li^2$, which fully confirms the hypothesis of the spatial expansion of energy and the validity of the relationship $w_m = \frac{\vec{B}\vec{H}}{2}$.

6.4 The theorems of generalized forces in magnetic fields

6.4.1 Energy balance in a system of *n* circuits carrying currents

Let us consider a system of n circuits carrying electric currents that have intensities $(i_1, i_2, \ldots, i_n)$. The circuits are connected to energy sources that have electromotive forces $(e_1, e_2, \ldots, e_n)$, as shown in figure 6.3. Given that both $(i_1, i_2, \ldots, i_n)$ and $(e_1, e_2, \ldots, e_n)$ are time dependent and the circuits can be considered mobile, the energy balance of the system shall be written as follows. The total energy produced by the generators in the time interval dt, i.e. $\sum_{k=1}^{n} e_k i_k dt$, must cover the energy losses due to the Joule–Lenz effect in the resistors $(R_1, R_2, \ldots, R_n)$ of the circuits, increase the magnetic energy located in the magnetic fields of the circuits, and compensate for the work done by magnetic forces acting on the circuits:

$$\sum_{k=1}^{n} e_k i_k dt = \sum_{k=1}^{n} R_k i_k^2 dt + dW_i + dW_m \tag{6.26}$$

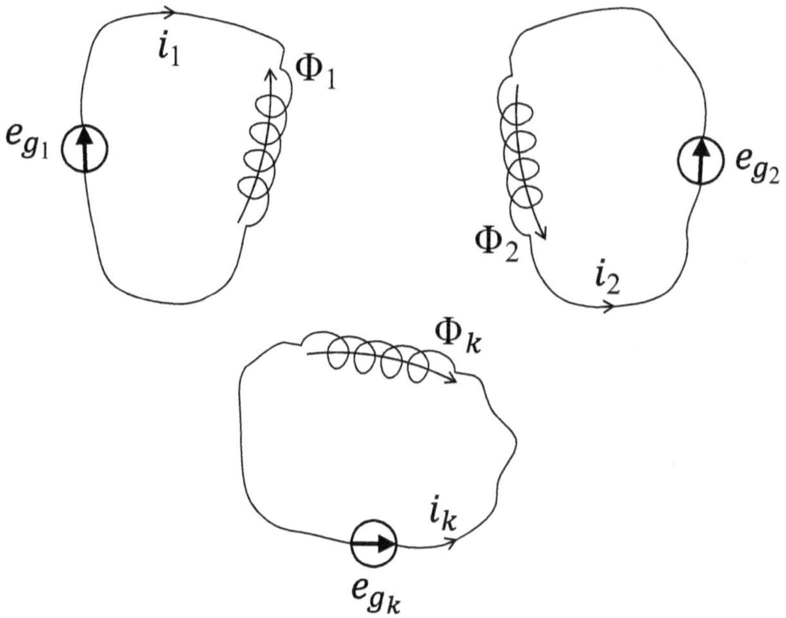

Figure 6.3. Electrical circuits, interacting by magnetic forces.

The mechanical work performed by the magnetic forces can be expressed according to the generalized forces X and the generalized coordinates x, respectively, as was done in the electric field, by $dL = dW_m = Xdx$. However, during the time dt (because i_k and e_k vary over time), the fluxes in the circuits are variable. Therefore, we may also need to add an induced electromotive voltage $e_k = -\frac{d\Phi_k}{dt}$ to the EMF e_{g_k} of each circuit, so that Ohm's law for each circuit becomes:

$$e_{g_k} + e_k = R_k i_k \text{ or } e_{g_k} - \frac{d\Phi_k}{dt} = R_k i_k \tag{6.27}$$

or $e_{g_k} = R_k i_k + \frac{d\Phi_k}{dt}$, which when substituted into (6.26) gives:

$$\sum_{k=1}^{n}\left(R_k i_k^2 + i_k \frac{d\Phi_k}{dt}\right)dt = \sum_{k=1}^{n} R_k i_k^2 dt + dW_i + dL$$

Hence, the variation in magnetic energy becomes:

$$dW_i = \sum_{k=1}^{n} i_k d\Phi_k - dL = \sum_{k=1}^{n} i_k d\Phi_k - \sum_{k=1}^{n} X_k dx_k \tag{6.28}$$

Thus, the variation in the magnetic energy of a system where the magnetic fluxes vary and the circuits move in a magnetic field depends on i, Φ, X, and x. As a result, we can calculate the magnetic energy W_i and also determine the generalized forces X.

6.4.2 Calculation of magnetic energy

If we consider a linear medium (with magnetic permeability μ independent of H) and immovable circuits, $dx = 0$, the magnetic energy is:

$$dW_i = \sum_{k=1}^{n} i_k d\Phi_k \tag{6.29}$$

In the case of a sufficiently slow variation of the currents (the quasi-stationary regime), the fluxes that cross the contours of the circuits are related to the currents by self-inductance and mutual inductances, according to the Maxwell relationship ($\Phi_k = \sum_j L_{kj} i_j$). If the parameter λ (defined as the ratio between the current intensity at the time t, namely i_j, and the final intensity i_f for which dW_i is calculated) is added, then: $i_j = \lambda i_f$ and $d\Phi_k = d\lambda \Phi_{k_{\text{final}}}$. Thus:

$$dW_i = \sum_k (\lambda i_{k_{\text{final}}}) \Phi_{k_{\text{final}}} d\lambda = \sum_k (i_k \Phi_k)_{\text{final}} \lambda d\lambda \tag{6.30}$$

The sum of the variations in energy from the initial moment when there was no magnetic field ($\lambda = 0$) until the final moment when i_k reaches the final value $i_{k_{\text{final}}}$ and $\lambda = 1$ gives the magnetic energy W_i of the magnetic field in the considered state:

$$W_i = \sum_k (i_k \Phi_k) \int_0^1 \lambda d\lambda = \frac{1}{2}\sum_{k=1}^{n} i_k \Phi_k \tag{6.31}$$

This is analogous to an expression that we encountered for a system of conductors in electrostatic equilibrium, where each conductor carries a charge q_k and is kept at the potential V_k ($W_e = \frac{1}{2}\sum_{k=1}^{n} q_k V_k$). If $\Phi_k = \sum_j L_{kj} i_j$, then for W_i, we obtain:

$$W_i = \frac{1}{2}\sum_{k=1}^{n}\sum_{j=1}^{n} L_{kj} i_j i_k \tag{6.32}$$

As an example, for two inductively coupled coils, the magnetic energy is:

$$W_m = \frac{1}{2}L_{11}i_1^2 + \frac{1}{2}L_{12}i_1i_2 + \frac{1}{2}L_{21}i_2i_1 + \frac{1}{2}L_{22}i_2^2$$

$$= \frac{1}{2}L_1i_1^2 + Mi_1i_2 + \frac{1}{2}L_2i_2^2$$

in which the three terms represent, in order: W_{m1}—the magnetic energy of the first coil (1); W_m—the energy of interaction between coil 1 and coil 2; and W_{m2}—the magnetic energy of the second coil (2). Generally, the interaction energy between a current i and a magnetic field is $W_{\text{int}} = i\Phi_{\text{ext}}$.

6.4.3 First theorem of generalized forces in magnetic fields

From the equation for the energy balance (6.28), if **the flux remains constant** ($d\Phi_k = 0$—that is, no energy is exchanged with the external sources/generators), it follows that mechanical work is performed on account of the decrease in the intrinsic magnetic energy of the system, and then we obtain:

$$(dW_i)_{\Phi_k} = -\sum_{k=1}^{n} X_k dx_k, \quad X_k = -\left(\frac{\partial W_i}{\partial x_k}\right)_{\Phi_k = \text{const}} \tag{6.33}$$

Statement:

The generalized force X_k corresponding to the generalized coordinate x is equal to the partial derivative of the magnetic energy (expressed as generalized fluxes and coordinates) in relation to the generalized coordinate, taken with a changed sign, for constant magnetic fluxes passing through circuits.

The relationship $(dW_i)_\Phi = -Xdx$ can be given the following interpretation: because there are no induced electromotive forces (Φ_k = constant), there is no exchange of energy between the sources and the field; the intrinsic magnetic energy of the field decreases as mechanical work is performed.

6.4.4 Second theorem of generalized forces in magnetic fields

Assuming that **the currents in the circuits remain constant**, the equation for the energy balance can be written as: $(dW_i)_{i_k} = \sum_k d(i_k\Phi_k) - Xdx$. However:

$$W_i = \frac{1}{2}\sum_{k=1}^{n} i_k\Phi_k \Rightarrow \sum_{k=1}^{n} i_k\Phi_k = 2W_i \Rightarrow (dW_i)_{i_k} = 2(dW_i)_{i_k} - Xdx$$

$$\Rightarrow (dW_i)_{i_k} = Xdx \tag{6.34}$$

and therefore:

$$X_k = \left(\frac{\partial W_i}{\partial x_k}\right)_{i_k = \text{const}} \tag{6.35}$$

Statement:
The generalized force X_k, corresponding to the generalized coordinate x_k, is equal to the partial derivative of the magnetic energy (expressed as a function of coordinates and currents) in relation to the generalized coordinate, at constant currents.
Relation (6.34) can be interpreted as follows: if the currents are kept constant by the external generators, these generators provide some energy for the magnetic field, resulting in its increase by x_k, while another portion of the energy is used to perform mechanical work by magnetic forces. Both theorems of generalized forces, i.e. equations (6.33) and (6.35), allow for the calculation of the generalized force.

Example 1 *Calculation of the load-bearing force of an electromagnet.*
The load-bearing force is the force necessary to detach the armature from the poles of an electromagnet. In figure 6.4, x is the thickness between intervals $\left(l_{01} = l_{02} = \frac{l_0}{2}\right)$.

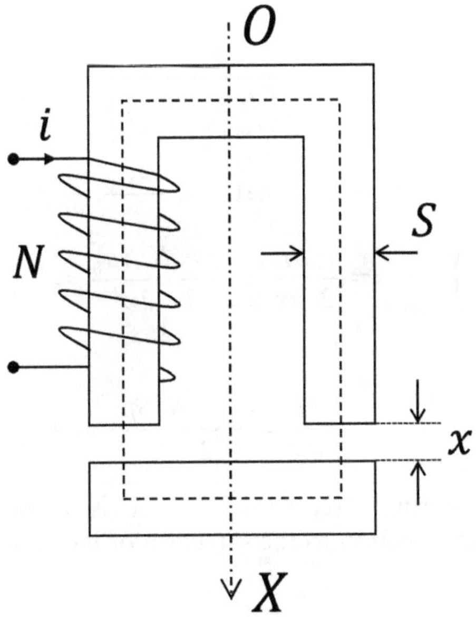

Figure 6.4. *Electromagnet used in an electromagnetic crane.*

Solution: *The energy stored in the magnetic field of the electromagnet is:* $W_i = \frac{1}{2}Li^2 = \frac{\Phi^2}{2L}$; therefore, *according to equation (6.33), the generalized force at a constant flux is:*

$$X_k = -\left(\frac{\partial W_i}{\partial x_k}\right)_{\Phi_k = \text{const}} = -\frac{\Phi^2}{2}\left(-\frac{1}{L^2}\right)\frac{\partial L}{\partial x} = \frac{\Phi^2}{2L^2}\frac{\partial L}{\partial x} = \frac{i^2}{2}\frac{\partial L}{\partial x}$$

However, $L = \frac{N\Phi}{i} = \frac{N}{i}\frac{\theta}{R_m} = \frac{N}{i}\frac{Ni}{R_m} = \frac{N^2}{R_m}$, where R_m is the reluctance. For a magnetic circuit of constant section S (considered to be a tube of magnetic field lines), $\Phi = \vec{B}\vec{S}$, the flux is considered constant, so that $B = \frac{\Phi}{S} = \mu H \Rightarrow H = \frac{\Phi}{\mu S}$. Using the law of magnetic circuits, i.e. $\int_\Gamma \vec{H}\vec{dl} = Ni = \theta = F_m$ (magnetomotive force), we obtain:

$$\oint_\Gamma \frac{\Phi}{\mu S} dl = Ni = \theta \Rightarrow \Phi \oint_\Gamma \frac{dl}{\mu S} = \theta$$

This leads to: $\Phi = \frac{\theta}{\oint_\Gamma \frac{dl}{\mu S}} = \frac{\theta}{R_m}$, in which $\theta = Ni = F_m$ and $R_m = \oint_\Gamma \frac{dl}{\mu S}$ is the reluctance. In this scenario, the reluctance is:

$$R_m = \oint_\Gamma \frac{dl}{\mu S} = \frac{l_{\text{iron}}}{\mu_0 \mu_r S} + \frac{2x}{\mu_0 S}$$

and then:

$$L = \frac{N\Phi}{i} = \frac{\text{NF}}{iR_m} = \frac{\text{NN}i}{i\left(\frac{l_{\text{iron}}}{\mu_0\mu_r S} + \frac{2x}{\mu_0 S}\right)} = \frac{N^2 S\mu_0}{\frac{l_{\text{iron}}}{\mu_r} + 2x}$$

As $\mu_r \gg 1$ and $x \ll l_{\text{iron}}$, the result is that $L \approx \frac{N^2 S\mu_0}{2x}$, $\frac{\partial L}{\partial x} = -\frac{N^2 S\mu_0}{2x^2}$. Therefore:

$$X = -\left(\frac{\partial W_i}{\partial x}\right)_{\Phi=\text{const}} = \frac{i^2}{2}\frac{\partial L}{\partial x} = -\frac{i^2}{2}\frac{N^2 S\mu_0}{2x^2} = -\frac{i^2 L}{2x} = -\frac{\Phi^2}{2Lx}$$

or:

$$X = \left(\frac{\partial W_i}{\partial x}\right)_{i=\text{const}} = \frac{i^2}{2}\frac{\partial L}{\partial x} = -\frac{i^2}{2}\frac{N^2 S\mu_0}{2x^2} = -\frac{i^2 L}{2x} = -\frac{\Phi^2}{2Lx}$$

The sign of the load-bearing force shows that X is actually an attractive force oriented in the opposite direction to the direction of increase in x.

Further reading

[1] Antohe Ş and Antohe V-A 2023 *Electrostatics. Formalism of the Electrostatic Field in Vacuum and Matter* 1st edn (Bristol: IOP Publishing) 266 pages

[2] Purcell E M and Morin D J 2013 *Electricity And Magnetism* 3rd edn (Cambridge: Cambridge University Press) 853 pages

[3] Jackson J D 1998 *Classical Electrodynamics* 3rd edn (New York: Wiley) 832 pages

[4] Griffiths D J 2017 *Introduction to Electrodynamics* 4th edn (Cambridge: Cambridge University Press)

[5] Clayton R, Keith W and Nasar S 1997 *Introduction to Electromagnetic Fields* (New York: McGraw-Hill)

[6] Sadiku M and Nelatury S 2021 *Elements of Electromagnetics (Oxford Series Electric Computer Engineer)* (Oxford: Oxford University Press)

[7] Maxwell J C 2003 *A Treatise on Electricity and Magnetism* **vol 1** (Mineola, NY: Dover) 506 pages

[8] Friesen D J 2023 *Electromagnetics for Practicing Engineers: Electrostatics and Magnetostatics* **vol 1** (Norwood, MA: Artech House) 234 pages

[9] Panofsky W K H and Phillips M 2012 *Classical Electricity and Magnetism* 2nd edn (North Chelmsford, MA: Courier Corporation) 512 pages

[10] Jones E T 2007 *Induction Coil: Theory and Applications* (Newberg, OR: Barclay Press) 276 pages

IOP Publishing

Electromagnetism and Special Methods for
Electric Circuits Analysis

Ştefan Antohe and Vlad-Andrei Antohe

Chapter 7

Magnetization state

This chapter discusses the magnetization state using an analogy with the polarization state studied in electrostatics. The study of the magnetization state opens perspectives going to understand the physical processes taking place in magnetic materials, from the phenomenological study of different magnetic materials, such as diamagnetic, paramagnetic, and ferromagnetic materials, to the microscopic approach to magnetization processes.

7.1 Magnetic properties of substances

Just as the electrostatic field can be produced either by charged bodies with electric charge or by electrically polarized bodies, it has been found experimentally that the stationary magnetic field is also determined by two distinct states of bodies:

- the state of conductors passing electric currents (the electrokinetic state);
- the state of magnetized bodies (the state of magnetization).

In the case of nonstationary electrokinetic states (those that vary in time), it has also been established that a magnetic field is produced by the variation of an electric field over time, as seen during the study of the magnetic circuit law. In the following, the magnetization state will be described in detail by analogy with the study of the polarization state of dielectrics.

7.1.1 The definition of the magnetization state

The existence of magnetized bodies was established experimentally; that is, bodies that have the ability to interact with each other and act on other bodies (Fe, Ni, Co, alloys, etc.) near them with forces and moments of a new type (different from those of an electrical nature). They are also subject to couples and forces produced by the magnetic fields created by electric currents.

doi:10.1088/978-0-7503-5854-5ch7

Definition: *a body is in a state of magnetization if it is subjected to supplementary ponderomotive actions such as forces and torques when it is located in a magnetic field, even though it does not carry an electric current.*

The simplest example of a magnetized body is the magnetic needle. As with electrostatic states, magnetization states can be permanent when they do not depend on an external magnetic field $\overrightarrow{B_0}$ or temporary when they depend on an external magnetic field.

7.1.2 Magnetic moment and intensity of magnetization

The state of polarization is characterized at the macroscopic scale by the electrical moment $\vec{p}$ and at the microscopic scale by the intensity of polarization $\vec{P}$. In an identical way, the state of magnetization of a body is characterized at the macroscopic scale by the primitive quantity $\vec{m}$, called the **magnetic moment**, and at the microscopic scale by the **intensity of magnetization** or simple magnetization, $\vec{M}$. Experimentally, it has been found that a small magnetized body is acted upon by an external magnetic field with ponderomotive actions of the torque and force types. The moment of torque exerted on a small magnetized body that has a magnetic moment $\vec{m}$ is:

$$\overrightarrow{M_c} = \vec{m} \times \overrightarrow{B_0} \tag{7.1}$$

and the force with which a nonuniform magnetic field acts on the magnetic moment $\vec{m}$ is:

$$\vec{F} = \nabla \left(\vec{m}\overrightarrow{B_0}\right) \tag{7.2}$$

The torque $\overrightarrow{M_c}$ tends to rotate the small magnetized body to align its moment $\vec{m}$ parallel to the magnetic field $\overrightarrow{B_0}$, and in the case of a nonuniform magnetic field, a resultant force is also exerted on it, which has the components:

$$\begin{cases} F_x = m_x \dfrac{\partial B_{0x}}{\partial x} + m_y \dfrac{\partial B_{0y}}{\partial x} + m_z \dfrac{\partial B_{0z}}{\partial x} \\ F_y = m_x \dfrac{\partial B_{0x}}{\partial y} + m_y \dfrac{\partial B_{0y}}{\partial y} + m_z \dfrac{\partial B_{0z}}{\partial y} \\ F_z = m_x \dfrac{\partial B_{0x}}{\partial z} + m_y \dfrac{\partial B_{0y}}{\partial z} + m_z \dfrac{\partial B_{0z}}{\partial z} \end{cases} \tag{7.3}$$

If the gradient of the nonuniform magnetic field is known and the components of the force, i.e. F_x, F_y, and F_z, can be experimentally measured, the components of the magnetic moment of the small magnetized body can be determined, solving the above system. The magnetic moment $\vec{m}$ is then a primitive vectorial quantity that characterizes the magnetized body.

Definition: *The magnetic moment of a small magnetized body is a primitive quantity that completely characterizes its magnetization state and determines the ponderomotive actions (couples and forces) with which an external magnetic field acts on it.*

In the general case, the magnetic moment $\vec{m}$ can have a permanent component $\vec{m_p}$ (independent of the external magnetic field) and a temporal component $\vec{m_t}$ (induced by the external magnetic field $\vec{B_0}$ that is canceled when the magnetic field is removed), so that:

$$\vec{m} = \vec{m_p} + \vec{m_t} \tag{7.4}$$

If the magnetized body has larger dimensions, its state of magnetization is characterized by the magnetic moment $\vec{m}$, which is the vector sum of the magnetic moments $\overrightarrow{\Delta m}$ of its various component parts:

$$\vec{m} = \sum \overrightarrow{\Delta m} \tag{7.5}$$

Similar to the intensity of polarization $\vec{P}$ that characterizes a polarized body, the magnetization intensity (magnetization) $\vec{M}$ is also introduced for magnetized bodies $\vec{M}$ as the magnetic moment per unit volume:

$$\vec{M} = \lim_{\Delta V \to 0} \frac{\overrightarrow{\Delta m}}{\Delta V} = \frac{\overrightarrow{dm}}{dV} \tag{7.6}$$

Depending on the magnetization, the magnetic moment of the magnetized body can be calculated by:

$$\vec{m} = \int_V \vec{M} dV \tag{7.7}$$

The unit of measurement for $\vec{m}$ is derived from:

$$\vec{F} = \nabla \left(\vec{m}\vec{B_0}\right) \Longrightarrow \langle \vec{m} \rangle = \frac{\langle F \rangle \langle L \rangle}{\langle B \rangle} = \frac{\langle I \rangle \langle B \rangle \langle L \rangle \langle L \rangle}{\langle B \rangle} = Am^2$$

and for magnetization, it is:

$$\langle \vec{M} \rangle = \frac{\langle \vec{m} \rangle}{\langle V \rangle} = \frac{Am^2}{m^3} = \frac{A}{m}$$

7.1.3 The equivalence between a small magnetized body and a current loop

As in the case of polarized dielectric bodies, where the equivalence between a small polarized body and an electric dipole has been studied, in the case of magnetized bodies, there is a theorem of equivalence between a small magnetized body and a current loop, which is stated as follows: *in the stationary regime, a small magnetized body with a magnetic moment $\vec{m}$ and a current loop that has the moment $\vec{m_b}$ are* *equivalent not only in terms of the forces and mechanical moments that are exerted on them by an external magnetic field but also from the point of view of the magnetic field they produce in vacuum, if the following condition is met:*

$$\vec{m} = \vec{m_b} = I\vec{S} \tag{7.8}$$

The demonstration of this theorem is equivalent to the one that was performed for electrostatics; specifically, it takes a small magnetized body that has a moment $\vec{m}$ and a current loop that has a moment $\vec{m_b} = I\vec{S}$ that are brought in turn to the same distance from a current loop $\overrightarrow{m_b{}'} = I'\vec{S'}$. Recalling the principle of action and reaction for forces and couples, it is found that the equivalence theorem applies under the conditions in which $\vec{m} = \vec{m_b}$.

7.1.4 Amperian currents (magnetization currents)

As a consequence of the equivalence between a small magnetized body and a current loop, a distribution of currents equivalent to the magnetization state of the magnetized body can be found. The current that flows through a small circular conducting wire equivalent to a small magnetized body that it replaces is called an Amperian current (or magnetization current) because Ampère hypothesized that, on the microscopic scale, this equivalence is actually realized. If we consider a uniformly magnetized small cylindrical body (with magnetization parallel to the axis of the cylinder) that has a cross-sectional area S and a height h (see figure 7.1), its magnetic moment is:

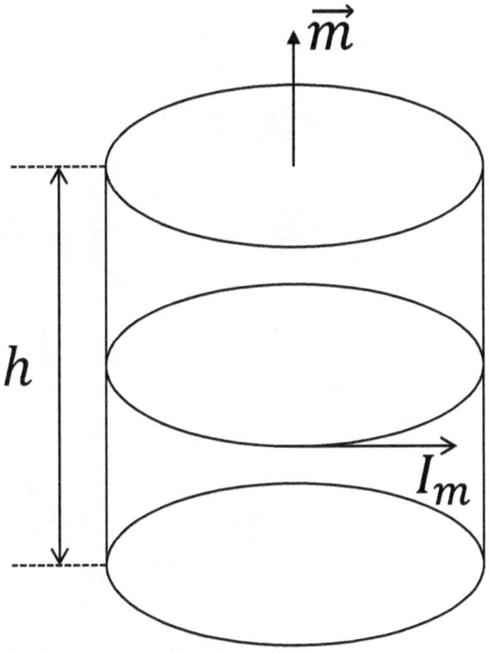

Figure 7.1. Uniformly magnetized cylindrical body that has a magnetic moment $\vec{m}$.

$$\vec{m} = \vec{M}V = MSh \tag{7.9}$$

which leads to:

$$I_m = Mh. \tag{7.10}$$

Generally, 'Amperian currents' refers to a distribution of conduction currents that generates a magnetic field in vacuum equivalent to that created by a magnetized body that has a magnetization $\vec{M}$. The distribution of the Amperian current can be volumetric or superficial.

a) In general, $I_m = \int_\Gamma \vec{M}\vec{dh}$ and $I_m = \int_{S_\Gamma} \vec{j_m}\,\vec{ds} = \int_{S_\Gamma} (\nabla \times \vec{M})\vec{ds}$, resulting in: $\vec{j_m} = \nabla \times \vec{M}$.

b) At the surface of a magnetic body, one can define a superficial distribution of currents whose linear density is equal to the tangential component of $\vec{M}$, hence: $\vec{j_{mt}} = \frac{\Delta I_m}{\Delta h} = \vec{M_t}$.

7.2 The relationship between the vectors $\vec{B}$, $\vec{H}$, and $\vec{M}$

7.2.1 Experimental measurement of magnetic flux density in a substance

One of Faraday's experiments with a dielectric placed between the plates of a capacitor, carried out to study the behavior of dielectrics in an electric field, can be paralleled here, as follows: let us consider a toroidal coil in which a current of intensity I (as shown in figure 7.2) has been established. If the space inside the coil is filled with vacuum, then the magnetic field induced, $\vec{B_0}$, can be calculated using Ampère's circuit law:

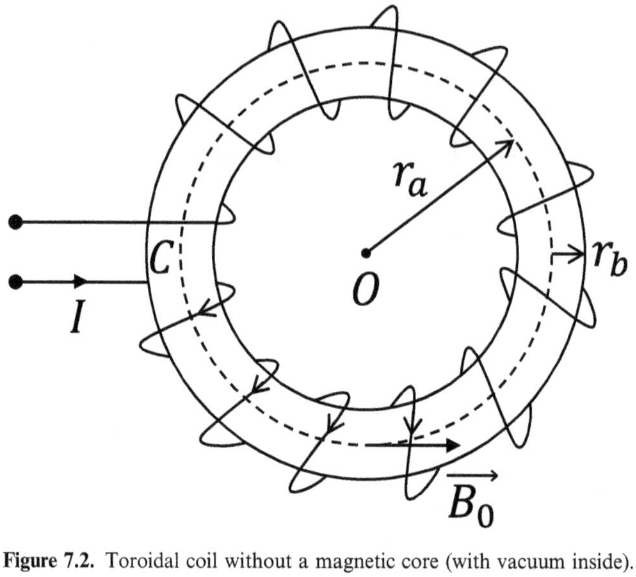

Figure 7.2. Toroidal coil without a magnetic core (with vacuum inside).

$$\int_C \overrightarrow{B_0} \overrightarrow{dl} = \mu_0 NI \tag{7.11}$$

where C is a closed path inside the coil of medium radius r_a, N is the number of turns in the coil, and μ_0 is the magnetic permeability of vacuum. At a distance r_a, $\overrightarrow{B_0}$ is constant and has the value:

$$\overrightarrow{B_0} = \frac{\mu_0 NI}{2\pi r_a} \overrightarrow{e_\theta} \tag{7.12}$$

If the condition $r_b << r_a$ is met, then $\overrightarrow{B_0}$ can be consistently considered to be a constant value inside the coil.

We now introduce the homogeneous and isotropic substance to be studied inside the coil. In this case, a direct relationship allowing the calculation of the magnetic field induction $\overrightarrow{B_0}$ can no longer be used, because the magnetic properties of the substance are not known. This is why we use an experimental measurement of $\overrightarrow{B}$. For this purpose, a new toroidal coil is used, having n turns that are wound between the circular wires of the first coil (as shown in figure 7.3). The magnetic flux passing through the secondary coil n turns (drawn with thick lines in figure 7.3) is given by:

$$\Phi = n\pi \, r_b^2 B \tag{7.13}$$

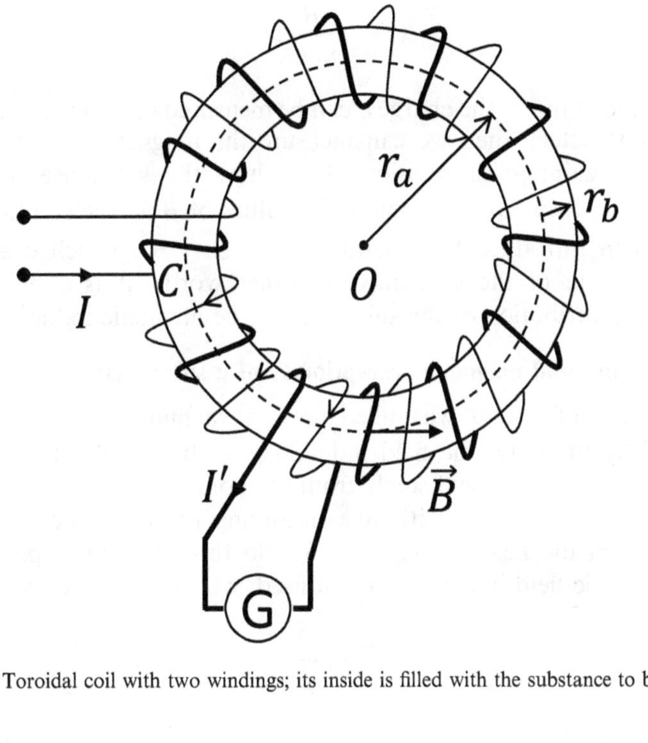

Figure 7.3. Toroidal coil with two windings; its inside is filled with the substance to be studied.

assuming that $\vec{B}$ has the same value at every point inside the coil. We observe a variation in the magnetic flux corresponding to a variation in the magnetic field induction $\overrightarrow{dB}$, given by:

$$d\Phi = n\pi\ r_b^2 dB \tag{7.14}$$

However, the variation of the magnetic flux through the secondary coil causes an induced voltage (electromotive force) in it, and if the secondary circuit is completed by a ballistic galvanometer, the value of the induced current can be measured:

$$I' = \frac{E}{R} = -\frac{1}{R}\frac{d\Phi}{dt} \tag{7.15}$$

where R is the electrical resistance of the secondary coil circuit. The charge carried by the current I' through the n turns of the secondary coil is:

$$q = \int_{t_0}^{t} I'dt = -\frac{1}{R}\int_{\Phi_0}^{\Phi} d\Phi = -\frac{\Delta\Phi}{R} \tag{7.16}$$

If, at the initial time, $\Phi_0 = 0$, the result is:

$$q = -\frac{\Phi}{R} = -\frac{n\pi\ r_b^2 B}{R} \tag{7.17}$$

leading to:

$$B = -\frac{R}{n\pi r_b^2}q \tag{7.18}$$

It is thus concluded that if the charge q can be measured (and this can be done with a ballistic galvanometer), then we can measure the magnetic induction (B) in any substance, thus gaining information about how the substance responds to an external magnetic field. Once we know the values of $\vec{B}$ in the substance and $\overrightarrow{B_0}$ in vacuum, we can introduce the quantity $\overrightarrow{M'} = \vec{B} - \overrightarrow{B_0}$, which characterizes the magnetization state of the substance; in other words, it is a physical quantity describing the contribution of the substance to the magnetic induction inside it.

7.2.2 Permeability and magnetic susceptibility of a substance

When the inside of a coil was filled with a vacuum, the magnetic field was characterized by the magnetic field induction $\overrightarrow{B_0}$. It is useful to characterize the properties of the magnetic field solely from the point of view of the electric current geometry that creates it (i.e. without considering the magnetic properties of the medium in which the field propagates). To do this, the vector physical quantity called the magnetic field intensity is often used, which is defined by:

$$\vec{H} = \frac{\vec{B}}{\mu_0} = -\frac{R}{n\pi r_b^2}q\ \frac{1}{\mu_0} = -\frac{Rq}{n\pi\mu_0 r_b^2} \tag{7.19}$$

It can be seen that the value of the magnetic field intensity differs from that of magnetic field induction, and also that the two vector physical quantities have different units. The unit of measurement for the intensity of the magnetic field results from Ampère's magnetic circuit law, which is written as:

$$\int_C \overrightarrow{B_0}\overrightarrow{dl} = \mu_0 I \text{ or } \int_C \mu_0 \overrightarrow{H}\overrightarrow{dl} = \mu_0 I \implies \int_C \overrightarrow{H}\overrightarrow{dl} = I \tag{7.20}$$

where the number of turns is considered to be $N = 1$. Relation (7.20) leads to the definition of the unit of magnetic field strength, which in the International System of Units (SI) is A m^{-1}. The expression for the intensity of the magnetic field can be calculated using Ampère's law if the core of the coil is a substance of any kind. Under the conditions of the experiment that was used to determine the induction of the magnetic field in the substance, we introduced the vector field $\overrightarrow{M'} = \overrightarrow{B} - \overrightarrow{B_0}$, which characterizes the magnetization state of the substance. With this, one can write:

$$\overrightarrow{B_0} = \overrightarrow{B} - \overrightarrow{M'} \tag{7.21}$$

or:

$$\overrightarrow{H} = \frac{\overrightarrow{B_0}}{\mu_0} = \frac{\overrightarrow{B}}{\mu_0} - \frac{\overrightarrow{M'}}{\mu_0} \tag{7.22}$$

Noting that $\frac{\overrightarrow{M'}}{\mu_0} = \overrightarrow{M}$ (which we can call **the intensity of magnetization, the vector of magnetic polarization,** or simply **the magnetization**), relation (7.22) becomes:

$$\overrightarrow{H} = \frac{\overrightarrow{B}}{\mu_0} - \overrightarrow{M} \text{ or } \overrightarrow{B} = \mu_0\overrightarrow{H} + \mu_0\overrightarrow{M} \tag{7.23}$$

Once we know the quantities $\overrightarrow{H}$, $\overrightarrow{B}$, $\overrightarrow{B_0}$, and $\overrightarrow{M}$, we can define several other quantities that characterize the magnetization state of the substance.

The magnetic permeability of the substance (μ) is defined (when the substance is homogeneous and isotropic) as the ratio of B to H:

$$\mu = \frac{B}{H} \tag{7.24}$$

We can now define the magnetic induction as:

$$\overrightarrow{B} = \mu\overrightarrow{H} \tag{7.25}$$

where μ is a scalar for homogeneous and isotropic media and a tensor for inhomogeneous and anisotropic media. In the case of vacuum,

$$\overrightarrow{B_0} = \mu_0\overrightarrow{H} \Rightarrow \quad \mu_0 = \frac{\overrightarrow{B_0}}{\overrightarrow{H}} \tag{7.26}$$

In general, to describe the magnetic properties of substances, the dimensionless quantity called ***relative magnetic permeability*** is used, i.e. $\frac{\mu}{\mu_0} = \frac{B}{B_0} = \mu_r$. This shows how many times B (in a substance) is larger than B_0. Often, in the case of homogeneous and isotropic substances, another scalar quantity is used for the characterization of magnetic properties. This quantity, called ***magnetic susceptibility***, is defined as the ratio of magnetization to the intensity of the magnetic field:

$$\chi_m = \frac{M}{H} \tag{7.27}$$

In the case of homogeneous and isotropic media, this is a scalar, but in general, it is a tensor. If $\vec{B} = \mu\vec{H} = \mu_0\vec{H} + \vec{M}\mu_0$, then $\mu_0(\mu_r - 1)H = M\mu_0$ and thus $\chi_m = (\mu_r - 1)$. Hence, $\vec{M'} = \mu_0\chi_m\vec{H}$.

7.2.3 Classification of magnetic substances

By defining the magnetic susceptibility of a homogeneous and isotropic substance as $\chi_m = \frac{M}{H}$, a classification of substances can be made in terms of their behavior in a magnetic field, taking into account the sign and value of this susceptibility:

- Substances for which $\chi_m < 0$ are called ***diamagnetic*** materials.
- Substances for which $\chi_m > 0$ but not too large (on the order of units or tens) are called ***paramagnetic*** materials.
- Substances for which $\chi_m > 0$ and very large (10^2–10^4) are called ***ferromagnetic*** materials.

The way in which a substance reacts when it is introduced into a magnetic field and how the magnetic field changes in a space containing the substance, compared to vacuum, can be inferred from the following experiment. Using a specially constructed electromagnet (dimensions: inner diameter $D_i = 0.1$ m; outer diameter, $D_e = 0.4$ m; length, $l = 0.4$ m; the space between the two coaxial cylinders is filled with a winding made from a copper tube through which water circulates at a flow rate of 130 l h^{-1} for cooling, and a power of 400 kW), a magnetic field of 3 T can be created (see figure 7.4). The magnetic field is practically uniform in the center and decreases to 1.8 T toward the ends of the solenoid, so that the gradient of its decrease is $dB_z/dz = 17$ T m^{-1}. The value of the force acting on the sample can be measured by means of a dynamometer. It is found that if the sample size is not too large, the force acting on it is directly proportional to its mass and the gradient of the magnetic field at the location where it is situated and does not depend on the shape of the sample. In order to describe this force using an expression of the type $\vec{F_Z} = m\frac{\partial B}{\partial z}\vec{k}$ (previously established as the component of the force acting on a magnetic moment $\vec{m}$ in an inhomogeneous magnetic field), in the particular case of $\vec{B}\,(0, 0, B_Z)$, it is natural to assume that the sample acquires a magnetic moment in the magnetic field that is directly proportional to the mass of the sample. In other words, the magnetic moment of the sample is equal, as will be seen later, to the vector sum of the

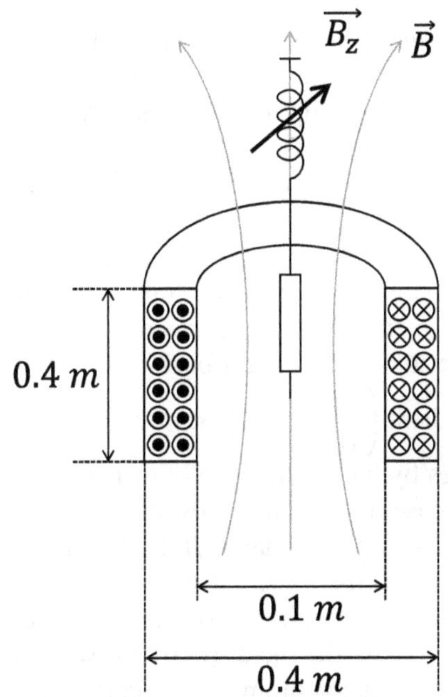

Figure 7.4. Specially designed electromagnet used to study different substances' behaviors in a magnetic field.

Table 7.1. Forces with which a magnetic field acts on substances introduced into the field.

Substance	H_2O	Cu	NaCl	N_2 liquid	Na	$CuCl_2$	O_2 liquid	Fe	Fe_3O_2
F **(N kg^{-1})**	−0.22	−0.026	−0.15	+40	+0.2	+2.8	+75	+4000	+1200
Category	Diamagnetic			Paramagnetic				Ferromagnetic	

magnetic moments of the sample's atoms. In order to compare the forces that the magnetic field exerts on different substances, they are related to the unit of mass and expressed in $N\,kg^{-1}$. Experiments show that for the great majority of pure substances, the magnetic force, although measurable, is extremely small (compared to the gravitational force). In table 7.1, the value of the magnetic force per unit of mass is given for several substances measured in a field of $B_z = 1.8$ T and a gradient of $dB_z/dz = 17\ \text{T m}^{-1}$. The second finding shows that for some substances, the force is vertically oriented downward and tends to move substances to regions with a more intense magnetic field, while for others, the force is vertically oriented upward and tends to repel the substance from the magnetic field. The relationship $\overrightarrow{F_Z} = m\frac{\partial B}{\partial z}\vec{k}$ shows that if $m > 0$, then the moment moves in the direction of the field gradient, that is, toward the region with a more intense magnetic field; and if $m < 0$, the magnetic moment moves toward weaker fields. In accordance with this relationship, it can be stated that in the magnetic field, substances attracted to regions with a more intense field acquire a magnetic moment $\vec{m}$ parallel to the applied magnetic field, whereas substances in the magnetic field propelled toward the weaker field region acquire a magnetic moment $\vec{m}$ antiparallel to the magnetizing field.

As shown in table 7.1, there are also several substances that are strongly propelled by the field toward the region with a more intense field, for example, N_2 and O_2. For this reason, a drop of liquid nitrogen placed at the bottom end of the solenoid does not fall but ascends to the center of the solenoid, since the magnetic force is stronger than the gravitational force. A small number of substances, of which iron is one, are driven by powerful forces two to three orders of magnitude larger than the gravitational force. Moreover, in these substances, it has been observed that if the value of B_z, and therefore that of the gradient $\left(\frac{\partial B}{\partial z}\right)$, decreases by half, the magnetic force acting on the unit of mass decreases by half in the case of iron and magnetite, while decreasing almost four times in the case of the other substances. This can only be possible if the force of the magnetic field is proportional to B_z in the case of iron but proportional to the square of B_z in the case of the other substances. The behavior of substances in a magnetic field, as described above, allows for their classification, as follows:

1. Substances rejected by the field (marked with a '−' sign in table 7.1, indicating that the magnetic moments of their atoms are oriented antiparallel to the magnetic field) are called ***diamagnetic*** materials. The forces exerted on them by the magnetic field are weak. This category includes Pb, quartz, sulfur, carbon, inorganic complexes, and all organic compounds.
2. Substances that are attracted toward the more intense zone of the magnetic field, marked with a '+' sign in table 7.1, are divided into two categories:
   (i) ***Paramagnetic*** materials form a category of substances that are poorly attracted by magnetic fields with a force proportional to the square of the magnetic field, B^2. The paramagnetic effect is temperature dependent, while the diamagnetic effect is independent of temperature. As the temperature is lowered, the paramagnetic effect increases. Precisely for this reason, the magnetic force acting on a unit mass of liquid oxygen (90 K) is remarkably high. It can be seen that it is not at all easy to explain the behavior of substances in magnetic fields, especially when it is found that Cu is diamagnetic and $CuCl_2$ is paramagnetic, while Na is paramagnetic and NaCl is diamagnetic.
   (ii) Substances that behave like iron are called ***ferromagnetic*** materials. In addition to the pure materials, i.e. Fe, Co, and Ni, their alloys with other elements and a series of crystalline compounds also belong to this category.

7.3 Microscopic approach to the magnetization of substances

7.3.1 Atomic magnetic moments

7.3.1.1 Orbital magnetic moment

According to the classic planetary model of the atom, the movement of an electron in a circular orbit around the nucleus corresponds to the angular momentum $\vec{L} = \vec{r} \times \vec{p}$ whose magnitude is $L = rp = rmv$, since $\vec{v} \perp \vec{r}$; see figure 7.5. However,

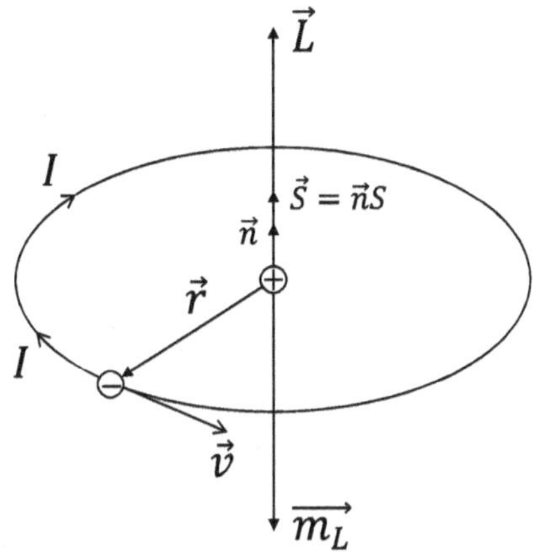

Figure 7.5. Orbital magnetic moment of the atom.

the movement of the electron in a circular orbit is equivalent to a circular current, oriented in the opposite direction to the tangential speed of the electron. As a result, from a classical point of view, the orbital motion of the electron can be associated with a classical magnetic moment whose value will be determined.

The intensity of the electric current produced by the movement of the electron in the circular orbit can be expressed as follows:

$$I = \frac{dq}{dt} = \frac{-e}{T} = -e\nu = -\frac{e\omega}{2\pi} = -\frac{e\nu}{2\pi r} \tag{7.28}$$

The circular electric current of intensity I produced by the orbital motion of the electron creates a magnetic moment $\overrightarrow{m_L}$, whose magnitude is $m_L = I \cdot S$, where $S = \pi r^2$ is the area of the electronic orbit of radius r, equivalent to a circular current:

$$m_L = IS = \frac{-e}{T}S = -\frac{e\omega}{2\pi}\pi r^2 = -\frac{emvr}{2m} = -\frac{e}{2m}L \tag{7.29}$$

This can be rewritten in vector form as:

$$\overrightarrow{m_L} = -\frac{e}{2m}\vec{L} = -\gamma\vec{L} \tag{7.30}$$

where $\gamma = \frac{e}{2m} = \frac{|\overrightarrow{m_L}|}{|\vec{L}|}$ (the ratio of the orbital magnetic moment of the electron to its angular momentum) is called *the orbital magnetomechanical ratio of the electron.* Due to the negative electric charge of the electron, the circular electric current generated by the orbital movement of the electron moves in the opposed direction to the movement of the electron; as a result, the orbital magnetic moment is opposed to the angular momentum. In establishing relation (7.30), a particular electronic orbit of circular shape was used, but it can be shown that even in the case of an elliptical orbit, and in general, the movement of an electron in a field of central forces (such as that generated by the nucleus) involves a relationship between $\overrightarrow{m_L}$ and $\vec{L}$ that is similar to (7.30).

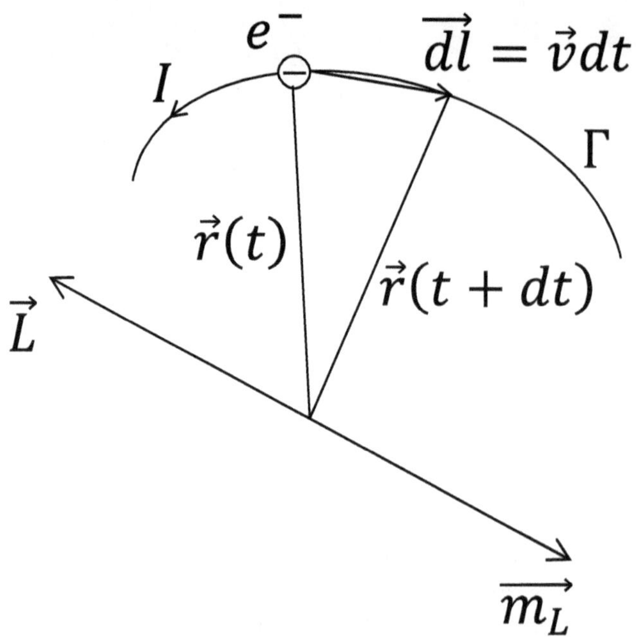

Figure 7.6. Orbit of a charged particle in motion in a field of central forces.

Let us consider a general plane orbit of radius $\vec{r}$ at time t (see figure 7.6). As is known from mechanics, movement in the central field conserves angular momentum; as a result, in the calculation of the magnetic moment presented below, the angular momentum $\vec{L} = \vec{r} \times \vec{p} = \text{constant}$ in the last integral can be moved in front of the integral:

$$\overrightarrow{m_L} = I\vec{S} = I\frac{1}{2}\int_{\Gamma} \vec{r} \times d\vec{l} = I\frac{1}{2}\int_0^T \vec{r} \times \vec{v}dt = -\frac{e}{T}\frac{1}{2}\int_0^T (\vec{r} \times \vec{v})dt$$

$$= -\frac{e}{2T}\int_0^T \frac{\vec{r} \times m\vec{v}}{m}dt = -\frac{e}{2m}\frac{1}{T}\int_0^T \vec{L}dt = -\frac{e}{2m}\vec{L}\frac{1}{T}T$$

$$= -\frac{e}{2m}\vec{L} = -\gamma\vec{L} \tag{7.31}$$

It can be seen, in addition, that $\overrightarrow{m_L} = -\frac{e}{2m}\vec{L}$ appears in the relationship; along with $\vec{L}$, these are the only universal constants, i.e. those that do not depend on the form of electronic orbits in atoms. Any electronic orbit in the atom is equivalent to a magnetic moment $\overrightarrow{m_L}$ that generates a magnetic field whose induction can be obtained using the relationships:

$$\vec{B} = -\mu_0 \nabla V_m \text{ with } V_m = \frac{I\Omega}{4\pi} = \frac{I}{4\pi}\int \frac{\vec{ds}\cdot\vec{r}}{r^3} = \frac{\vec{m}\cdot\vec{r}}{4\pi r^3}$$

$$\Rightarrow \vec{B} = \nabla \times \vec{A}, \text{ where } \vec{A} = \frac{\mu_0}{4\pi}\frac{\vec{m}\times\vec{r}}{r^3}$$

If $V_m = \frac{\vec{m}\cdot\vec{r}}{4\pi r^3}$, then $\vec{B} = -\mu_0 \nabla V_m = \frac{\mu_0}{4\pi r^3}\left[\frac{3(\vec{m}\vec{r})\vec{r}}{r^2} - \vec{m}\right]$. These relationships can now be used to elucidate certain manifestations of bodies in the state of magnetization. However, the angular momentum of the orbital motion of electrons in atoms is quantized; that is, it cannot have a continuous spectrum of values but only a selection of discrete values given by a formula established in quantum theory:

$$|\vec{L}| = \hbar\sqrt{l(l+1)}, \quad l = 0, 1, 2, \ldots \tag{7.32}$$

This is called an orbital quantum number, where $\hbar = \frac{h}{2\pi}$ is the reduced Planck constant. The relationship between $\vec{m_L}$ and $\vec{L}$ leads to the conclusion that the magnitude of the orbital magnetic moment is also quantized:

$$|\vec{m_L}| = \frac{e\hbar}{2m}\sqrt{l(l+1)}, \quad l = 0, 1, 2, \ldots \tag{7.33}$$

The universal constant:

$$\mu_B = \frac{e\hbar}{2m} = 9.29 \times 10^{-24}\ \text{A m}^2 \tag{7.34}$$

is called the ***Bohr magneton*** and has been chosen as the unit of measurement for magnetic moments.

7.3.1.2 Electronic spin and the magnetic moment of electronic spin

In addition to orbital motion, the electron may also have an additional degree of freedom similar to the rotation around its own axis. This degree of freedom is associated with an angular momentum, called the spin angular momentum, whose expression, as established in quantum theory, is:

$$|\vec{S}| = \hbar\sqrt{s(s+1)} \tag{7.35}$$

Here, s is called the spin quantum number; it takes a single value $s = 1/2$ and should not be confused with the spin angular momentum, which according to (7.35) has the value:

$$|\vec{S}| = \hbar\frac{\sqrt{3}}{2} \tag{7.36}$$

However, as the electron is a particle that carries an electric charge, its spin angular momentum can also be associated with a spin magnetic moment. The relationship between the spin angular momentum (i.e. $\vec{S}$) and the spin magnetic moment (i.e. $\vec{m_S}$) is assumed by analogy to be the same as in the case of orbital moments:

$$\vec{m_S} = -\frac{e}{2m}\vec{S} \tag{7.37}$$

By contrast, experimental measurements of the value of the magnetomechanical ratio, i.e. $\frac{|\overrightarrow{m_S}|}{|\vec{S}|}$, have shown that it is twice as large as in the case of orbital motion, a phenomenon known in the research literature as the *'spin magnetic anomaly'*. Therefore, the correct relationship between $\overrightarrow{m_S}$ and $\vec{S}$ is:

$$\overrightarrow{m_S} = -2\frac{e}{2m}\vec{S} = = -2\gamma\vec{S} \tag{7.38}$$

7.3.1.3 The total angular momentum and the magnetic moment of the electron

The total angular momentum of the electron is equal to the sum of the kinetic moments $\vec{L}$ and $\vec{S}$ ($\vec{J} = \vec{L} + \vec{S}$). The value of the total angular momentum, by analogy with (7.33) and (7.35), is quantized:

$$|\vec{J}| = \hbar\sqrt{j(j+1)} \tag{7.39}$$

where j is the total quantum number of the electron and can take the following values:

$$\begin{cases} j = l + s = l + \dfrac{1}{2} \\ j' = l - s = l - \dfrac{1}{2} \end{cases} \tag{7.40}$$

The total magnetic moment of the electron is equal to the vector sum of the orbital and spin magnetic moments, $\overrightarrow{m_L}$ and $\overrightarrow{m_S}$:

$$\overrightarrow{m_{LS}} = \overrightarrow{m_L} + \overrightarrow{m_S} = -\gamma(\vec{L} + 2\vec{S}) \tag{7.41}$$

It can be seen that the total magnetic moment $\overrightarrow{m_{LS}}$ is not oriented in the opposite direction to the total angular momentum $\vec{J}$. Indeed, taking into account (7.38), one can write:

$$\overrightarrow{m_{LS}} = -\gamma(\vec{L} + \vec{S} + \vec{S}) = -\gamma(\vec{J} + \vec{S}) \tag{7.42}$$

Thus, $\overrightarrow{m_{LS}}$ is opposed to the resultant of the vectors $\vec{J}$ and $\vec{S}$. This is due to the spin anomaly. A representation of the vectors is given in figure 7.7. In order to determine

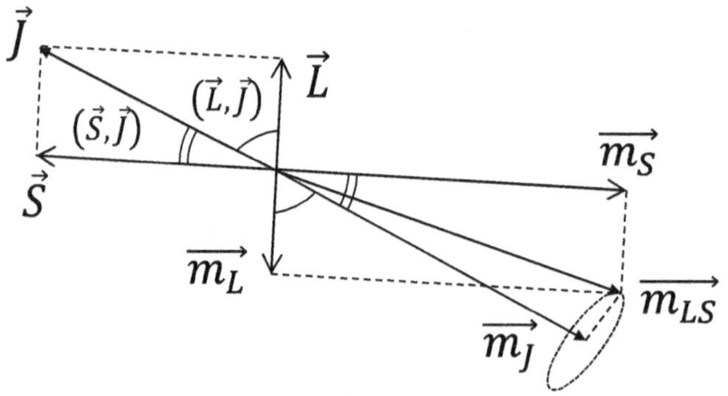

Figure 7.7. Illustration of the kinetic and magnetic moments in an atom.

the value of the projection of the total magnetic moment $\overrightarrow{m_{LS}}$ in the direction $\vec{J}$, denoted by $\overrightarrow{m_j}$, we use the vector sum. This allows us to establish the relationship between the total magnetic moment and the total angular momentum of an electron in an atom. The projection of the resulting magnetic moment $\overrightarrow{m_{LS}}$ in the direction of $\vec{J}$, namely ($\overrightarrow{m_j}$), is given by:

$$m_j = m_L \cos(\vec{L}, \vec{J}) + m_s \cos(\vec{S}, \vec{J})$$

$$= m_L \frac{J^2 + L^2 - S^2}{2\text{JL}} + m_S \frac{J^2 + S^2 - L^2}{2\text{JS}}$$

$$= -\gamma L \frac{J^2 + L^2 - S^2}{2\text{JL}} - 2\gamma S \frac{J^2 + S^2 - L^2}{2\text{JS}}$$

$$= -\frac{\gamma}{2J}\hbar^2[j(j+1) + l(l+1) - s(s+1) + 2j(j+1) + 2s(s+1) - 2l(l+1)]$$

$$= -\frac{\gamma J}{2J^2}\hbar^2[2j(j+1) + j(j+1) + s(s+1) - l(l+1)]$$

$$= -\frac{\gamma J\hbar^2}{\hbar^2}\left[1 + \frac{j(j+1) + s(s+1) - l(l+1)}{2j(j+1)}\right] = -\gamma gJ \tag{7.43}$$

Hence, $m_j = -\gamma gJ$, where $g = 1 + \dfrac{j(j+1) + s(s+1) - l(l+1)}{2j(j+1)}$ is called **Landè's factor**. We can now discuss two cases:

(a) When the atom has only an orbital magnetic moment ($\vec{J} = \vec{L}$ and $\vec{S} = 0$), then:

$$g_J = g_L = 1 \tag{7.44}$$

(b) When the orbital magnetic moment is zero ($\vec{L} = 0$) and the atom has only a spin magnetic moment ($\vec{J} = \vec{S}$), then:

$$g_J = J_S = 2 \tag{7.45}$$

In general, the atom possesses both an angular momentum and an angular spin momentum, so *Landè's factor* takes values between two and zero; the latter value corresponds to the situation where the contributions of orbital and spin motion cancel each other.

7.3.1.4 Nuclear magnetic moment

Along with the orbital magnetic moment and the spin moment, the magnetic moment of the nucleus also contributes to the total magnetic moment of the atom. Any nucleus is characterized by an angular momentum, whose value is:

$$|\vec{I}\,| = \hbar\sqrt{I(I+1)},\, I = 0,\, 1,\, 2,\, \ldots \tag{7.46}$$

and, correspondingly, a nuclear magnetic moment:

$$\overrightarrow{m_I} = g_N \frac{e}{2m_N}\vec{I} = g_N \gamma_N \vec{I} \tag{7.47}$$

where I is the nuclear quantum number and γ_N is the nuclear magnetomechanical ratio. The magnitude of the nuclear magnetic moment is quantized and is given by:

$$\overrightarrow{m_I} = g_N \frac{e}{2m_N}\hbar^2\sqrt{I(I+1)} = g_N \mu_N \sqrt{I(I+1)} \tag{7.48}$$

where:

$$\mu_N = \frac{e\hbar^2}{2m_p} = 5 \cdot 10^{-27}\ \mathrm{A\ m^{-1}} \tag{7.49}$$

This value, called the ***nuclear Bohr magneton***, is much lower than the Bohr magneton of the electron, due to the mass of the proton, i.e. $m_p = 1840\ m_e$. Given the above, we can calculate the total kinetic moment of the atom, $\vec{J} + \vec{I} = \vec{F}$, and the total magnetic moment:

$$\overrightarrow{m_F} = \overrightarrow{m_j} + \overrightarrow{m_N} \tag{7.50}$$

In reality, these combinations may or may not occur, depending on many factors. In quantum mechanics, it has been demonstrated that not only are the kinetic moments quantified, but also their projections in a direction in space. If we choose this direction to be the OZ axis of the coordinate system, the relationships quantifying the projections of the kinetic moments can be written as:

$$\begin{cases} L_z = m_l\hbar,\ m_l = 0,\ \pm1, \pm2, \pm3,\ \ldots\ , \pm l \\ S_z = m_s\hbar,\ m_s = \pm\dfrac{1}{2} \\ J_z = m_j\hbar,\ m_j = 0, \pm1, \pm2, \pm3,\ \ldots\ , \pm j \\ I_z = m_I\hbar,\ m_I = 0, \pm1, \pm2, \pm3,\ \ldots\ , \pm I \end{cases} \tag{7.51}$$

where m_l, m_s, m_j, and m_I are the magnetic quantum numbers, i.e. the orbital, spin, total, and nuclear numbers, respectively.

As will be discussed later, if the atoms are placed in a magnetic field oriented in the OZ direction, then the projections of the magnetic moments in the direction of the field are precisely quantified by these numbers, which is why they are called magnetic quantum numbers.

7.3.1.5 Action of the magnetic field on magnetic moments

(A) If, in a homogeneous magnetic field of induction $\vec{B}$, a magnetic moment $\vec{m}$ is inserted that makes an angle of θ with the field, a torque of forces is exerted on it that tends to orient it parallel to the magnetic field. The torque moment is given by the vector product of the magnetic moment and the magnetic field induction:

$$\overrightarrow{M_c} = \vec{m} \times \vec{B} \tag{7.52}$$

This torque tends to orient the magnetic moment parallel to the magnetic field (in the stable equilibrium position, where $\theta = 0$ and $\overrightarrow{M_c} = 0$). However, if the magnetic field acts on a magnetic moment with a torque of $\overrightarrow{M_c}$, it follows that the magnetic moment in the magnetic field is characterized by a potential energy. Consider the variation in the potential energy of a magnetic moment $\vec{m}$ between the states in the field characterized by the angles θ_1 and θ_2 between the magnetic moment and the magnetic field of induction $\vec{B}$. This variation is equal to the mechanical work performed to rotate the moment $\vec{m}$ between them:

$$\Delta U = L = \int_{\theta_1}^{\theta_2} M_c(\theta)d\theta = \int_{\theta_1}^{\theta_2} mB \sin\theta d\theta$$

$$= -mB\cos\theta_2 - (-mB\cos\theta_1) = U_2 - U_1$$

Thus, the potential energy of a magnetic moment $\vec{m}$, located in a magnetic field of induction $\vec{B}$, is given by:

$$U = -\vec{m}\cdot\vec{B} = -mB\cos\theta = -m_zB \tag{7.53}$$

where m_z is the projection of the magnetic moment on the magnetic induction $\vec{B}$ considered to lie on the OZ axis of a Cartesian coordinate system. One can observe that this is in perfect analogy with the potential energy of an electric dipole that has a moment of $\vec{p}$, placed in an electric field $\vec{E}$. In the expression of potential energy, $\vec{m}$ can be: $\overrightarrow{m_L}$, $\overrightarrow{m_S}$, $\overrightarrow{m_j}$, or $\overrightarrow{m_N}$. Using the most general expression for the total magnetic moment of the electron in the atom, i.e. $\overrightarrow{m_j} = -\gamma g_j\vec{J}$, we obtain the following for the potential energy of interaction between a magnetic moment and a magnetic field:

$$U = -\overrightarrow{m_j}\cdot\vec{B} = -\gamma g_j\vec{J}\vec{B} = -\gamma g_j JB\cos\theta = -\gamma g_j J_z B \tag{7.54}$$

If it is taken into account that J_z is quantized by $J_z = \hbar m_j$, it follows that the potential energy of the magnetic moment in a magnetic field is also quantized, being given by the relation:

$$U = -g_j\gamma\hbar Bm_j = -g_j\beta_B Bm_j \tag{7.55}$$

where $\beta_B = \gamma\hbar$, and $m_j = 0, \pm1, \pm2, \pm3, \ldots, \pm j$ is the total quantum magnetic number.

(B) If the magnetic moment is placed in a homogeneous magnetic field, it only experiences a torque, which tends to rotate it parallel to the magnetic field. However, if the magnetic moment is introduced into an inhomogeneous magnetic field, then other forces can be exerted on it in addition to couples. It is known that in conservative fields, one can find the force exerted on a body if one knows the potential energy of the body in that field, using:

$$\vec{F} = -\nabla U \tag{7.56}$$

Although the magnetic field is not a conservative field, the relation (7.56) can still be applied; in this case, if we limit ourselves to a small region of the field where the field lines seem to be open, such as that between the poles of an electromagnet, then according to (7.56) and (7.53), we obtain:

$$\vec{F} = -\nabla U = -\nabla(-\vec{m} \cdot \vec{B}) = \nabla(m_x B_x + m_y B_y + m_z B_z) \tag{7.57}$$

Thus, if the components of the magnetic moment are independent of x, y, and z, then according to relation (7.57), the force exerted on the magnetic moment has the following components:

$$\begin{aligned} \vec{F} &= \left(m_x \frac{\partial B_x}{\partial x} + m_y \frac{\partial B_y}{\partial x} + m_z \frac{\partial B_z}{\partial x}\right)\vec{i} \\ &+ \left(m_x \frac{\partial B_x}{\partial y} + m_y \frac{\partial B_y}{\partial y} + m_z \frac{\partial B_z}{\partial y}\right)\vec{j} \\ &+ \left(m_x \frac{\partial B_x}{\partial z} + m_y \frac{\partial B_y}{\partial z} + m_z \frac{\partial B_z}{\partial z}\right)\vec{k} \\ &= F_x\vec{i} + F_y\vec{j} + F_z\vec{k} \end{aligned} \tag{7.58}$$

If we consider a magnetic field that has only the nonzero component of the gradient of B_z, along the OZ axis, then the force exerted on the magnetic moment is:

$$\vec{F_z} = m_z \frac{\partial B_z}{\partial z}\vec{k} \tag{7.59}$$

In other words, the magnetic moment is subjected to a force equal to the product of the magnetic moment and the gradient of the magnetic field and is oriented in the direction of the field gradient. If $m_z > 0$, the force is oriented above the magnetic field, and if $m_z < 0$, the force is oriented below the field. So, an inhomogeneous magnetic field applied to a magnetic moment exerts both a torque of forces that tend to rotate it to become parallel to the field and a force that gives rise to a translational movement of the moment in the direction of the field gradient. If the interaction energy is expressed in terms of the magnetic flux Φ established by a circuit, then expressions (7.53) and (7.57) take another form:

$$U = -\vec{m}\vec{B} = -I\vec{S}\vec{B} = -I\Phi \tag{7.60}$$

and respectively:

$$\vec{F} = -\nabla U = \nabla (I\Phi) = I\left(\frac{\partial \Phi}{\partial x}\vec{i} + \frac{\partial \Phi}{\partial y}\vec{j} + \frac{\partial \Phi}{\partial z}\vec{k}\right) \tag{7.61}$$

7.3.1.6 Potential energy of interaction between two magnetic moments

In the case of the electrostatic field, we consider the energy of interaction between two electric dipoles with moments denoted by $\vec{p_1}$ and $\vec{p_2}$. We start from the expressions for the potential energy and the electric field of an electric dipole that has a dipole moment of $\vec{p}$, given by: $U_p = -\vec{p}\,\vec{E}$ and $\vec{E} = \frac{1}{4\pi\varepsilon_0 r^3}\left[\frac{3(\vec{p}\cdot\vec{r})\vec{r}}{r^2} - \vec{p}\right]$. By analogy, in the case of two magnetic moments $\vec{m_1}$ and $\vec{m_2}$, one can write the potential energy of interaction between them. If a magnetic moment $\vec{m_1}$ is present at a point O in space (see figure 7.8), then at a point P_1 sufficiently far from it, the magnetic moment creates a field given by:

$$\vec{B_1} = \frac{\mu_0}{4\pi r_{12}^3}\left[\frac{3(\vec{m_1}\cdot\vec{r_{12}})\vec{r_{12}}}{r_{12}^2} - \vec{m_1}\right] \tag{7.62}$$

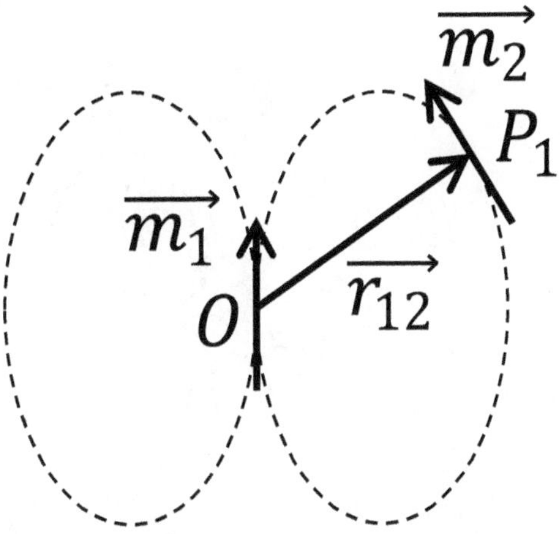

Figure 7.8. Magnetic interaction between two magnetic moments.

If the magnetic moment $\vec{m_2}$ is placed in this field, it follows that the potential energy of interaction between them is:

$$U = -\vec{m_2}\cdot\vec{B_1} = -\frac{\mu_0}{4\pi r_{12}^3}\left[\frac{3(\vec{m_1}\cdot\vec{r_{12}})(\vec{m_2}\cdot\vec{r_{12}})}{r_{12}^2} - (\vec{m_1}\cdot\vec{m_2})\right] \tag{7.63}$$

that is, the potential energy of the magnetic interaction is:

$$U = \frac{\mu_0}{4\pi r_{12}^3}\left[\overrightarrow{m_1} \cdot \overrightarrow{m_2} - \frac{3(\overrightarrow{m_1} \cdot \overrightarrow{r_{12}})(\overrightarrow{m_2} \cdot \overrightarrow{r_{12}})}{r_{12}^2}\right] \quad (7.64)$$

An example of interaction between magnetic moments is the hyperfine interaction, which is a magnetic-type interaction between an electron belonging to an atom and the atom's own nucleus. This interaction comprises several terms; however, the term corresponding to the interaction between the spin magnetic moment of the electron and the nuclear magnetic moment is of a dipolar nature. Indeed, the distance between the electron and the nucleus is large enough (an order of magnitude of $\sim 10^{-10}$ m; see figure 7.9) that we can express the magnetic field created by the spin magnetic moment of the electron at the point where the nucleus is located as the magnetic field of a magnetic moment at a great distance from the electron: $\overrightarrow{B_S} = \frac{\mu_0}{4\pi r^3}\left[\frac{3(\overrightarrow{m_S} \cdot \vec{r})\vec{r}}{r^2} - \overrightarrow{m_S}\right]$. Thus, the potential energy of the nuclear magnetic moment $\overrightarrow{m_N}$ in this field is:

$$U_{\text{HF}} = \frac{\mu_0}{4\pi r^3}\left[\overrightarrow{m_S} \cdot \overrightarrow{m_N} - \frac{3(\overrightarrow{m_S} \cdot \vec{r})(\overrightarrow{m_N} \cdot \vec{r})}{r^2}\right] \quad (7.65)$$

This term of the interaction energy between the atomic electron and the nucleus is called *the energy of the hyperfine interaction.*

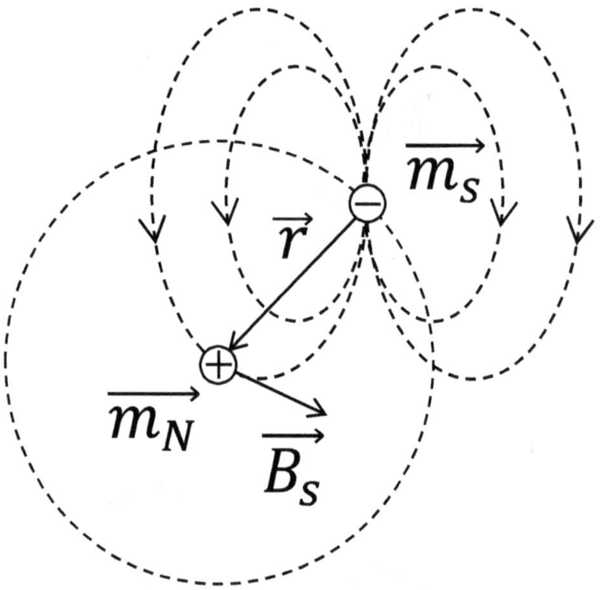

Figure 7.9. Dipolar interaction between the spin magnetic moment of the electron and the nuclear magnetic moment of the atom.

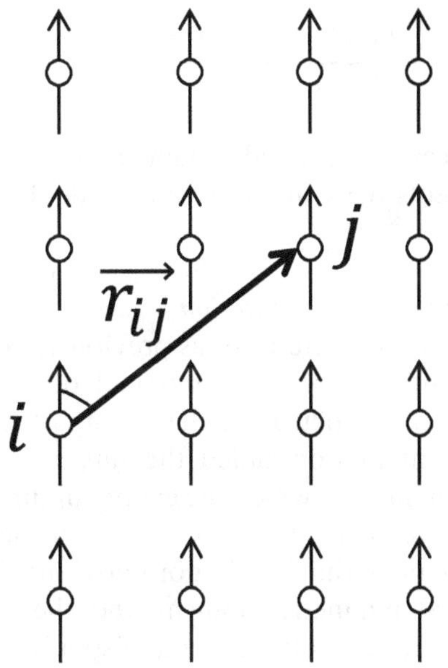

Figure 7.10. Crystalline network whose nodes have nonzero atomic magnetic moments.

Let us now consider a crystalline substance whose atoms have nonzero magnetic moments (see figure 7.10). If there were a network vacancy at node j, the magnetic field created at this node j by a magnetic moment $\overrightarrow{m_i}$ located at node i of the lattice would be:

$$\overrightarrow{B_i} = \frac{\mu_0}{4\pi r_{ij}^3}\left[\frac{3(\overrightarrow{m_i}\cdot\overrightarrow{r_{ij}})(\overrightarrow{r_{ij}})}{r_{ij}^2} - \overrightarrow{m_i}\right] \tag{7.66}$$

If the network contained N atoms, then the field created by all these atoms at node j would be:

$$\overrightarrow{B} = \sum_{i=1}^{N-1}\overrightarrow{B_i} = \frac{\mu_0}{4\pi}\sum_{i=1}^{N-1}\left[\frac{3(\overrightarrow{m_i}\cdot\overrightarrow{r_{ij}})(\overrightarrow{r_{ij}})}{r_{ij}^5} - \frac{\overrightarrow{m_i}}{r_{ij}^3}\right] \tag{7.67}$$

where $\overrightarrow{r_{ij}}$ is the relative position vector between nodes i and j. If, in this field, there is an atom j with a magnetic moment $\overrightarrow{m_j}$, the potential energy of interaction between them is:

$$U_j = -\overrightarrow{m_j}\cdot\overrightarrow{B_i} = \frac{\mu_0}{4\pi}\sum_{i=1}^{N-1}\left[\frac{\overrightarrow{m_i}\cdot\overrightarrow{m_j}}{r_{ij}^3} - \frac{3(\overrightarrow{m_i}\cdot\overrightarrow{r_{ij}})(\overrightarrow{m_j}\cdot\overrightarrow{r_{ij}})}{r_{ij}^5}\right] \tag{7.68}$$

and the potential energy of the magnetic interaction of the entire crystal lattice can be obtained by summing this relationship according to i and j, so that:

$$U = \frac{\mu_0}{4\pi}\sum_{i<j=1}^{N-1}\left[\frac{\vec{m_i}\cdot\vec{m_j}}{r_{ij}^3} - \frac{3(\vec{m_i}\cdot\vec{r_{ij}})(\vec{m_j}\cdot\vec{r_{ij}})}{r_{ij}^5}\right] \tag{7.69}$$

The condition $i < j$ is necessary to avoid including the energy of interaction between a pair of magnetic moments twice and to also exclude the interaction of a moment with itself.

7.3.1.7 Energy of the spin–orbit interaction

The electron in the atom is characterized, as previously shown, by both an orbital magnetic moment $\vec{m_L}$ and the magnetic moment of spin $\vec{m_S}$. As a result, it is expected that the question of calculating the energy of interaction between these moments will arise, an interaction called the *spin–orbit interaction.* Of course, taking into account the manner in which the energy of dipolar interaction has been written so far, there is a tendency to proceed here in the same way. However, a major difficulty arises, namely that we do not know the distance between the two moments (i.e. the orbital magnetic moment and the spin magnetic moment). Indeed, if the location of the magnetic moment of spin is certain (i.e. the location of the electron), then the location of the orbital magnetic moment is uncertain, because we only obtain magnitude and orientation from the definition of the magnetic moment of a current loop but not its application point (as is known from vector algebra, the application point of the vector product of two vectors is undetermined). It is customary for the application point to be graphically represented in the center of the circular orbit. It can be there, but it can also be anywhere within the limits of the electric current distribution that creates it. For this reason, in order to find the expression for the energy of the spin–orbit interaction, another approach is used, as described below. Previously, it was noted that the interaction between the orbital and spin motions of the atomic electron takes place within the electric field of the nucleus. Therefore, we can attempt to express the energy of the spin–orbit interaction using the electric field of the nucleus. As previously demonstrated, a particle charged with a charge q, moving at a low speed $\vec{v}$, creates, at a distance $\vec{r}$ with respect to the charge, a magnetic field whose induction is:

$$\vec{B} = \frac{\mu_0}{4\pi}\frac{q(\vec{v}\times\vec{r})}{r^3} \tag{7.70}$$

and the electric field created by this charge at a distance $\vec{r}$ is:

$$\vec{E} = \frac{q\vec{r}}{4\pi\varepsilon_0 r^3} \tag{7.71}$$

Thus, the magnetic field created at point P can be expressed in terms of the electric field, as follows:

$$\vec{B} = \mu_0\varepsilon_0(\vec{v}\times\vec{E}) = \frac{(\vec{v}\times\vec{E})}{c^2} \tag{7.72}$$

Building on these observations, let us analyze the magnetic field created by the atomic nucleus in the electron reference system. In the reference frame *RS* bound to the nucleus (see figure 7.11), the electron is in rotational motion around the nucleus at a speed $\vec{v}$. However, in the reference frame *R′S′* bound to the electron, the nucleus is seen to move in the opposite direction with a velocity $\vec{v'} = -\vec{v}$, also on a circular trajectory of radius $\vec{r}$. As a result, a nucleus with a charge of *Ze* moving at low speed creates a magnetic field at a distance $\vec{r}$ from it. The induction of the field is given by:

$$\vec{B'} = \frac{\mu_0}{4\pi} Ze \frac{\vec{v'} \times \vec{r}}{r^3} \tag{7.73}$$

and the electric field created by the nucleus at the distance $\vec{r}$, where the electron is located, is:

$$\vec{E'} = \frac{Ze\vec{r}}{4\pi\varepsilon_0 r^3} \tag{7.74}$$

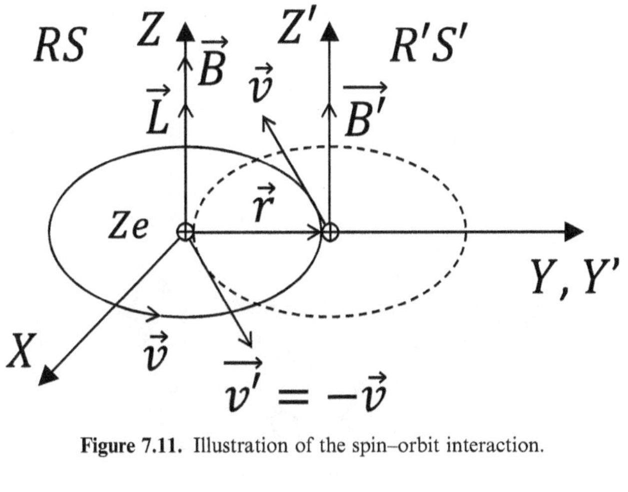

Figure 7.11. Illustration of the spin–orbit interaction.

Therefore:

$$\vec{B'} = -\frac{\mu_0}{4\pi} Ze \frac{\vec{v} \times \vec{r}}{r^3} = -\frac{\mu_0}{4\pi} \frac{Ze}{m} \frac{m\vec{v} \times \vec{r}}{r^3} = +\frac{\mu_0 Ze}{4\pi m} \frac{\vec{L}}{r^3} \tag{7.75}$$

The electron itself, with a magnetic moment of spin $\vec{m_S} = -2\gamma\vec{S}$, is located in the magnetic field associated with the orbital movement of the electron, so that the spin–orbit interaction energy is:

$$U = -\vec{m_S} \cdot \vec{B'} = \frac{2\mu_0 Ze\gamma}{4\pi m} (\vec{L} \cdot \vec{S}) = \lambda(\vec{L} \cdot \vec{S}) \tag{7.76}$$

So, the spin–orbit interaction energy depends on the relative orientation of the vectors $\vec{L}$ and $\vec{S}$. The proportionality constant $\lambda = \frac{2\mu_0 Ze\gamma}{4\pi m} = \frac{2Ze\gamma}{4\pi\varepsilon_0 c^2 m}$ is called *the spin–orbit interaction constant*. In an external magnetic field, a system such as a crystal formed from atoms or ions possessing orbital magnetic moments, spin magnetic moments, and nuclear moments will possess all the interaction energy types listed above. The sum of the energies of these interactions gives the expression for the classical Hamiltonian of the system, from which we must start in order to write the expression for the Hamiltonian operator frequently used in quantum mechanics.

7.4 The precession of a magnetic moment in magnetic field

Suppose that particles possessing a magnetic moment do not have links that prevent their movement around their axis of rotation. Assuming that the magnetic orbital moment of the electron makes an angle α with the induction of a magnetic field $\vec{B}$ (see figure 7.12), this will act on the magnetic moment $\overrightarrow{m_L}$ with the torque $M_C = m_L B \sin\alpha$, which tends to rotate it along $\vec{B}$. The torque moment is

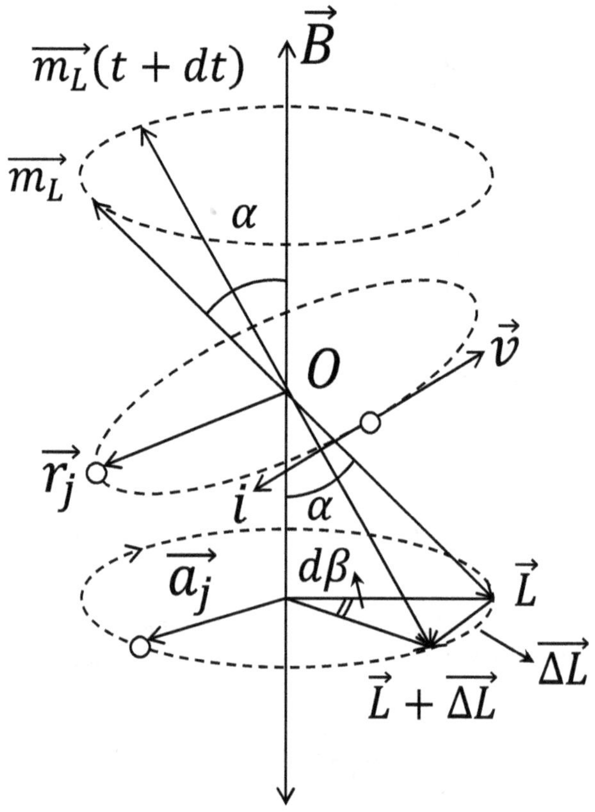

Figure 7.12. Precessional movement of an electronic orbit in a magnetic field.

perpendicular to the plane formed by $\vec{B}$ and $\overrightarrow{m_L}$ ($\overrightarrow{M_C} = \overrightarrow{m_L} \times \vec{B}$). If this moment acts for a period of time dt, then the following variation of angular momentum occurs:

$$\overrightarrow{\Delta L} = \overrightarrow{M_C} dt \tag{7.77}$$

However, $\Delta L = L \sin \alpha \; d\beta$ (see figure 7.12), because its projection $\vec{L}$ on the normal plane to $\vec{B}$ is rotated at the same time as $d\beta$, $L \sin \alpha \; d\beta = m_L B \sin \alpha dt$, resulting in:

$$\frac{d\beta}{dt} = \frac{m_L B}{L} = \frac{eB}{2m} = \Delta\omega_L \tag{7.78}$$

where $\overrightarrow{\Delta\omega_L}$ is the variation in the angular velocity of the electron in orbit, when the magnetic field is applied. Thus, in the presence of the magnetic field, the angular momentum $\vec{L}$ starts a precessional movement around the magnetic field $\vec{B}$ with an angular speed of precession of $\overrightarrow{\Delta\omega_L} = \frac{e\vec{B}}{2m}$. The magnitude of the angular momentum does not change, but it rotates around the axis of the field with $\overrightarrow{\Delta\omega_L}$. In a weak magnetic field, the electron describes the same orbit as in the absence of the field, but this orbit is related to a system of axes that performs a precessional movement relative to the magnetic field. Therefore, the action of the magnetic field is reduced to a precessional movement of the electron's orbit with an angular velocity of $\overrightarrow{\Delta\omega_L}$. The precessional motion of the electron's orbit with the angular velocity $\overrightarrow{\Delta\omega_L}$ can be associated with an orbital magnetic moment $\overrightarrow{\Delta m_L}$, since the precessional movement of the electron's orbit is equivalent to a current of intensity $\Delta I = -\frac{e}{T} = -\frac{e\Delta\omega_L}{2\pi}$. The magnetic moment associated with the precessional movement of an electron j is:

$$\overrightarrow{\Delta m_L} = -\frac{e\Delta\omega_L}{2\pi}\pi \overline{r}_j^2 = -\frac{e^2}{4m}\overline{r}_j^2 \vec{B} \tag{7.79}$$

Thus, when the magnetic field $\vec{B}$ is applied, each electron orbit acquires an additional orbital magnetic moment oriented in the opposite direction to that of the applied magnetic field, and this induced magnetic moment is nonzero only as long as $\vec{B}$ is nonzero. The radii of the orbits of different electrons in an atom may differ, and the planes of the orbits may not always be normal to $\vec{B}$. If we orient $\vec{B}$ along the OZ axis and the electron's orbit is not in the XOY plane, then simultaneously with the movement of the electron in its orbit, the orbit itself performs a precessional movement around the direction of the magnetic field. The trajectory of the compound motion of the electron is a circle of radius a_j in a plane perpendicular to $\vec{B}$. For this reason, for electron j we must replace $\overline{r}_j^2$ in (7.79) with $a_j^2 = x_j^2 + y_j^2$. Also, in order to obtain the induced magnetic moment of an atom, we have to take the sum of all the induced magnetic moments of all the electrons. The average magnetic moment induced by the precessional motion of all the orbits of an atom is therefore:

$$\langle \overrightarrow{\Delta m_L} \rangle = -\frac{e^2\overrightarrow{B}}{4m}\left\langle \sum_{j=1}^{z}\left(x_j^2 + y_j^2\right)\right\rangle = -\frac{e^2\overrightarrow{B}}{4m}\left[\sum_{j=1}^{z}\left(\left\langle x_j^2 + y_j^2\right\rangle\right)\right] \tag{7.80}$$

However, for an isotropic distribution of electric charges in an atom, $\langle x_j^2\rangle + \langle y_j^2\rangle = \frac{2}{3}\bar{r}_j^2$, due to the fact that the three spatial coordinates are equivalent, i.e. $\langle x_j^2\rangle = \langle y_j^2\rangle = \langle z_j^2\rangle = \frac{1}{3}\bar{r}_j^2$. Defining the double average of the squares of the radii of electronic orbits in an atom as $\langle \bar{r}_j^2\rangle = \frac{\sum_{j=1}^{z} r_j^2}{Z}$ leads to:

$$\langle \overrightarrow{\Delta m_L} \rangle = -\frac{e^2 Z}{6m}\overrightarrow{B}\,\langle \bar{r}^2\rangle \tag{7.81}$$

which expresses the average magnetic moment induced by the precessional motion of all the electronic orbits in an atom.

The magnetic moment of an atom is given by the vector sum of the magnetic moments of its electrons. The magnetic moment of the electron has three components: (i) its own magnetic moment of spin; (ii) its orbital magnetic moment; and, in the presence of an external magnetic field, (iii) its variation of the orbital magnetic moment due to orbital precession around the magnetic field. The magnetization vector $\overrightarrow{M}$ of a substance is equal to the vector sum of the magnetic moments of the atoms contained in the unit volume:

$$\overrightarrow{M} = \sum_{i=1}^{n}\overrightarrow{m_i} \tag{7.82}$$

where n is the number of atoms in the unit volume and $\overrightarrow{m_i}$ the magnetic moment of atom i.

7.5 Diamagnetism

Diamagnetic bodies are characterized by very low and negative magnetic susceptibility. When they are introduced into an external magnetic field, they generate their own magnetic field, which is oriented in the opposite direction to that of the external magnetic field; hence, this is similar to an induction process. When these bodies are introduced into an inhomogeneous magnetic field, they are repelled toward regions where the magnetic field is minimal. Diamagnetism is due both to atomic electrons that perform orbital movements, also called ***bound electrons***, characterized by Heitler–London functions, as well as to ***free electrons***, characterized by Bloch functions.

7.5.1 Diamagnetism of bound electrons

If one takes into account the orbital motion of electrons, as previously found, the orbits of electrons in an external magnetic field undergo precessional movement, and this generates an induced magnetic moment, the expression of which was previously established to be:

$$\langle \overrightarrow{\Delta m_L} \rangle = -\frac{e^2 Z}{6m} \vec{B} \langle \bar{r}^2 \rangle \tag{7.83}$$

The magnetization $\vec{M}$ of the substance is then:

$$\vec{M} = -\frac{ne^2 Z}{6m} \langle \bar{r}^2 \rangle \vec{B} \tag{7.84}$$

where n is the number of atoms per unit volume. The magnetic susceptibility is obtained using relation (7.84) for magnetization:

$$\chi_m = -\frac{\mu_0 M}{B} = -\frac{\mu_0 n e^2 Z}{6m} \langle \bar{r}^2 \rangle \tag{7.85}$$

i.e. the Langevin relationship or equation, which gives a negative value for χ_m.

The phenomenon of ***Larmor precession*** affects all electronic orbits, regardless of whether or not their magnetic moments are compensated in the atom; it causes all bodies to exhibit atomic diamagnetism. In the case of paramagnetic bodies, the diamagnetism is masked by the permanent magnetic moments that orient themselves in an applied magnetic field. Electronic orbits do not depend on temperature, while the magnetic susceptibility resulting from the precessional motion is also independent of temperature. The diamagnetism due to Larmor precession, and therefore the magnetic susceptibility resulting from this movement, depend on Z. The determination of its value allows interesting conclusions to be drawn regarding the study of the atom, as well as various chemical compounds, especially organic ones.

In the case of diamagnetic metals such as Ag, Au, Cu, Zn, Cd, Hg, Bi, etc it has been found that the total magnetic susceptibility corresponding to the metallic state is higher than that which would result from the Larmor precession movement. This is the case because metal crystals contain both bound electrons in atoms (causing atomic diamagnetism) and quasi-free electrons. The latter are responsible for this increase in susceptibility.

7.5.2 Diamagnetism of conduction electrons

If free electrons in a crystal move in a uniform magnetic field, the overall movement, as previously derived, follows a helicoidal trajectory resulting from translational movement along the magnetic field and periodic circular motion in a plane perpendicular to the magnetic field. The disordered, unquantized motion of a free electron in a crystal due to thermal agitation is quantized in the presence of an external magnetic field. The quantization energy has the form:

$$W = \mu_B B(2n + 1) \tag{7.86}$$

where μ_B is the Bohr magneton. Taking into account the expression for the energy of a magnetic moment placed in an external magnetic field of induction $\vec{B}$, ($W = -\vec{m}\vec{B}$), it follows that the presence of the magnetic field induces a magnetic moment. If the concentration of free electrons is n, the susceptibility is:

$$\chi_d = -\frac{n\mu_B}{3kT} \tag{7.87}$$

Since this is negative, the susceptibility is of the diamagnetic type but dependent on temperature. The temperature dependence of χ_d is due to the temperature dependence of the concentration of free electrons in a non-degenerate semiconductor (subject to Boltzmann statistics) appearing in relation (7.87). However, in metal crystals in which the electron concentration (on the order of $n = 10^{28}$–10^{29} m^{-3}) is independent of temperature, Fermi–Dirac statistics are used, which give the law of distribution of electrons over the energy levels of the allowed band. For common temperatures, i.e. $T < T_F$, the electrons in the immediate vicinity of the Fermi level pass from the occupied levels into the lowest free ones. The number of electrons participating in this process is only a fraction of the total number of free electrons, a fraction proportional to the T/T_F ratio. In this situation, i.e under the action of the magnetic field, out of the total number of free electrons, only the electrons grouped in the immediate vicinity of the Fermi level change their energy. In other words, only nT/T_F of the total number of valence electrons change their energy, and in this case, the diamagnetic susceptibility is:

$$\chi_d = -\frac{n\mu_B}{3kT_F} \tag{7.88}$$

which is independent of temperature.

7.5.3 Diamagnetism in superconductors

As discovered by Kamerlingh Onnes in mercury (Hg), the electrical resistivity of Hg becomes practically zero at a temperature of around 4 K, which is known as the superconducting state. However, superconductivity is not specific to Hg but has been observed in other materials (i.e. Al, Sn, Zn, Cd) at certain very low temperatures, specific to each one, called critical temperatures (T_C). This state, according to the quantum theory of superconductivity, is caused by the association of the quasi-free conduction electrons in pairs, in which the electrons have antiparallel spins. The existence of this state of superconductivity indicates the possibility of obtaining magnetic fields of very high intensity in solenoids using coils formed from such superconducting wires carrying very intense electric currents. From the magnetic point of view, superconductors are ideal diamagnetic materials, presenting high diamagnetic susceptibility. However, due to this diamagnetism, which can occur only under the action of an external magnetic field, the state of superconductivity can be destroyed. Indeed, if one considers a solenoid consisting of a superconducting material carrying a high-intensity current, it generates a very intense magnetic field that crosses its superconducting wires. The superconducting wires, being diamagnetic, give rise to a magnetic field in the opposite direction to that of the magnetizing field, and at a certain critical value (H_C), they cancel it, reducing the magnetic flux in the solenoid to zero. Due to these diamagnetic properties of superconductors, which can cancel the state of superconductivity for certain field values, the intensity of the magnetic fields obtained using solenoids with superconducting circular wires cannot increase infinitely.

7.6 Paramagnetism

Paramagnetic substances have positive magnetic susceptibility and orient their permanent magnetic moments in the direction of a magnetizing field, generating their own magnetic field in the same direction as the external field. Paramagnetic bodies are drawn into a nonuniform magnetic field, i.e. toward regions of maximum intensity. Paramagnetism is due to both bound and conduction electrons. In the case of paramagnetic materials, susceptibility is not a constant quantity but depends on the temperature according to the law $\chi = C/T$, established by Pierre Curie, where the constant C depends on the nature of the paramagnetic substance.

7.6.1 Paramagnetism of bound electrons

Taking into account both orbital motion and spin motion, all the constituent electrons of an atom present both an orbital magnetic moment and a magnetic moment of spin. In an atom that constitutes a well-defined ensemble, each of these two groups of moments creates a nonzero magnetic moment, although these can cancel out overall due to the spin–orbit interaction. As a result, the atom as a whole may or may not have an overall magnetic moment. Atoms with fully occupied internal electronic shells and an even number of electrons have no overall magnetic moment. This is because their orbital and spin magnetic moments completely cancel each other out. As a result, only the following types of atoms exhibit magnetic properties:

- Atoms and molecules possessing an odd number of electrons, so that the total spin of the system cannot be zero (Na atoms, organic free radicals, etc);
- Atoms and free ions with an incomplete inner electronic layer, such as the transition elements (Cr, Mn, Fe, Co, Ni, etc.) which have an incomplete 3*d* layer, rare earths, actinides, etc;
- A small number of compounds with even numbers of electrons comprising molecular oxygen and double organic radicals.

The susceptibility of paramagnetic substances is determined using the Langevin relationship, while the variation of magnetization intensity versus the intensity of the magnetic field is given by Curie's law. To establish these laws, we consider a paramagnetic substance containing n atoms per unit of volume; each atom of the substance has a magnetic moment $\vec{m}$. The problem will be approached in two ways, i.e. a classical approach and a simplified quantum approach.

In the classical treatment, magnetization occurs in the presence of an external magnetic field as a result of the orientation of magnetic moments in the direction of the field. However, the orientation process is opposed by the movement of thermal agitation, which tends to disorient the magnetic moments. At a given temperature, a stationary state is created in which a nonzero overall magnetic moment is present per unit of volume, oriented in the direction of the field, so that the substance exhibits magnetization. The interaction energy between the magnetic field and the magnetic moment $\vec{m}$ is:

$$W = -\vec{m}\vec{B} = -mB\cos\theta \tag{7.89}$$

In the classical treatment, the magnetic moment of the atom can make any angle with the direction of the magnetic field. As a result, by calculating the average value of the projections of the magnetic moments in the direction of the magnetic field, we can obtain the following expression for the intensity of magnetization using the Boltzmann distribution function:

$$M = n\langle m\cos\theta\rangle = \frac{nm\int_{\pi}^{0} e^{\frac{mB\cos\theta}{kT}}\cos\theta\sin\theta d\theta}{\int_{\pi}^{0} e^{\frac{mB\cos\theta}{kT}}\sin\theta d\theta} \tag{7.90}$$

Using the new variables $x = \frac{mB}{kT}$ and $=\cos\theta$, this takes the form of:

$$M = \begin{cases} nm\dfrac{\int_{-1}^{1} ue^{xu}du}{\int_{-1}^{1} e^{xu}du} = nm\dfrac{d}{dx}\left(\ln\int_{-1}^{1} e^{xu}du\right) = \\ nm\dfrac{d}{dx}\left(\ln\dfrac{e^{x}-e^{-x}}{x}\right) = nm\left(\coth x - \dfrac{1}{x}\right) = nm\mathcal{L}(x) \end{cases} \tag{7.91}$$

where $\mathcal{L}(x) = \coth x - \frac{1}{x}$ is known as the ***Langevin function***. A similar situation was already seen in the case of the microscopic approach to the orientational polarization of dielectrics. In the case where $x \ll 1$ (weak fields and high temperatures), $\mathcal{L}(x) \cong \frac{x}{3}$ and therefore:

$$M = \frac{nm^2B}{3kT} = \frac{C}{T}B \tag{7.92}$$

The constant $C = \frac{nm^2}{3k}$ is called the ***Curie constant***. By replacing $\vec{B} = \mu_0\vec{H}$ in (7.92), we obtain:

$$\vec{M} = \frac{\mu_0 nm^2}{3kT}\vec{H} = \chi_{\text{param}}\vec{H} \tag{7.93}$$

Thus, for χ_{param}, a positive and temperature-dependent value is obtained:

$$\chi_m = \frac{\mu_0 nm^2}{3kT} \tag{7.94}$$

Figure 7.13 shows the dependence of the magnetization of the substance (expressed as the mean value of the projection of the magnetic moments of the G_d^{3+} ions in the direction of the magnetic field, in Bohr magnetons) as a function of B/T, for gadolinium octahydrate sulfate, at $T = 1.3$ K. It can be seen that the magnetization

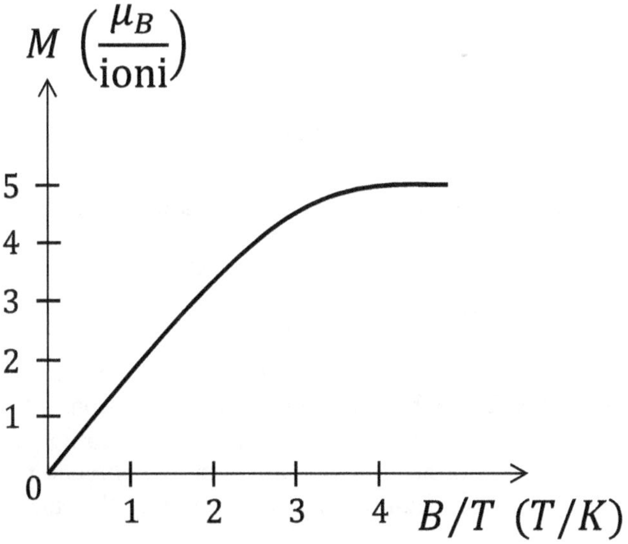

Figure 7.13. Dependence of the magnetization of gadolinium octahydrate sulfate, at $T = 1.3$ K, as a function of B/T.

shows a linear dependence on B in the region starting from the origin. In this part of the curve, the condition $x \ll 1$ (i.e. weak fields and high temperatures) is fulfilled. When the magnetic induction B reaches a certain value, saturation of the magnetization occurs. In this part of the magnetization curve, the Langevin function $\mathcal{L}(x)$ accurately represents the relationship between the magnetic field and the magnetization M. The magnetization reaches saturation at $B = 5$ T. This saturation is explained by the alignment of all the magnetic moments of the gadolinium ions in the direction of the magnetic field.

From the point of view of quantum mechanics, we consider a paramagnetic body containing n atoms per unit volume, each with an elementary magnetic moment $\overrightarrow{m_i}$ produced by the spin–orbit interaction. The total magnetic moment of the body is $M = n\langle m \rangle$, while the mean value $\langle m \rangle$ is obtained from the relation:

$$\langle m \rangle = \sum_{i=1}^{n} m_i P_i \tag{7.95}$$

where m_i is the magnetic moment corresponding to a given state and P_i is the probability of that state. If w_i is the energy of the atom corresponding to the state in which the magnetic moment is $\overrightarrow{m_i}$, the probability of being in this state is given by:

$$P_i = Ae^{-\frac{w_i}{kT}} \tag{7.96}$$

The constant A is found from the normalization condition $\sum_{i=1}^{n} P_i = 1$, which implies there is certainty that at least one of the states is occupied:

$$\sum_{i=1}^{n} P_i = 1 \quad \Rightarrow \quad A\sum_{i=1}^{n} e^{-\frac{w_i}{kT}} = 1 \quad \Rightarrow \quad A = \frac{1}{\sum_{i=1}^{n} e^{-\frac{w_i}{kT}}}$$

Therefore,

$$P_i = \frac{e^{-\frac{w_i}{kT}}}{\sum_{i=1}^{n} e^{-\frac{w_i}{kT}}} \tag{7.97}$$

The total magnetic moment of the atom is quantized and is given by the relationship:

$$m_i = \hbar\gamma g_j\sqrt{j(j+1)} = g_j\mu_B\sqrt{j(j+1)} \tag{7.98}$$

where $g_j = 1 + \frac{j(j+1)+s(s+1)-l(l+1)}{2j(j+1)}$ is the gyromagnetic factor. Under the influence of a magnetic field $\vec{B}$, the projection of the total magnetic moment $\vec{M}$ in the direction of the field is given by the sum of the projections of the elementary magnetic moments $\vec{m_i}$. However, each elementary magnetic moment $\vec{m_i}$ can orient itself in $2j+1$ ways, the projection of which in the direction of the field is: $m_{iz} = g_j m_j \mu_B$, where m_j is the quantum magnetic number that takes the values:

$$m_j = 0, \pm 1, \pm 2 \pm \ldots \pm j \quad (2j+1) \text{ values} \tag{7.99}$$

In this situation, the corresponding magnetic energy is:

$$w_i = -\vec{m_i}\vec{B} \tag{7.100}$$

which, if $\vec{B}$ is taken along the OZ axis, reduces to:

$$w_i = -m_{iz}B = -g_j\mu_B B m_j \tag{7.101}$$

Once we have an expression for energy and we know that the average of a quantity is calculated using the relation $\langle\alpha\rangle = \sum \alpha P_i(\alpha)$, we can obtain the following relation for the value of the total magnetic moment projected in the direction of the magnetic field:

$$M = n\langle m\rangle = n\frac{\sum_{m_j=-j}^{j} g_j m_j \mu_B e^{\frac{g_j m_j \mu_B B}{kT}}}{\sum_{m_j=-j}^{j} e^{\frac{g_j m_j \mu_B B}{kT}}} \tag{7.102}$$

or:

$$M = n g_j \mu_B \frac{\sum_{m_j=-1}^{j} m_j e^{\frac{g_j m_j \mu_B B}{kT}}}{\sum_{m_j=-j}^{j} e^{\frac{g_j m_j \mu_B B}{kT}}} \tag{7.103}$$

In weak fields, the condition $\frac{g_j m_j \mu_B B}{kT} \ll 1$ is preserved; as a result, we can develop the exponential series and stop at the first-order terms of the expansion:

$$M = ng_j\mu_B \frac{\sum_{m_j=-j}^{j} m_j\left(1 + \frac{g_j m_j \mu_B B}{kT}\right)}{\sum_{m_j=-j}^{j}\left(1 + \frac{g_i m_j \mu_B B}{kT}\right)} \tag{7.104}$$

At low magnetic fields and high temperatures, the term $\frac{g_i m_j \mu_B B}{kT}$ can be neglected in the denominator, being smaller than 1, $\sum_{m_j=-1}^{j}(\frac{g_i m_j \mu_B B}{kT}) = 0$ (because $\sum_{m_j=-1}^{j}(m_j) = 0$), and we obtain:

$$M = \frac{ng_j\mu_B}{\sum_{m_j=-j}^{j}(1)}\left(\sum_{m_j=-j}^{j} m_j + \frac{g_j\mu_B B}{kT}\sum_{m_j=-1}^{j} m_j^2\right)$$

$$= ng_j\mu_B \frac{\left[\frac{g_i m_j \mu_B B}{kT} 2\frac{j(j+1)(2j+1)}{6}\right]}{2j+1}$$

In other words:

$$\vec{M} = \frac{ng_j^2\mu_B^2 j(j+1)}{3kT}\vec{B} \tag{7.105}$$

which takes into account that:

$$\begin{cases} \sum_{m_j=-j}^{j} m_j = 0;\ \sum_{m_j=-j}^{j} 1 = 2j+1 \\ \sum_{m_j=-j}^{j} m_j^2 = 2\frac{j(j+1)(2j+1)}{6} = \frac{j(j+1)(2j+1)}{3} \end{cases}$$

Thus, the susceptibility is:

$$\chi_p = \frac{n\mu_0 g_j^2\mu_B^2 j(j+1)}{3kT} \cong \frac{C}{T} \tag{7.106}$$

In this scenario, we have again obtained a Curie-type law, where:

$$C = \frac{n\mu_0 g_j^2\mu_B^2 j(j+1)}{3k} \tag{7.107}$$

If spin–orbit coupling is broken, then the $g_j^2 j(j+1)$ term comes from the sum of the term $g_L^2 l(l+1)$ (due to orbital motion, with $g_L = 1$) and the term $g_S^2 s(s+1)$ (due to spin movement, with $g_S = 2$). Thus, in this case, χ is:

$$\chi_p = \frac{n\mu_0\mu_B^2[l(l+1)+4s(s+1)]}{3kT} \tag{7.108}$$

and therefore:

$$C = \frac{n\mu_0\mu_B^2[l(l+1)+4s(s+1)]}{3k} \tag{7.109}$$

Therefore, in the case of either the presence of spin–orbit coupling or its absence, a relationship analogous to the Langevin equation is obtained for paramagnetic susceptibility. According to Langevin, relations (7.106) and (7.108) must be interpreted as follows: at a temperature of 0 K, the elementary magnetic moments are distributed in a disordered manner and are easily oriented parallel to even a weak magnetic field; however, as the temperature increases, the thermal agitation of these moments increases, and it becomes increasingly difficult for the moments to align with the magnetic field.

7.6.2 The paramagnetism of conduction electrons

The paramagnetism of conduction electrons is due only to the spin movement, since the displacement of the quasi-free electrons between the nodes of a crystalline lattice, represented by a modulated plane wave and representing an open rectilinear trajectory, cannot generate a magnetic field. The spin of each conduction electron is associated with its own magnetic moment that has a permanent character and can be oriented by the action of the magnetic field. If the number of conduction electrons in a unit volume is denoted by N and it is taken into account that, under the action of the magnetic field, the magnetic spin moment can have only two orientations (i.e. $m_s = \pm\frac{1}{2}$, $g_s = 2$), then the potential energy takes the form: $W = \pm\mu_B B$. Thus, due to the orientation of the atomic magnetic moment, i.e. either in the direction of the field or in the opposite direction, the initial energy level W is split into two levels, separated from each other by the energy interval $\Delta W = 2\mu_B B$.

On the basis of reasoning similar to that in the previous subsection, the magnetic moment for all conduction electrons is given by the following equation:

$$M = Ng_s\mu_B \frac{\sum\limits_{m_s=-\frac{1}{2},\frac{1}{2}} m_s e^{\frac{g_s m_s \mu_B B}{kT}}}{\sum\limits_{m_s=-\frac{1}{2},\frac{1}{2}} e^{\frac{g_s m_s \mu_B B}{kT}}} = N2\mu_B \frac{\frac{1}{2}e^{\frac{\mu_B B}{kT}} - \frac{1}{2}e^{-\frac{\mu_B B}{kT}}}{e^{\frac{\mu_B B}{kT}} + e^{-\frac{\mu_B B}{kT}}} = \\ = N\mu_B \frac{e^{\frac{\mu_B B}{kT}} - e^{-\frac{\mu_B B}{kT}}}{e^{\frac{\mu_B B}{kT}} + e^{-\frac{\mu_B B}{kT}}} = N\mu_B \tanh\left(\frac{\mu_B B}{kT}\right) \tag{7.110}$$

Starting from the condition $\frac{\mu_B B}{kT} \ll 1$, developing the hyperbolic tangent in series, and retaining the first term of the expansion, we obtain:

$$M = \frac{N\mu_B^2 B}{kT} \tag{7.111}$$

This is a relationship of the Curie type, in which the paramagnetic susceptibility is:

$$\chi = \frac{\mu_0 N \mu_B^2}{kT} \tag{7.112}$$

This time, we again find the Curie law previously established using quasi-classical considerations.

In the case of conductors, it has been found that the magnetic susceptibility due to conduction electrons is independent of temperature. According to Pauli's exclusion principle, the same energy level allows the existence of only two spin values. Therefore, the magnetic moments must also be in opposite directions. Under the action of the external magnetic field, only one of the two electrons orients its magnetic moment in the direction of the external field; the second cannot do so, since it would end up in the same energy state, and we would then have two electrons with the same quantum spin numbers. Because of this, one of the two electrons, namely the one that does not have its magnetic spin moment oriented in the direction of the magnetic field, must move to an unoccupied energy level to be able to orient its own magnetic moment. In other words, under the influence of the magnetic field, some of the valence band electrons in the immediate vicinity of the Fermi level must move to the free levels above the Fermi level to orient their magnetic spin moments along $\overrightarrow{B}$, thus modifying their energy. This process does not affect the totality of N conduction electrons but only a fraction $\frac{NT}{T_F}$ of them. In this situation, the magnetic susceptibility is:

$$\chi_P = \frac{\mu_0 N \mu_B^2}{kT_F} \tag{7.113}$$

which is independent of temperature.

7.7 Ferromagnetism

Diamagnetic and paramagnetic substances are magnetized and therefore possess a magnetic moment only under the action of an external magnetic field; they retain their magnetization as long as this field acts, exhibiting perfect reversibility. Unlike these substances, ferromagnetic materials exhibit spontaneous magnetization. In other words, they have a nonzero magnetic moment even in the absence of an external magnetic field. Permanent magnetization suggests that the magnetic moments of the constituent atoms have a natural tendency to align due to interactions between them. Upon subjecting such a ferromagnetic body to the action of a magnetizing field $\overrightarrow{H}$, it is found that the magnetization intensity $\overrightarrow{M}$ increases greatly, even for low values of the magnetizing field, which implies a very high magnetic susceptibility, on the order of 10^2–10^4, compared to a value of 10^{-5} for paramagnetic bodies. The intensity of magnetization quickly reaches a

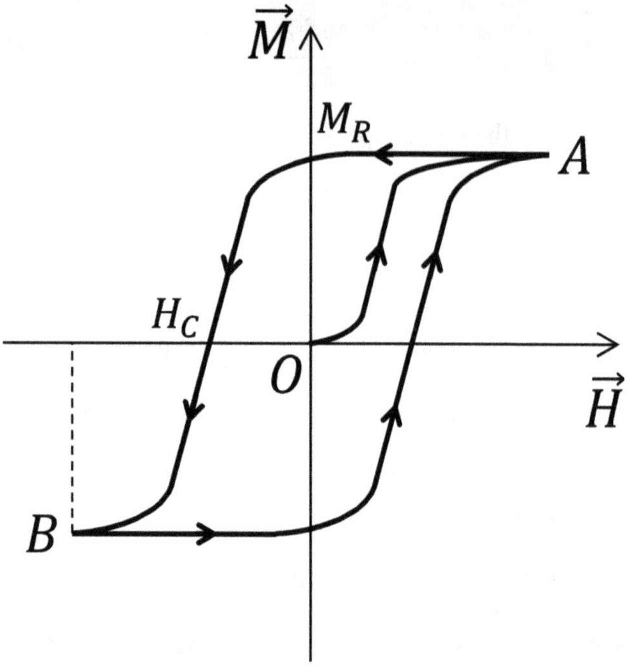

Figure 7.14. Dependence of magnetization on the intensity of the magnetic field.

saturation value (see figure 7.14). Upon decreasing the intensity of the magnetizing field, a decrease in the magnetization intensity of the ferromagnetic body is also observed, but it no longer passes through the values reached in the first magnetization process (see the *OA* curve in figure 7.14). This means that when the value of the magnetizing field returns to zero, the ferromagnetic body still shows remanent magnetization represented by the OM_R segment. To cancel this remanent magnetization, an opposing magnetizing field represented by OH_C, called a *'coercive field'*, must be applied. Due to this remanent magnetization, a certain magnetic energy is required for the magnetization process of a ferromagnetic body.

7.7.1 Weiss's theory for magnetic domains and the Curie–Weiss law

Starting from the observation that a ferromagnetic substance exhibits permanent magnetization in the absence of an external field, Weiss was the first to attempt an explanation of ferromagnetism by proposing that this permanent magnetization is due to a natural tendency of the magnetic moments of the atoms of ferromagnetic substances to align due to the interaction between them. This intrinsic property of the substance is associated with an internal field. Weiss considered that in limited regions of a ferromagnetic medium, of microscopic dimensions, the elementary magnetic moments are ordered parallel to a certain direction and oriented in the same direction. Weiss called these small regions, in which saturation magnetization is achieved by spontaneous magnetization, *magnetic domains*. In the absence of an external field, the orientation of these magnetic domains is random, which means that the ferromagnetic medium as a whole does not exhibit detectable magnetic properties. When the body is subjected to the action of a magnetizing field $\vec{H}$, these areas orient themselves and, due to their own spontaneous magnetization, cause the magnetization intensity of the body to increase rapidly, even for weak values of the magnetizing field. When these areas have been fully oriented, the saturation magnetization intensity is obtained. If the action of the field ceases, the magnetic domains, due to their large dimensions (10^{-2}–10^{-1} mm, can be seen with an optical microscope), cannot become immediately and completely disoriented, which explains the presence of remanent magnetization. The 'friction' suffered by the magnetic domains in their orientation or disorientation process, due to the 'viscosity' of the medium, justifies the hysteresis cycle and the consequent loss of energy through hysteresis. According to Weiss's hypothesis, the existence of magnetic domains, which determine the total magnetization $\vec{M}$, is attributed to the presence of an internal magnetic field $\vec{H_I}$ (*molecular field*) proportional to the total magnetization $\vec{M}$ called the *Weiss field*:

$$\vec{H_I} = \lambda\vec{M} \tag{7.114}$$

where λ is a constant specific to a given substance and is called *the Weiss field constant*. The molecular field $\vec{H_I}$ acts at very small distances and represents the mediated action of atomic magnetic moments on a single atomic moment. In this case, the magnetization of ferromagnetic bodies introduced into an external field is due to the action of a resultant field $\vec{H_R}$ formed by the inner and outer field $\vec{H}$, i.e. the sum:

$$\vec{H_R} = \vec{H} + \lambda\vec{M} \tag{7.115}$$

As a result, the magnetization of the substance is:

$$\vec{M} = \chi(\vec{H} + \lambda\vec{M}) \tag{7.116}$$

The following approximations can be made: (i) the three vectors appearing in relation (7.116) are aligned and have the same physical meaning, and (ii) the susceptibility has the expression given by Langevin for a paramagnetic substance, i.e. $\chi = \frac{C}{T}$ (it is assumed that the temperature is high enough that the substance behaves like a paramagnet). Applying these approximations, we can write relation (7.116) as $\vec{M} = \frac{C}{T}(\vec{H} + \lambda\vec{M})$ or again, $M(1 - \frac{C\lambda}{T}) = \frac{C}{T}H$, and finally:

$$\chi = \frac{M}{H} = \frac{C}{T - C\lambda} \tag{7.117}$$

Thus, the magnetic susceptibility of the considered substance depends on temperature. The law:

$$\chi = \frac{C}{T - C\lambda} = \frac{C}{T - T_c} \tag{7.118}$$

is called **the *Curie–Weiss law***. The quantity $T_c = C\lambda$ is called **the *Curie temperature***. For $T = T_c$, it is observed that $\chi \to \infty$, i.e. χ has nearly the maximum value. Therefore, ferromagnetism has an existence bounded by a maximum temperature, which is the Curie temperature itself. This temperature is the boundary between the ferromagnetic phase and the paramagnetic phase. At $T \leqslant T_c$, the substance is ferromagnetic, while at $T \geqslant T_c$, it is paramagnetic. According to figure 7.15(a), in the case of paramagnetic substances, the dependence $\chi(T)$ is a hyperbola with asymptotes at the coordinate axes. In the case of ferromagnetic substances, we are again dealing with a hyperbola (when the substance is in the paramagnetic phase at $T > T_c$), but the asymptotes are, on the one hand, the OT axis, and on the other hand, to the right of the $T = T_c$ line parallel to the magnetic susceptibility or χ_m axis ($T = T_c$ is shown as a dashed line in figure 7.15(b)). The Curie temperatures for various ferromagnetic substances are given in table 7.2. The experimental determination of the Curie temperature (T_c), together with the theoretical value of the Curie constant, allows us to find the value of λ. Upon introducing a ferromagnetic substance into an external magnetic field whose induction increases, it has been found that at a certain induction value, the magnetization of the substance reaches

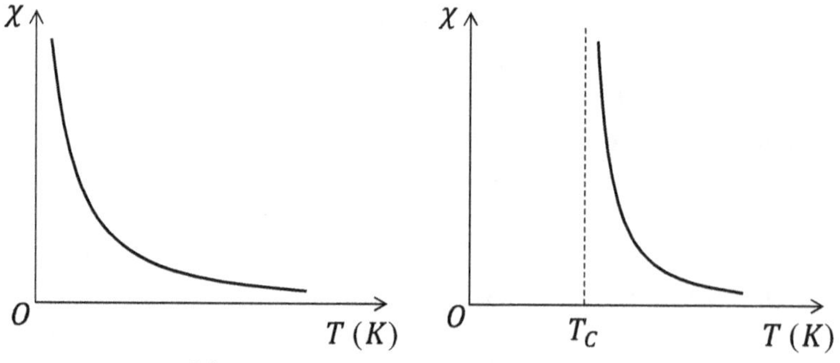

Figure 7.15. Temperature dependence of the magnetic susceptibility of (a) a paramagnetic substance and (b) a ferromagnetic substance.

Table 7.2. Typical values of the Currie temperature for several common materials.

Substance	T_C (K)
Gadolinium (Gd)	278.5
Nickel (Ni)	627.2
Iron (Fe)	1043
Cobalt (Co)	1388

saturation. The variation of the ratio between saturation magnetization at a given temperature and saturation magnetization at 0 K, in the range from zero to T_c, is shown in figure 7.16.

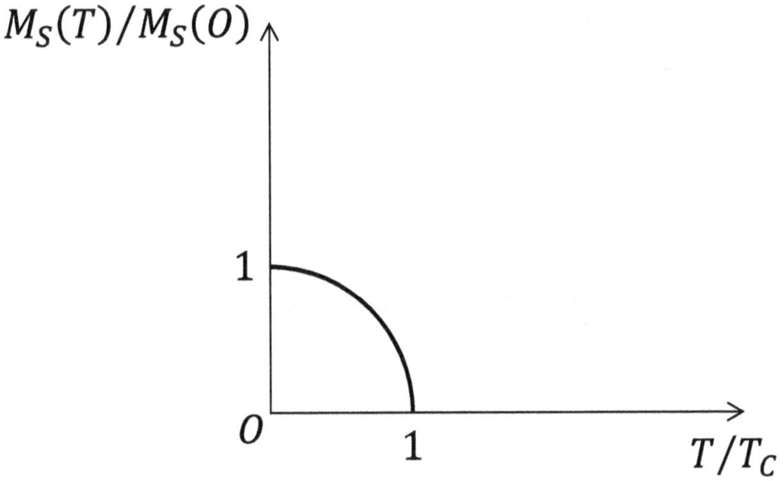

Figure 7.16. Graphical representation of the dependence $M_S(T)/M_S(O) = f(T/T_C)$.

Table 7.3 shows the values of the saturation magnetization at 0 K and at room temperature (RT) for different ferromagnetic substances. With magnetization at saturation, it is possible to find the λ constant of the internal field $\overrightarrow{H_I}$ for a given substance at a given temperature.

Table 7.3. Magnetizations of several elements at $T = 0$ K and RT.

	M_S $(\times 10^3)$ A m^{-1}	
Substance	0 K	RT
Nickel (Ni)	510	485
Iron (Fe)	1740	1700
Cobalt (Co)	1446	1400

7.7.2 Ferromagnetic domains and the formation of magnetic domains

Based on Weiss's hypothesis of the presence of magnetic domains, one can represent the distribution of magnetic domains in the absence of an external magnetic field, as shown in figure 7.17. Under the action of an external magnetic field $\overrightarrow{H_{ext}}$, the magnetization of the substance in the direction of the field increases due to two processes:

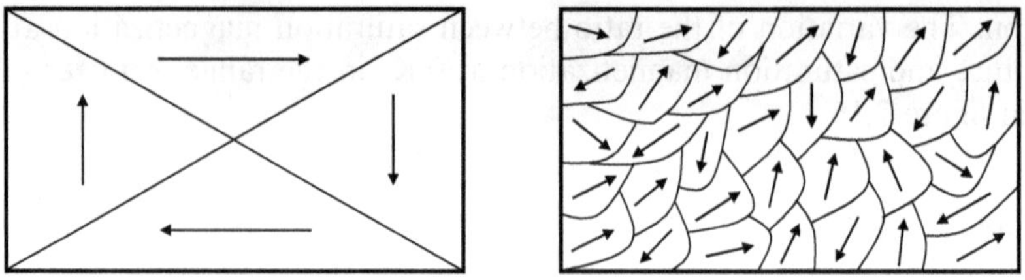

$H_{ext} = 0\,; M = 0$ $\qquad$ $H_{ext} = 0\,; M = 0$

(a) $\qquad$ (b)

Figure 7.17. Magnetic monodomains in: (a) a single crystal and (b) a polycrystal.

(a) At low values of the applied external field, the favorably oriented domains increase in volume to the detriment of the unfavorably oriented ones (see figure 7.18(a)).
(b) At high values of $\overrightarrow{H_{ext}}$, we observe rotation in the direction of the applied field in all domains except those that were already oriented in the direction of the applied field (see figure 7.18(b)).

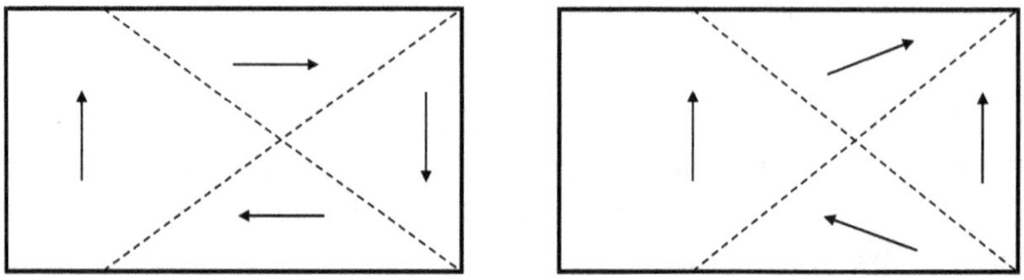

$H_{ext} = 0\,; M \neq 0$ $\qquad$ $H_{ext} \neq 0\,; M \neq 0$

(a) $\qquad$ (b)

Figure 7.18. (a) Expansion of the monodomains whose magnetic moments are favorably oriented with respect to the magnetic field. (b) The rotation of the magnetic moments of the monodomains to become parallel to the field.

On the curve that shows the dependence of magnetization $\overrightarrow{M}$ on the intensity of the external magnetic field $\overrightarrow{H_{ext}}$ (see figure 7.19), three sections can be distinguished:

I. A reversible shift of the domain boundaries (called Bloch walls).
II. An irreversible displacement of the domain boundaries.
III. A rotation of the domains in the direction of the field.

After this analysis of the phenomena that take place in a ferromagnet, as in the case of the other two forms of magnetism, questions still remain, namely, which magnetic moments are due to ferromagnetism? and what is the mechanism of their orientation within the magnetic monodomains? As for the first problem, it has been answered since 1915, when Einstein and Haas experimentally highlighted the spin anomaly.

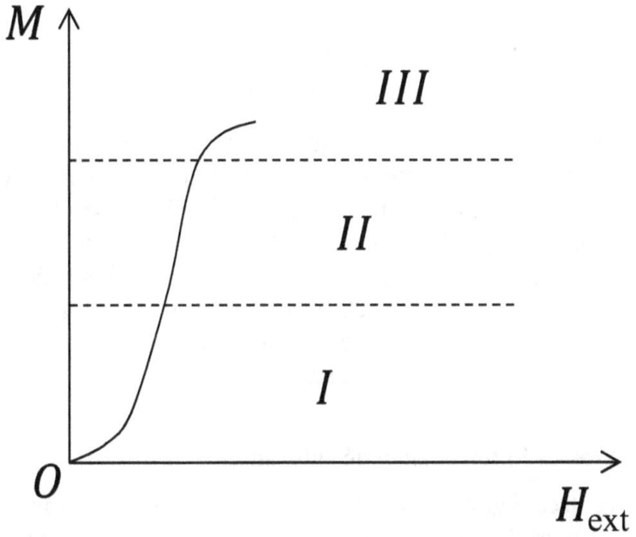

Figure 7.19. Dependence of magnetization on the strength of the magnetizing field in the case of a ferromagnetic substance.

Thus, it was found that the magnetic moments of spin are responsible for the phenomenon of ferromagnetism. In this way, monodomains arise through the parallel orientation of a large number of such elementary spin magnetic moments. However, the problem arises of determining the nature of the forces that, at short distances, act in the sense of ordering the spin magnetic moments in the same direction (forming monodomains) but, at long distances, act in the sense of disorienting them. The existence of magnetic domains with spontaneous magnetization results in the appearance of an inner magnetic field in the ferromagnetic body $\overrightarrow{H_I} = \lambda\overrightarrow{M}$. Given that ferromagnetism is determined only by the elementary spin magnetic moments, the Weiss field constant can be determined from the defined expression for the Curie constant, $C = \frac{n\mu_0\mu_B^2[l(l+1) + 4s(s+1)]}{3k}$ (which in the case of decoupled moments $\vec{L}$ and $\vec{S}$ becomes $C = \frac{n\mu_0\mu_B^2[4s(s+1)]}{3k}$). Thus:

$$C\lambda = T_c \Longrightarrow \frac{1}{\lambda} = \frac{C}{T_c} = \frac{n\mu_0\mu_B^2[4s(s+1)]}{3kT_c} \tag{7.119}$$

Given this result, the inner Weiss field is:

$$\overrightarrow{H_I} = \frac{3kT_c}{4n\mu_0 s(s+1)\mu_B^2}\overrightarrow{M} \tag{7.120}$$

If we consider that each elementary magnetic moment is equal to one Bohr magneton, for n monoelectronic atoms, we obtain $M = n\mu_B^2$, while the value of the Weiss field between two such elementary magnets is:

$$H_I = \frac{3kT_c}{4\mu_0 s(s+1)} \tag{7.121}$$

At the Curie temperature, this field can reach values much higher than those produced by purely magnetic interactions, which occur to a very small extent because they manifest themselves at long distances. Because of this, the physical origin of the Weiss field has been attributed to the exchange interaction, i.e. the exchange energy of parallel or antiparallel spin systems. This follows from Pauli's exclusion principle, according to which, in quantum mechanics, it is generally not possible to change the relative direction of two spins without changing the distribution of electric charges in the region considered.

7.7.3 Hysteretic losses in ferromagnetic substances

As shown above (see figure 7.14), the graphical representation of $\vec{B} = \vec{B}(\vec{H})$ presents a hysteresis cycle. Starting from the origin and increasing the intensity of the magnetic field $\vec{H}$, the magnetic induction $\vec{B}$ increases and reaches a saturation value $\vec{B_S}$. As $\vec{H}$ returns to zero, a residual magnetic induction $\vec{B_R}$ becomes apparent (see figure 7.20). To reduce the magnetic induction $\vec{B}$ to zero, we must apply an opposing magnetic field, called the coercive field $\vec{H_C}$. Increasing the magnetizing field further gives rise to the saturation of $\vec{B}$ in the 3rd quadrant. As $\vec{H}$ returns to

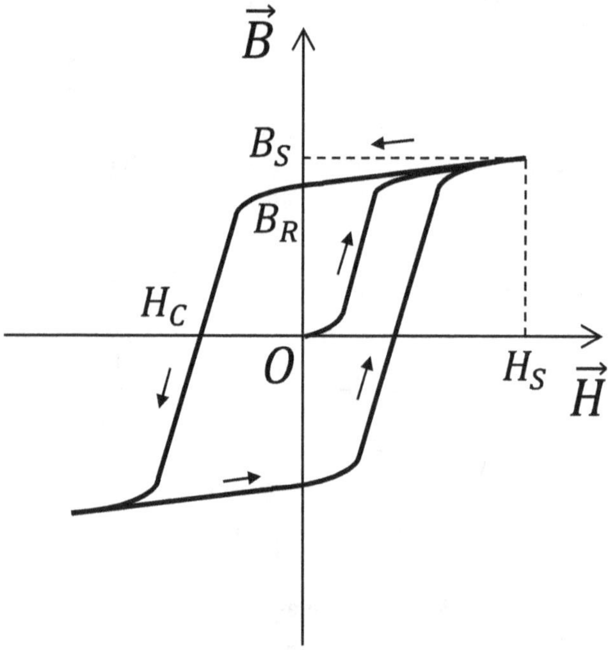

Figure 7.20. Magnetic hysteresis curve.

zero and then moves to higher values in the positive direction, the hysteresis curve ends at the maximum value B_S of magnetic induction (its saturation value) in the first quadrant. At any point of the hysteresis curve, the normal permeability is given by $\mu = \frac{B}{H}$, and at the points with maximum induction values, the maximum permeability is given by $\mu_m = \frac{Bm}{H}$. When a hysteresis curve shows symmetry with respect to its origin, it is called a normal hysteresis cycle. Permanent magnets have high values of B_R and H_C. Soft ferromagnetic substances used in transformer laminations and coil cores have low values of B_R and H_C. The area bounded by the hysteresis cycle is a measure of the energy consumed per unit volume of substance when a cycle is completed. This energy consists of the hysteresis and heat losses. The energy lost can be calculated by starting from the energy stored in the magnetic field and expressed using both the induction of the magnetic field $\vec{B}$ and its intensity $\vec{H}$. The energy transferred from a circuit carrying an electric current to the magnetic field is:

$$W_M = \int_{\Phi_0}^{\Phi} I d\Phi \tag{7.122}$$

where I is the intensity of the instantaneous current established in the circuit, and Φ is the magnetic flux that crosses the surface defined by the circuit. In order to express the energy W_M in terms of the physical quantities $\vec{B}$ and $\vec{H}$, we will consider a toroidal coil (see figure 7.21) whose core is made from a homogeneous and isotropic ferromagnetic substance. The radius of the torus and thus the coil (or a turn) complies with the condition $r_a \gg r_b$.

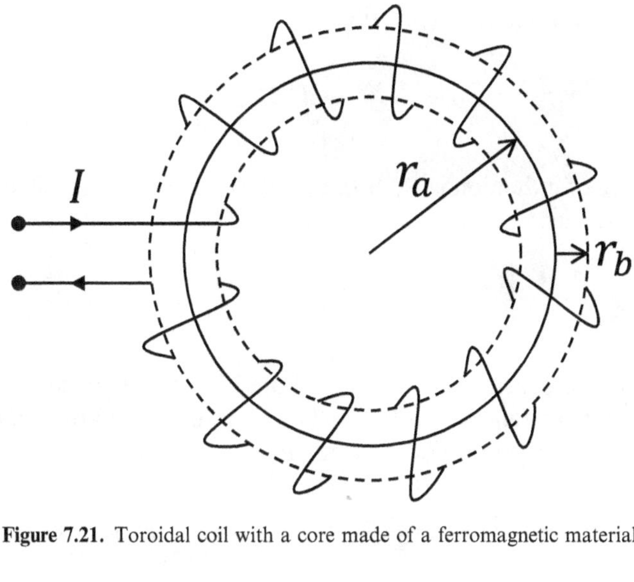

Figure 7.21. Toroidal coil with a core made of a ferromagnetic material.

The circulation of a vector $\vec{H}$ on a contour C that has the radius of r_a (mean radius of torus) is:

$$\oint \vec{H} \vec{dl} = NI \tag{7.123}$$

and

$$H = \frac{NI}{2\pi r_a} \tag{7.124}$$

The induction flux of the magnetic field in the coil is:

$$\Phi = N\pi\, r_b^2 B \tag{7.125}$$

where N is the number of turns and r_b is the radius of the core cross-section. Differentiating equation (7.125) yields:

$$d\Phi = N\pi\, r_b^2 dB \tag{7.126}$$

With these results, the energy stored in the magnetic field of the coil can be given as:

$$dW_M = \frac{2\pi r_a H}{N} N\pi\, r_b^2 dB \tag{7.127}$$

The energy stored in a magnetic field whose induction varies from an initial value B_0 to a final value B_f is:

$$W_M = 2\pi^2 r_a r_b^2 \int_{B_0}^{B_f} HdB = V \int_{B_0}^{B_f} HdB \tag{7.128}$$

where V is the volume occupied by the ferromagnetic substance. The energy transferred to a unit volume of the substance (or the density of the energy transferred to the magnetic field) has the expression:

$$w_M = \frac{W_M}{V} = \int_{B_0}^{B_f} HdB \tag{7.129}$$

If the initial value of induction is $B_0 = 0$, and B_f is denoted by B, the integral (7.129) can also be written as:

$$w_M = \int_0^B HdB \tag{7.130}$$

which represents the density of the energy stored in the magnetic field in the space occupied by the ferromagnetic substance. Taking into account the relation $B = \mu H$, the calculation of the integral (7.130) leads to:

$$w_M = \mu \int_0^B HdH = \frac{\mu H^2}{2} = \frac{1}{2}\frac{B^2}{\mu} = \frac{1}{2}\vec{H}\vec{B} \tag{7.131}$$

In the graphical representation of the dependence $\vec{B}(\vec{H})$, the energy density stored in the magnetic field is depicted by the closed surface between the curve and the B

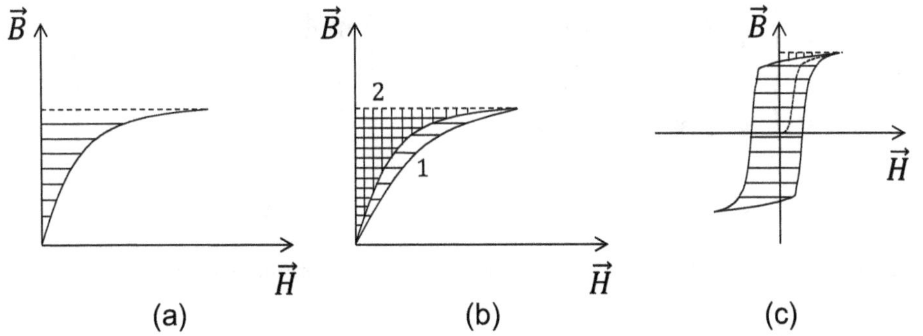

Figure 7.22. (a) The energy stored in the magnetic field. (b) The cross-hatched region represents the energy returned by the substance to the source. (c) The area of the hysteresis curve is proportional to the energy losses due to hysteresis.

axis (see figure 7.22(a)). This represents the energy expended by the source to overcome the self-induced electromotive force (EMF) caused by the increased current in the windings of the toroidal coil. When H returns to zero, the current from the turns of the toroidal coil decreases, and the self-induced EMF opposes the decrease in the inductive flux. In other words, the self-induced EMF has the same sense as the EMF of the external source, so the substance returns energy to the source, this energy being represented by the (cross-hatched) area between curve 2 and the B axis in figure 7.22(b). Thus, the energy supplied by the source at the time of magnetization is not equal to that provided by the substance at demagnetization; the amount lost through the Joule effect is precisely the difference between these two energies. Following the whole cycle, the surface area delimited by the plot represents the energy lost by heating the ferromagnetic substance, and this constitutes the hysteresis loss (see figure 7.22(c)). For the energy density consumed in a hysteresis cycle, Steinmetz's empirical relationship is valid:

$$w_{M\text{cycle}} = \eta B_m^{\beta} \tag{7.132}$$

where $\beta = 1.6$ and η takes different values for different substances. Tabulating the B_m and η values allows the hysteresis losses of ferromagnetic substances to be calculated.

7.7.4 Antiferromagnetism

The antiferromagnetic state differs from the ferromagnetic state in that the electron spins are oriented antiparallel in pairs (see figure 7.23). This state occurs in a number of substances, such as Mn, Cr, MnO, MnO_2, CVr_2O_2, FeO, NiO, etc. Antiferromagnetic substances exhibit a different magnetic susceptibility variation curve than that of ferromagnetic substances (see figure 7.24) in the sense that at a specific temperature, called the ***Néel temperature***[1] (T_N), the susceptibility reaches a

[1] Named after a discovery made by ***Louis Néel***, a French physicist who received the Nobel Prize in 1970.

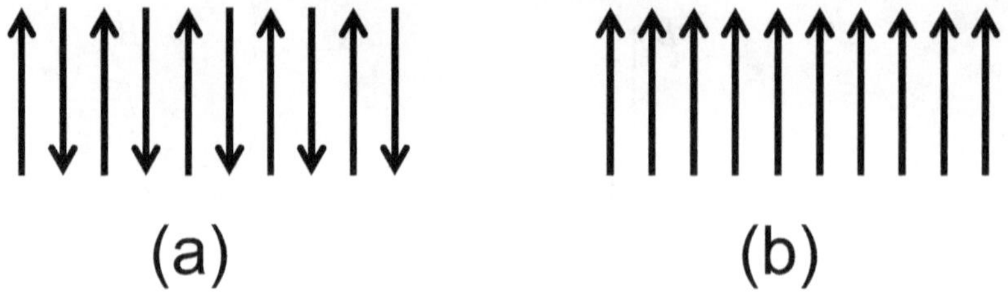

Figure 7.23. Orientation of the magnetic spin moments in: (a) an antiferromagnet and (b) a ferromagnet.

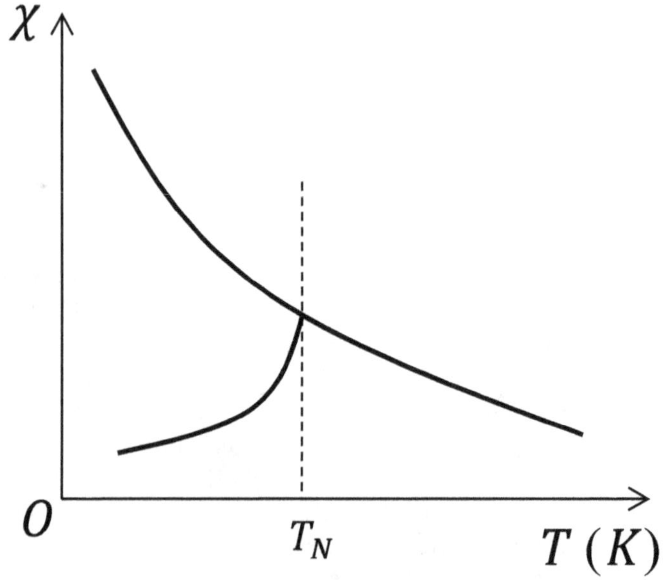

Figure 7.24. Temperature dependence of the magnetic susceptibility of an antiferromagnetic substance.

maximum, decreasing when $T > T_N$. The body has antiferromagnetic properties at $T < T_N$, because at temperatures less than the Néel temperature, the spins lock in the antiparallel positions (within a pair, the spins are parallel but have opposing directions). In the case of antiferromagnetic bodies, it is considered that there are two sublattices formed by ions with nonzero magnetic moments but oriented in opposite directions. At 0 K, the magnetic moments of one sublattice are parallel to each other, having the same sense, while the magnetic moments of the second are also parallel to each other, but are oriented in the opposite direction to those of the first sublattice. When the temperature increases, this ideal state is destroyed and the two sublattices are destroyed. In this situation, at a certain temperature, although the number of moments oriented in one direction continues to remain equal to that of the moments oriented in the opposite direction, there are no longer the same two-way relations between the orientations of the magnetic moments on one hand and their membership of a sublattice, on the other hand. Because of this, at temperatures above this critical temperature (the Néel temperature), the antiferromagnetic body exhibits paramagnetic properties. Theoretically, we now consider that the antiferromagnetic behavior could be due to the participation of oxygen in the electronic exchange mechanism.

7.7.5 Ferrimagnetism

Ferrimagnetic bodies are characterized by a crystal lattice where the two sublattices can be occupied by two kinds of ions of different magnetic moments, which in an ordered state give a total nonzero magnetization, even though the magnetic moments of the two sublattices are oriented in opposite directions. For this reason, Néel considered that ferrimagnetism is nothing more than uncompensated anti-ferromagnetism (see figure 7.25). Each type of sublattice can be occupied, depending on the temperature, by different proportions of the respective ion species. In addition, each ion species has, on a case-by-case basis, a greater or lesser probability of presenting a magnetic moment of one sign or another. Therefore, the problem of ferrimagnetism as a whole appears quite complicated. The best-known ferrimagnetic materials are ferrites, predominantly ionic crystals, in which conduction electrons exist in small numbers, resulting in high values of resistivity. Ferrites are generally classified into *soft ferrites* and *hard ferrites*. Soft ferrites are oxides that have the general formula MFe_2O_4, where M is a bivalent metal ion such as Mg, Cu, Mn, Ni, or Zn. Hard magnetic ferrites have the general chemical formula $MO_6Fe_2O_3$, where M is a metal ion of Ba, Pb, or Sn. Garnets constitute another type of ferrimagnetic substance. Their chemical composition is $5Fe_2O_3{-}3M_2O_3$, where the metal ion (M) is yttrium (Y) or a rare earth element. Such ferrimagnetic properties are also exhibited by perovskites with the formula $Fe_2O_3M_2O_3$, where the metal ion is the same as in garnets.

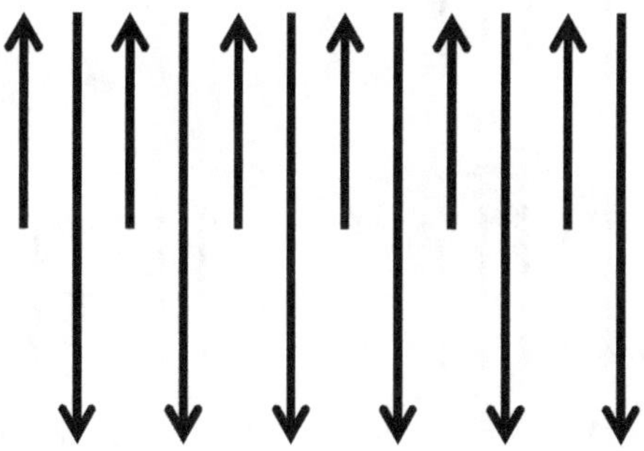

Figure 7.25. Two sublattices which have different magnetic moments that are oriented antiparallel.

From a phenomenological point of view, the process of partial compensation of the magnetic moments of the two sublattices oriented in opposite directions results from considering the chemical formula of ferrites. For example, consider a spinel ferrite such as magnetite, in which the metal ion (M) is replaced by an Fe^{2+} ion. The chemical formula for magnetite spinel ferrite has the form $FeOFe_2O_3$

[($Fe^{2+}Fe_2^{3+}$)OH], where the ferrous ion Fe^{2+} is found in the state $^5D_4$ with spin 2, and the ferric ion Fe^{2+} is found in the state $^6D_{5/2}$ with spin 5/2 and contributes to magnetization according to $5\mu_B$. Due to the fact that the spin magnetic moment of the O^{2-} is zero, since the 2*p* layer is complete, and considering that the spins of both types of iron ions are parallel, the chemical formula of magnetite leads to:

$$\left(2S_{\mathrm{Fe}}^{3+} + S_{\mathrm{Fe}}^{2+}\right)g\mu_B = \left(2\frac{5}{2} + 2\right)2\mu_B = 14\mu_B$$

which, taking $g = 2$ for the magnetic moment corresponding to the chemical formula of magnetite, gives a value of $14\mu_B$. This is in total contradiction to the experimental results, where the approximate value of $4.1\mu_B$ was obtained. According to Néel, this discordance is due to the fact that not all the spin magnetic moments are parallel. Considering the spin orientation model proposed by Néel (see figure 7.26), it follows that the spin magnetic moments of the Fe^{3+} ions compensate each other, leaving uncompensated only the spin magnetic moments of Fe^{2+} ions, whose magnetic moment according to the above relation results in $4\mu_B$. This model corresponds to reality, since structural analyses performed using x-rays and neutrons have shown that the elementary cell of magnetite contains 56 ions, of which 32 are O^{2-} ions and 24 are Fe cations, namely 16 Fe^{3+} cations and 8 Fe^{2+} cations. Fe^{3+} ions are located in the interstices of the O^{2-} sublattice; half are in tetrahedral positions (*A*) and have 4 O^{2-} neighbors, while the other half are in octahedral positions and have 6 O^{2-} neighbors. The Fe^{2+} ions are all located in octahedral (*A*) positions.

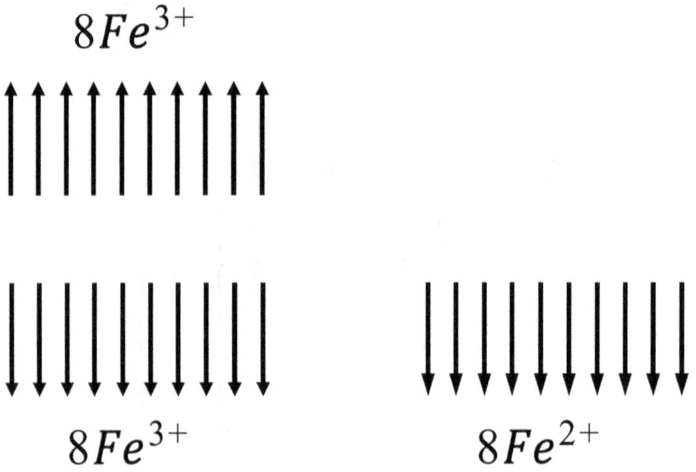

Figure 7.26. Spin sublattices of a ferrimagnetic substance of iron ions oriented in parallel.

Because they present high electrical resistance and offer the possibility of preparing mixed structures with various properties, ferrites are widely used in technology (for example, in transformer cores), but they also have semiconducting properties that find applications in computing elements, magnetic memories, etc.

Further reading

[1] Antohe Ş and Antohe V-A 2023 *Electrostatics. Formalism of the Electrostatic Field in Vacuum and Matter* 1st edn (Bristol: IOP Publishing) 266 pages

[2] Purcell E M and Morin D J 2013 *Electricity and Magnetism* 3rd edn (Cambridge: Cambridge University Press) 853 pages

[3] Jackson J D 1998 *Classical Electrodynamics* 3rd edn (New York: Wiley) 832 pages

[4] Griffiths D J 2017 *Introduction to Electrodynamics* 4th edn (Cambridge: Cambridge University Press)

[5] Bleaney B I and Bleaney B 1965 *Electricity and Magnetism* 3rd edn vol 1 (Oxford: Oxford University Press)

[6] Feynman R P, Leighton R B and Sands M 2011 *Feynman Lectures on Physics, Vol. II: The New Millennium Edition: Mainly Electromagnetism and Matter (Feynman Lectures on Physics)* (New York: Basic Books) 589 pages

[7] Secăreanu I, Ruxandra V, Gherbanovschi N, Logofătu M, Cazan-Corbasca M and Antohe Ş 1984 *Problems of Electricity and Magnetism (Culegere de probleme de Electricitate şi Magnetism)* (Bucharest: University of Bucharest Publishing House)

[8] Jiles D 2015 *Introduction to Magnetism and Magnetic Materials* 3rd edn (Routledge: CRC Press, Taylor and Francis Group) 626 pages

[9] Grosso G and Parravicini G P 2013 *Solid State Physics* 2nd edn (Cambridge, MA: Academic) 872 pages

[10] 2010 *Magnetic Properties of Solids (Material Science and Technologies)* ed K B Tamayo (New York: Nova Science) 342 pages

IOP Publishing

Electromagnetism and Special Methods for
Electric Circuits Analysis

Ştefan Antohe and Vlad-Andrei Antohe

Chapter 8

Transient phenomena in electrical circuits

The transient regime typically represents the state of an electric circuit before a permanent regime is reached or when a circuit passes from one permanent regime to another. Thus, the transient regime occurs within a determined time interval after an electrical circuit containing real reactive elements is connected/disconnected to/from an electromotive voltage source.

8.1 DC and AC electrical circuits in the transient regime

8.1.1 DC circuit containing a series-connected resistance and capacitance (RC) in the transient regime

8.1.1.1 Charging of a real capacitor by an electromotive voltage source

In order to connect and disconnect the source from the circuit, the design in figure 8.1 is used. If switch K is moved to position 1, then, since it is coupled to the source, we witness the process of charging the capacitor. In this case, a current of intensity i passes through resistor R at time t. Applying Kirchhoff's second theorem to the circuit leads to:

$$\mathcal{E} = Ri + U_c \tag{8.1}$$

The quantity $U_R = Ri$ represents the voltage drop across the resistor, and U_c is the voltage drop across the capacitor at time t. The relationship between the current i passing through the capacitor and the charge Q on the capacitor at time t can be found by applying the continuity equation for the surface S in figure 8.1:

$$i = \iint \vec{j}\,\vec{ds} = \frac{dQ}{dt} \tag{8.2}$$

doi:10.1088/978-0-7503-5854-5ch8

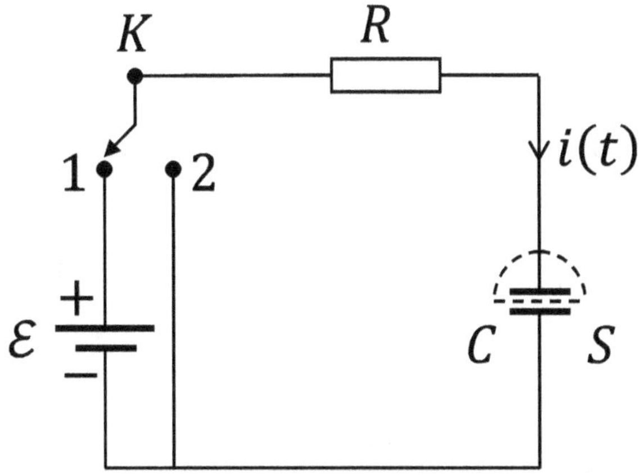

Figure 8.1. Circuit for charging and discharging a capacitor.

The positive sign of the charge differential on the right-hand side of this equation indicates that the charge inside the surface S increases over time (i.e. the capacitor is charging). The voltage drop across the capacitor is:

$$U_c = \frac{Q}{C} \tag{8.3}$$

Using relations (8.2) and (8.3), equation (8.1) becomes:

$$\mathcal{E} = R\frac{dQ}{dt} + \frac{Q}{C} \tag{8.4}$$

This is a first-order differential equation that can be solved by separating the variables:

$$\frac{dQ}{\mathcal{E}C - Q} = \frac{dt}{RC} \tag{8.5}$$

If, at time $t = 0$, when switch K is moved to position 1, the charge on the capacitor plates was equal to zero, and at time t the charge is $Q(t)$, then the integration of equation (8.5) leads to:

$$\int_0^{Q(t)} \frac{dQ}{\mathcal{E}C - Q} = \int_0^t \frac{dt}{RC} \implies \ln\frac{\mathcal{E}C - Q(t)}{\mathcal{E}C} = -\frac{t}{RC}$$

$$\Rightarrow Q(t) = \mathcal{E}C\left(1 - e^{-\frac{t}{RC}}\right) \tag{8.6}$$

Therefore, if at $t = 0$, $Q(0) = 0$, then at time $t \to \infty$ the charge on the capacitor reaches its maximum value $Q(\infty) = \mathcal{E}C$, where $\mathcal{E}$ is the electromotive voltage of the source. Between moments $t = 0$ and $t \to \infty$, the capacitor charge increases exponentially from 0 to $\mathcal{E}C$, as shown in figure 8.2. At time $t = 0$, the rate of increase of the electric charge on the capacitor is:

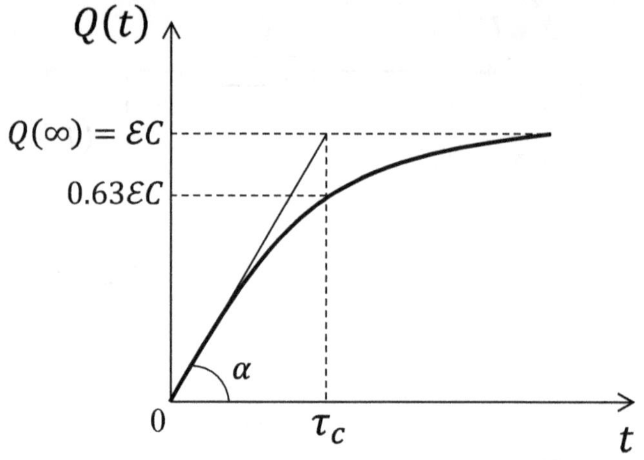

Figure 8.2. Time dependence of electric charge on the capacitor while charging.

$$\left.\frac{dQ}{dt}\right|_{t=0} = \frac{\mathcal{E}C}{RC} = \frac{\mathcal{E}}{R} = \frac{Q(\infty)}{RC} = \tan\alpha_0 \tag{8.7}$$

By integrating equation (8.7), we obtain the equation for the tangent to the curve (8.8) passing through the origin:

$$Q(t) = \frac{Q(\infty)}{RC}t \tag{8.8}$$

which shows that if the capacitor's charge were to always increase at its initial (origin) growth rate, it would acquire the maximum value $Q(\infty) = \mathcal{E}C$ in the time $\tau_c = RC$ determined by the values of the resistor and capacitance of the capacitor forming the circuit. *Time τ_c represents the time constant of the circuit, defined here as the time after which the charge on the capacitor would reach its maximum value under conditions in which it would increase at the same rate of variation as at the origin.* After this time, the electric charge on an actual capacitor to be charged is:

$$Q(\tau_c) = Q(\infty)\left(1 - e^{-\frac{t}{RC}}\right) = 0.63Q(\infty) \tag{8.9}$$

Thus, τ_c can also be defined as the time interval after which the electric charge on capacitor C acquires 0.63 of the maximum charge it can reach at a voltage of $\mathcal{E}$.

The current intensity in the circuit at a given time t is:

$$I(t) = \frac{dQ}{dt} = \frac{\mathcal{E}}{R}e^{-\frac{t}{\tau_c}} = I_0e^{-\frac{t}{\tau_c}} \tag{8.10}$$

where $I_0 = \frac{\mathcal{E}}{R}$ is the maximum value of the current in the circuit at $t = 0$. The current intensity in the circuit decreases exponentially over time, as shown in figure 8.3. After the time $t = \tau_c$, the current reaches the value $I(\tau_c) = \frac{I_0}{e} = 0, 37I_0$. *In other*

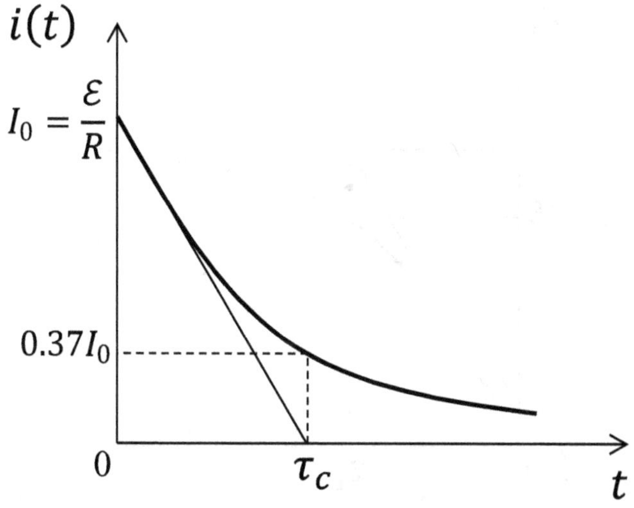

Figure 8.3. Dependence of current on time when charging a capacitor.

words, τ_c is the time after which the intensity reaches 0.37 *of* I_0. *Here, the equation for the tangent of the curve of I passing through the origin is:*

$$\frac{dI}{dt} = -\frac{I_0}{RC}e^{-\frac{t}{\tau_c}}\bigg|_{t=0} = -\frac{I_0}{RC} \tag{8.11}$$

The equation for the variation of the current I with this slope is:

$$\int_{I_0}^{I(t)} dI = -\frac{I_0}{RC}\int_0^t dt \Longrightarrow I = I_0\left(1 - \frac{t}{\tau_c}\right) \tag{8.12}$$

Thus, at $t = \tau_c$, it follows that $I(t) = 0$. As a result, the time constant τ_c can also be defined as the time interval after which the current would become zero if it decreased at the same rate of decrease as that observed from the origin.

The voltage at the terminals of the capacitor in the circuit is:

$$U_C = \frac{Q}{C} = \frac{\mathcal{E}C}{C}\left(1 - e^{-\frac{t}{\tau_c}}\right) = \mathcal{E}\left(1 - e^{-\frac{t}{\tau_c}}\right) \tag{8.13}$$

and this increases from zero at $t = 0$ to $U_{C\,\max} = \mathcal{E}$ when $t \to \infty$, according to the graph in figure 8.4. Therefore, charging a capacitor from a voltage source with an electromotive force (EMF) $\mathcal{E}$ takes place as follows: at the moment we connect the source, capacitor C is not charged, and the voltage across it is zero. The voltage drop across it subsequently increases, reaching a maximum value equal to the source voltage $\mathcal{E}$ At the moment we connect the source, all the voltage of source $\mathcal{E}$ is applied to the resistor R, determining the maximum current:

$$I_0 = \frac{\mathcal{E}}{R} \tag{8.14}$$

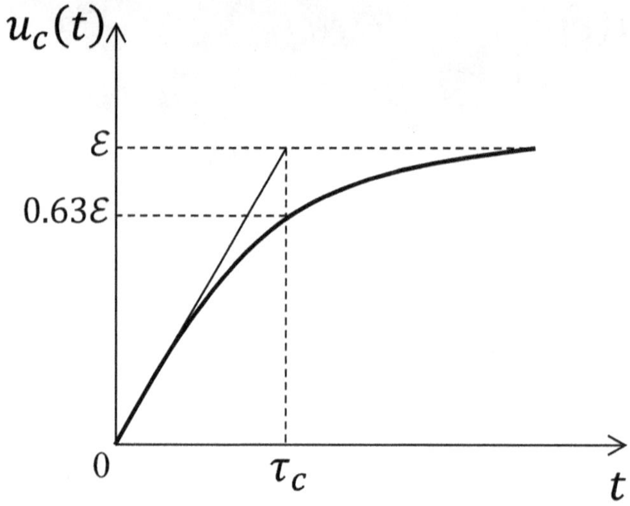

Figure 8.4. Time dependence of the voltage drop across a capacitor while it charges through a load resistance.

Over time, the capacitor begins to charge, and a potential difference develops between its plates, which increases over time. According to Kirchhoff's second law, the greater the voltage drop across the capacitor, the smaller the voltage drop across the resistor, and therefore the lower the current passing through the resistor $I_0 = \frac{U_R}{R}$. Finally, when the stationary state is reached, the electric charge on the capacitor's plates and the voltage between them acquire their maximum values of $Q(\infty) = \mathcal{E}C$ and $U_{C\,\max} = \mathcal{E}$, respectively. At this point, both the voltage drop across the resistor and the current in the circuit are zero, i.e. $I = \frac{U_R}{R}$, $U_R = 0$, and $I = 0$.

During the charging process of the capacitor, the total electric charge passing through the battery is :

$$Q(\infty) = \mathcal{E}C \tag{8.15}$$

The total energy consumed by the battery in transporting this electric charge is:

$$W = Q(\infty)\mathcal{E} = C\mathcal{E}^2 \tag{8.16}$$

The energy stored in the capacitor at the end of its charging process is:

$$W_C = \frac{1}{2}C\mathcal{E}^2 \tag{8.17}$$

The energy dissipated through resistor R during the charging process is:

$$W_R = \int_0^\infty Ri^2 dt = \int_0^\infty RI_0^2 e^{-\frac{2t}{\tau_C}} dt = \frac{1}{2}C\mathcal{E}^2 \tag{8.18}$$

It should be noted that regardless of the resistance and capacitance values, in the process of charging the capacitor through the resistor at a constant voltage, half of the source energy is lost in the resistor, and half is stored in the capacitor, with the capacitor's charging efficiency being:

$$\eta = \frac{W_C}{W} = \frac{1}{2} \tag{8.19}$$

8.1.1.2 Discharge of a real capacitor disconnected from an electromotive voltage source

The discharge of a capacitor through resistor R occurs when we move the switch in figure 8.1 to position 2. In this case, Kirchhoff's second theorem can be written for a given time t as follows:

$$Ri + \frac{Q}{C} = 0 \quad \text{or} \quad R\frac{dQ}{dt} + \frac{Q}{C} = 0 \tag{8.20}$$

The integration of equation (8.20) leads to:

$$\int_{Q_0}^{Q(t)} \frac{dQ}{Q} = -\int_0^t \frac{dt}{RC} \tag{8.21}$$

Thus:

$$\log \frac{Q(t)}{Q_0} = -\frac{t}{RC} \Longrightarrow Q(t) = Q_0 e^{-\frac{t}{\tau_C}} \tag{8.22}$$

The voltage across the terminals of the capacitor at time t is:

$$U_c = \frac{Q(t)}{C} = \frac{Q_0}{C} e^{-\frac{t}{\tau_C}} = \frac{C\mathcal{E}}{C} e^{-\frac{t}{\tau_C}} = \mathcal{E} e^{-\frac{t}{\tau_C}} \tag{8.23}$$

Thus, it also decreases exponentially over time.

The current intensity passing through resistance R is:

$$I(t) = \frac{dQ(t)}{dt} = -\frac{Q_0}{RC} e^{-\frac{t}{\tau_C}} = -\frac{C\mathcal{E}}{RC} e^{-\frac{t}{\tau_C}} = -I_0 e^{-\frac{t}{\tau_C}} \tag{8.24}$$

The minus sign in equation (8.24) indicates that the sense of the current flowing through R during the capacitor's discharge has the opposite sense to that of the current passing through the circuit during the process of charging the capacitor. Figure 8.5 shows the variations over time of the charge on the plates of the capacitor, the voltage across the terminals of the capacitor, and the current passing through the resistor during the process of discharging the capacitor. Transient phenomena occurring when charging and discharging a capacitor by means of a resistor can be visualized on an oscilloscope using the circuit shown in figure 8.1, in which K is an electronic switch capable of performing at least ten switching operations per second. Applying the voltage U_c or U_R to OY and using the oscilloscope's time base horizontally shows that τ_c is less than the time between two switching transitions of K.

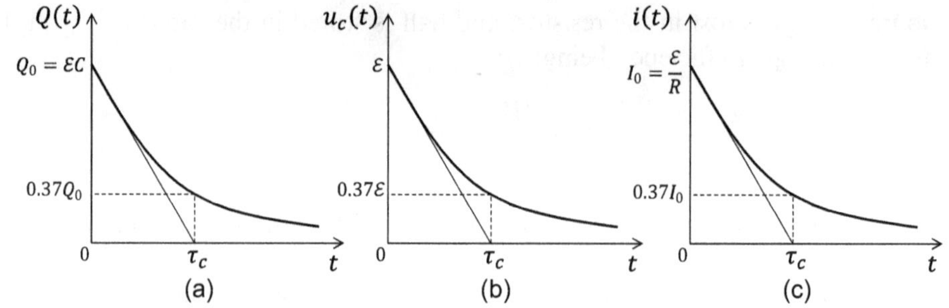

Figure 8.5. Time dependence of the electric charge on the capacitor's plates (a), the voltage across its plates (b) and the current flowing through the series resistance during the discharge of the capacitor (c).

8.1.2 *RC* series AC circuit in the transient regime

Let us consider an *RC* series circuit powered by a sinusoidal voltage source at a voltage of:

$$u = \sqrt{2}\,U \sin(\omega t + \gamma) \tag{8.25}$$

Considering that the circuit is closed at $t = 0$ when $u = \sqrt{2}\,U \sin\gamma$, Kirchhoff's second theorem applied to the circuit at moment t leads to:

$$Ri + \frac{Q}{C} = \sqrt{2}\,U \sin(\omega t + \gamma) \tag{8.26}$$

However:

$$i(t) = \frac{dq(t)}{dt} = \frac{d(Cu_C)}{dt} = C\frac{du_C}{dt} \Longrightarrow RC\frac{du_C}{dt} + u_C = \sqrt{2}\,U \sin(\omega t + \gamma)$$

or:

$$\frac{du_C}{dt} + \frac{u_C}{RC} = \sqrt{2}\frac{U}{RC}\sin(\omega t + \gamma) \tag{8.27}$$

which gives the complete solution for u_C, consisting of two solutions: u_1, associated with the homogeneous equation, and u_2, associated with the right-hand side of the equation (see figure 8.6):

$$u(t) = u_1(t) + u_2(t) = Ae^{-\frac{t}{\tau_C}} + \sqrt{2}u_C \sin\left[\omega t + \gamma - \left(\frac{\pi}{2} - \varphi\right)\right] \tag{8.28}$$

The steady-state component $u_2(t)$ has a frequency equal to that of the main voltage, and the effective voltage value is deduced based on Ohm's law in the AC RC circuit.

$$u_c = X_c I = X_c \frac{U}{\sqrt{R^2 + \frac{1}{C^2\omega^2}}} \tag{8.29}$$

With regard to the phase shift between this voltage u_c and the applied voltage $u(t)$, it is $\frac{\pi}{2} - \varphi$. Indeed, according to figure 8.7:

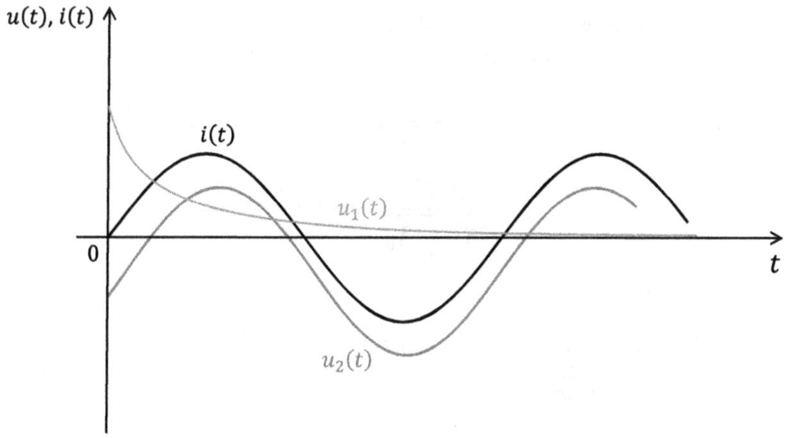

Figure 8.6. Transient voltage component $u_1(t)$ and steady-state component $u_2(t)$ in the transient regime in an AC RC circuit.

$$\varphi = \operatorname{arctg}\left(-\frac{X_c}{R}\right) = \operatorname{arctg}\left(-\frac{1}{RC\omega}\right) = -\operatorname{arctg}\frac{1}{RC\omega} \tag{8.30}$$

Thus,

$$u_2(t) = \sqrt{2}\frac{X_c U}{\sqrt{R^2 + \frac{1}{C^2\omega^2}}} \sin\left[\omega t + \gamma - \left(\frac{\pi}{2} - \varphi\right)\right] \tag{8.31}$$

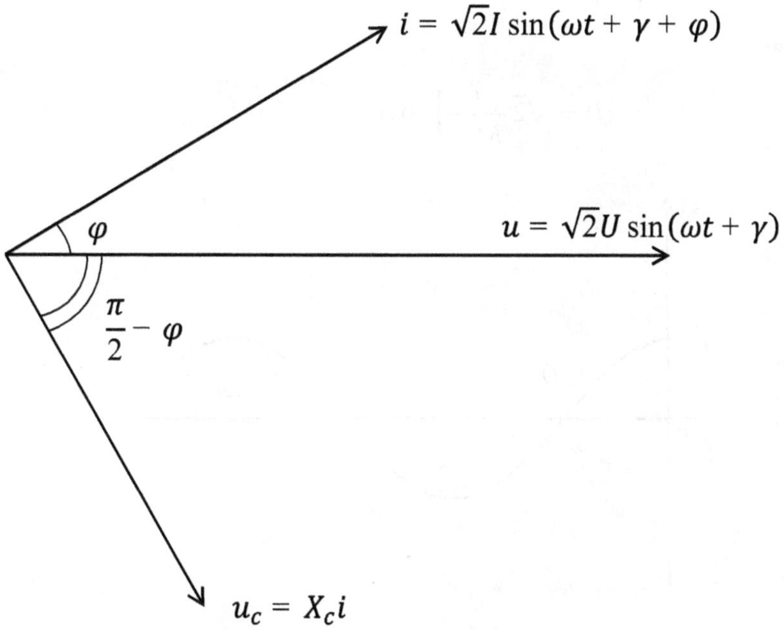

Figure 8.7. Phase shift between the applied voltage and the voltage across the capacitor terminals.

The constant A in the general solution for $u(t)$ given by equation (8.28) is determined by considering the initial conditions, i.e. those at $t = 0$, $u(t) = 0$:

$$\begin{cases} u(t) = Ae^{-\frac{t}{T_c}} + \sqrt{2}\dfrac{X_c U}{\sqrt{R^2 + X_c^2}} \sin\left(\omega t + \gamma + \varphi - \dfrac{\pi}{2}\right) \\ u(t) = Ae^{-\frac{t}{T_c}} - \sqrt{2}\dfrac{X_c U}{\sqrt{R^2 + X_c^2}} \cos(\omega t + \gamma + \varphi) \end{cases} \tag{8.32}$$

At $t = 0$, $u(t) = 0$, yielding:

$$0 = A - \sqrt{2}\frac{X_c U}{\sqrt{R^2 + \frac{1}{C^2\omega^2}}} \cos(\gamma + \varphi) \Longrightarrow A = \sqrt{2}\frac{X_c U}{\sqrt{R^2 + \frac{1}{C^2\omega^2}}} \cos(\gamma + \varphi)$$

and then:

$$u(t) = \sqrt{2}\frac{X_c U}{\sqrt{R^2 + \frac{1}{C^2\omega^2}}}\left[\cos(\gamma + \varphi)\;\; e^{-\frac{t}{T_c}} - \cos(\omega t + \gamma + \varphi)\right] \tag{8.33}$$

The current in the circuit is (see figure 8.8):

$$i(t) = \frac{dQ}{dt} = C\frac{du}{dt} =$$

$$= C\sqrt{2}X_c I\left[-\frac{1}{T_c}\cos(\gamma + \varphi)e^{-\frac{t}{T_c}} + \omega \sin(\omega t + \gamma + \varphi)\right] \tag{8.34}$$

$$= \sqrt{2}I\left[-\frac{1}{RC\omega}\cos(\gamma + \varphi)e^{-\frac{t}{T_c}} + \sin(\omega t + \gamma + \varphi)\right]$$

where, again:

$$i_1(t) = \sqrt{2}\frac{I}{RC\omega}\left[\cos(\gamma + \varphi)\;\; e^{-\frac{t}{T_c}}\right] \tag{8.35}$$

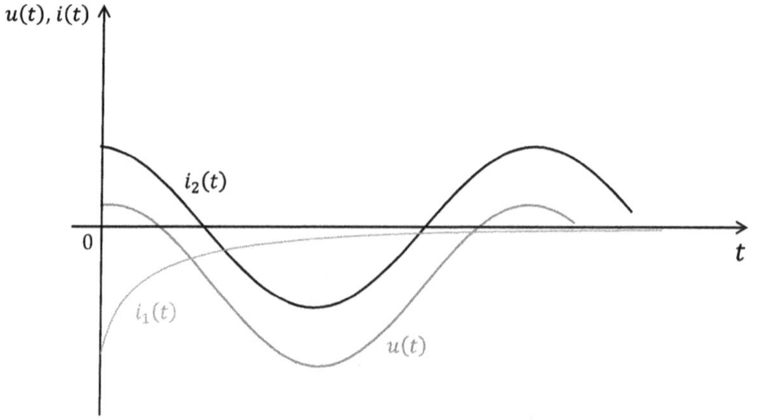

Figure 8.8. Transient component of the current $i_1(t)$ and the steady-state component $i_2(t)$for the transient regime in an AC RC circuit.

is the transient component and:

$$i_2(t) = \sqrt{2}I \sin(\omega t + \varphi + \gamma) \tag{8.36}$$

is the steady-state component. It is found that if the voltage is applied in such a way that $\gamma + \varphi = \pm\frac{\pi}{2}$, the transient components of voltage $u_1(t)$ and current $i_1(t)$ are zero, and the initial current i is equal to the steady-state current.

8.1.3 DC circuit containing a series-connected resistance and inductance (RL) in the transitory regime

In this section, we consider how the intensity of an electric current reaches a stationary value in a circuit containing a series-connected resistance R and a coil with an inductance L when the circuit is powered by a source at a constant electromotive voltage $\mathcal{E}$ (see figure 8.9). After we close the switch K (so that K is in position 1) at an initial time $t = 0$, the current intensity begins to increase from zero (its value before the power circuit is closed) to the maximum stationary value of $I_0 = \frac{\mathcal{E}}{R}$ determined by the total resistance R and the EMF $\mathcal{E}$ of the source. If the coil were removed from the circuit in figure 8.9, then the battery would give rise to a current of $I_0 = \frac{\mathcal{E}}{R}$ flowing through resistance R. In the presence of the coil, one should take into account the phenomenon of self-induction that manifests itself when the coil carries a time-varying current. In this case, when we close the circuit (by moving K to position 1), the current intensity varies, increasing over time, i.e. $\frac{dI}{dt} > 0$. Therefore, according to the law of self-induction, an EMF E_i is induced in the coil:

$$E_i = -L\frac{dI}{dt} \tag{8.37}$$

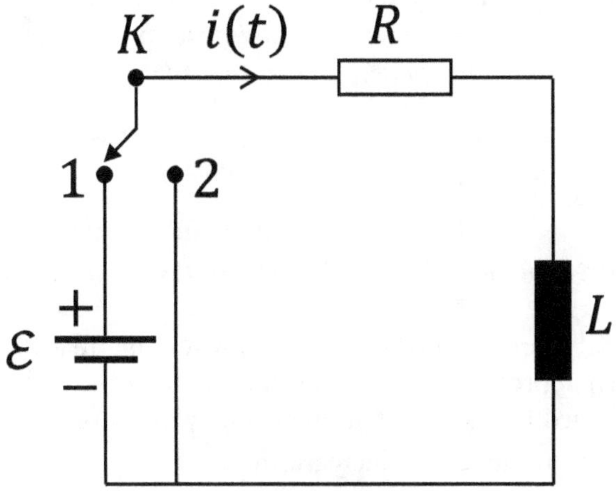

Figure 8.9. Circuit containing an inductance and a resistance connected in series, as used in the study of the transient regime.

In the presence of the coil, the circuit in figure 8.9 can be considered to contain two voltage sources with EMFs $\mathcal{E}$ and E_i connected in series and generating a current i passing through the resistor with resistance R. Applying Kirchhoff's second theorem to the circuit leads to:

$$\mathcal{E} + E_i = Ri \quad \text{or} \quad \mathcal{E} = Ri + L\frac{di}{dt} \tag{8.38}$$

This equation is integrated by the separation of variables, taking as limits $i_0 = 0$, i, the baseline $t = 0$, and time t. This leads to:

$$\int_0^i \frac{di}{\frac{\mathcal{E}}{R} - i} = \frac{R}{L}\int_0^t dt \Longrightarrow i = I_0\left(1 - e^{-\frac{t}{\tau_L}}\right) \tag{8.39}$$

where $I_0 = \frac{\mathcal{E}}{R}$. Here, $\tau_L = \frac{L}{R}$ represents the time constant of the circuit. The rate of increase in current is also a function of time, given by:

$$\frac{di}{dt} = \frac{I_0}{\tau_L}e^{-\frac{t}{\tau_L}} = \frac{\mathcal{E}}{L}e^{-\frac{t}{\tau_L}} \tag{8.40}$$

Thus, $\left.\frac{di}{dt}\right|_{t=0} = \frac{\mathcal{E}}{L}$. If the current were to increase continuously at the same rate as at the baseline (i.e. from the origin), then the current would depend on time according to a linear equation, which shows that it would reach a maximum after the time: $i = \frac{\mathcal{E}}{L}t$. If:

$$t = \tau_L \Longrightarrow i = \frac{\mathcal{E}}{L}\frac{L}{R} = \frac{\mathcal{E}}{R} = I_0 \tag{8.41}$$

Therefore, the time constant of the circuit τ_L *can be defined as the period of time after which the current intensity in the circuit would reach its maximum value* $\frac{\mathcal{E}}{R} = I_0$ *if it were to continuously increase at the same rate of increase as at the baseline (from the origin). On the other hand, after* $t = \tau_L$, *the current in the circuit reaches the value:*

$$i(\tau_L) = I_0(1 - e^{-1}) = 0.63I_0 \tag{8.42}$$

So, in other words, τ_L can also be defined differently, namely as *the time interval after which the current in the circuit reaches* 0.63 *of the maximum value* $I_0 = \frac{\mathcal{E}}{R}$ *in the steady-state*.

A plot (graphical representation) of $i(t)$ is shown in figure 8.10. The gradual (i.e. non-instantaneous) approach to the steady-state value of the current intensity in the RL circuit is explained by the fact that immediately after we close the circuit, when the current intensity in the circuit increases at the maximum rate ($\left.\frac{dI}{dt}\right|_{t=0} = \frac{\mathcal{E}}{L}$), an electromotive voltage is induced in the coil that has the maximum value of:

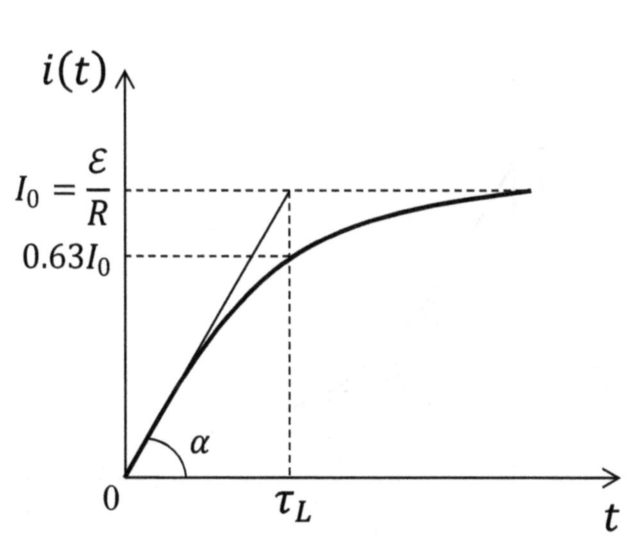

Figure 8.10. Dependence of current on time when an RL series circuit is closed.

$$E_{i\,\max} = -L\frac{dI}{dt}\bigg|_{t=0} = -L\frac{\mathcal{E}}{L} = -\mathcal{E} \tag{8.43}$$

This is equal to the external voltage but has the opposite sign, which completely cancels the effect of the source in the circuit. Only in the moments after closing the circuit, when the rate of variation of current intensity decreases according to the law (8.40), does the induced EMF value E_i become lower than that of the external source. At this point, an electric current with an intensity of $i(t) = \frac{\mathcal{E} - E_i}{R}$ starts to pass through the circuit. Finally, when the rate of current variation is zero, the induced EMF is zero, and all of the source voltage drop occurs at resistor R, causing the current intensity to reach a maximum of $I_0 = \frac{\mathcal{E}}{R}$.

Disconnecting the source from the circuit involves moving switch K to position 2. Kirchhoff's second theorem for this circuit leads to:

$$Ri + L\frac{di}{dt} = 0 \tag{8.44}$$

At $t = 0$, $i = I_0 = \frac{\mathcal{E}}{R}$, while at $t \to \infty$, $i = 0$; therefore:

$$\int_{i_0}^{i(t)} \frac{di}{i} = -\frac{R}{L}\int_0^t dt \Longrightarrow \ln\frac{i(t)}{I_0} = -\frac{R}{L}t \Longrightarrow i(t) = I_0 e^{-\frac{t}{\tau_L}} \tag{8.45}$$

A graphical representation of $i(t)$ is shown in figure 8.11. In this case, the time constant of the circuit has the following meaning: if the rate of decrease in current remains constant and equal to the initial rate (at $t = 0$, $\frac{di}{dt} = -\frac{\mathcal{E}}{R}\frac{R}{L}e^{-\frac{t}{\tau_L}} = -\frac{\mathcal{E}}{L}$), then the equation for the tangent to the curve is:

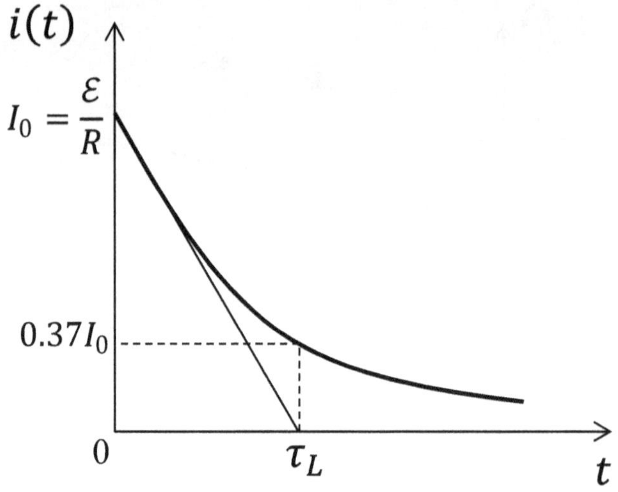

Figure 8.11. Time dependence of the current intensity in an RL series circuit in the transient regime when the circuit is opened.

$$i = I_0 - \frac{\mathcal{E}}{L}t = I_0\left(1 - \frac{t}{\tau_L}\right) \tag{8.46}$$

which shows that τ_L is the time after which the current in the circuit becomes zero if it decreases at the same rate as at the initial moment $\frac{\mathcal{E}}{L}$. After $t = \tau_L$, the current is $i(\tau_L) = \frac{I_0}{e} = 0.37I_0$; therefore, in other words, τ_L is the time after which the current reduces to 0.37 of its initial value I_0. Thus, the presence of the coil in the circuit has the effect that the current intensity does not instantly reach zero at the moment when the source is removed, but only after some time, during which all the energy stored in the coil dissipates as heat in the resistor.

8.1.4 Parallel LC circuit in the transient regime

Let us consider a circuit consisting of an ideal coil (without ohmic resistance) that has an inductance L and an ideal capacitor (that has zero conductance) with a capacitance of C. This is an ideal LC circuit whose behavior is desired in many practical applications (see figure 8.12). By moving the switch to position 1, we can charge the capacitor from the battery (whose EMF is $\mathcal{E}$) with the charge $Q_0 = \mathcal{E}C$; as a result, the energy stored in it is $W = \frac{1}{2}\frac{Q_0^2}{C}$. When we move the switch from position 1 to position 2, the capacitor, which is charged at $t = 0$, begins to discharge through the coil, thus passing an electric current through it (figure 8.12(a)). Since there is no energy loss through the Joule effect ($R = 0$), all energy with which the capacitor was charged is transferred to the coil. Therefore, in the second phase (figure 8.12(b)), when the capacitor is completely discharged, the current intensity passing through

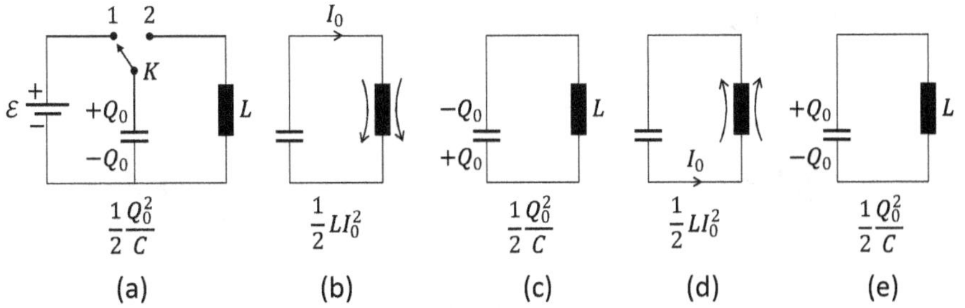

Figure 8.12. Reversible energy transfer between the electric field of a capacitor and the magnetic field of a coil.

the coil has the maximum value I_0, and therefore also the energy in the coil has the maximum value $\frac{1}{2}LI_0^2$. According to the law of conservation of energy, the following equality must apply:

$$\frac{1}{2}\frac{Q_0^2}{C} = \frac{1}{2}LI_0^2 \tag{8.47}$$

However, at any chosen point in time, the sum of the energy stored in the capacitor's electric field and that stored in the coil's magnetic field remains constant, having the value:

$$\frac{1}{2}\frac{Q^2}{C} + \frac{1}{2}Li^2 = \text{constant} = \frac{1}{2}\frac{Q_0^2}{C} \tag{8.48}$$

Due to the EMF induced in the coil, the capacitor is recharged in the third phase (figure 8.12(c)) but with reverse polarity. It is then discharged again through the coil in the fourth phase (figure 8.12(d)), recharging the capacitor again with the original polarity in the fifth phase (figure 8.12(e)). The process continues. In carrying out this process, the capacitor's electric charge Q and the current intensity in the circuit i periodically vary over time, i.e. they oscillate. These oscillations are accompanied by the mutual and periodic transformation of energy in the circuit from electrical to magnetic form. The oscillations described above are called electromagnetic oscillations. There is a close analogy between energy transformations in an LC circuit and those that occur in the case of mechanical oscillations of a mass m suspended by an elastic spring.

Mathematically, an analysis of the series LC circuit confirms the existence of oscillations of the electric charge and current passing through the circuit, as follows. At a given time t, after the capacitor (which initially had a charge Q_0) has been discharged through inductance L, Kirchhoff's second theorem for the circuit in figure 8.12 states that:

$$\frac{Q}{C} + L\frac{di}{dt} = 0 \tag{8.49}$$

Taking into account that $|i| = \frac{dQ}{dt}$, equation (8.49) can be rewritten in the form:

$$\frac{d^2Q}{dt^2} + \frac{Q}{LC} = 0 \tag{8.50}$$

which is a linear, homogeneous, second-order differential equation. If

$$\omega_0^2 = \frac{1}{LC} \text{ and } \omega_0 = \frac{1}{\sqrt{LC}} \tag{8.51}$$

denotes the *circuit's own frequency of oscillation* ω_0, then equation (8.50) becomes:

$$\frac{d^2Q}{dt^2} + \omega_0^2 Q = 0 \tag{8.52}$$

Its characteristic equation is $\lambda^2 + \omega_0^2 = 0$, which has the solutions:

$$\lambda = \pm i\omega_0 \tag{8.53}$$

According to the theory of linear and homogeneous differential equations with constant coefficients, the most general solution of the equation is:

$$Q(t) = C_1 e^{i\omega_0 t} + C_2 e^{-i\omega_0 t} \tag{8.54}$$

where C_1 and C_2 are the integration constants, generally complex numbers. The constants C_1 and C_2 are deduced directly from the initial conditions that Q and $\frac{dQ}{dt}$ must satisfy, so that, at $t = 0$:

$$\begin{cases} Q(0) = Q_0 & \Longrightarrow \quad Q_0 = C_1 + C_2 \\ \left.\dfrac{dQ}{dt}\right|_{t=0} & = 0 = i\omega_0 C_1 - i\omega_0 C_2 \end{cases} \Longrightarrow \begin{cases} 2C_1 = Q_0 \\ C_1 = C_2 \end{cases} \Longrightarrow C_1 = \frac{Q_0}{2}$$

Thus,

$$Q(t) = Q_0 \frac{1}{2}\left(e^{i\omega_0 t} + e^{-i\omega_0 t}\right) = Q_0 \cos \omega_0 t \tag{8.55}$$

Therefore, the electric charge contained in the capacitor has a periodic dependence on time, oscillating with a frequency and a period of:

$$\omega_0 = \frac{1}{\sqrt{LC}} \text{ and } T = 2\pi\sqrt{LC} \tag{8.56}$$

respectively. Since we know the electric charge contained in the capacitor at all times, we can calculate the voltage drops across it and the current flowing through the circuit as follows:

$$u_c(t) = \frac{Q}{C} = \frac{Q_0}{C} \cos \omega_0 t = \mathcal{E} \cos \omega_0 t \tag{8.57}$$

$$
\begin{aligned}
i(t) &= -\frac{dQ(t)}{dt} = \mathcal{E}\omega_0 \sin \omega_0 t = \mathcal{E}\frac{1}{\sqrt{LC}} \sin(\omega_0 t) \\
&= I_0 \cos\left(\frac{\pi}{2} - \omega_0 t\right) = I_0 \cos\left(\omega_0 t - \frac{\pi}{2}\right)
\end{aligned} \tag{8.58}
$$

where $I_0 = \frac{Q_0}{\sqrt{LC}}$ is the maximum current intensity and is realized at $\omega t = \frac{\pi}{2}$, that is, the exact time when the capacitor charge is zero. $Q(t)$ and $u_c(t)$ oscillate in phase, whereas $i(t)$ is delayed in respect to $Q(t)$ and $u_c(t)$ by $\frac{\pi}{2}$. The amplitudes of these quantities remain constant. The total energy stored in the electric field between the capacitor's plates and in the magnetic field in the coil is:

$$
\begin{aligned}
&\frac{1}{2}\frac{Q^2}{C} + \frac{1}{2}Li^2 = \frac{1}{2C}Q_0^2\cos^2\omega_0 t + \frac{1}{2}L\frac{Q_0^2}{LC}\sin^2\omega_0 t \\
&= \frac{1}{2}\frac{Q_0^2}{C} = \frac{1}{2}LI_0^2 = \text{constant}
\end{aligned}
$$

Thus, the law of conservation of energy is confirmed. Undamped electrical oscillations arise in this circuit.

8.1.5 RLC circuit in the transient regime

Let us consider a series RLC circuit coupled to a voltage source that has an EMF of $\mathcal{E}$, as shown in figure 8.13. When switch K is moved to position 1, capacitor C is charged with a charge of $Q_0 = \mathcal{E}C$. When switch K is moved to position 2, the capacitor discharges through resistor R and coil L. The electrical energy originally stored in the capacitor no longer remains constant over time but dissipates through the Joule effect into resistance R, which carries a current of intensity i generated by the variation in the capacitor's charge. As a result, the electrical oscillations that settle in the circuit are no longer undamped, as in the previous case, but are extinguished over time. To establish the differential equation describing the electrical oscillations in the series RLC circuit, Kirchhoff's second theorem will be written for the circuit in figure 8.13. At time t, at which the capacitor's upper plate is charged with the charge $+Q$, the voltage drop across the capacitor is $u_c = +\frac{Q}{C}$, and

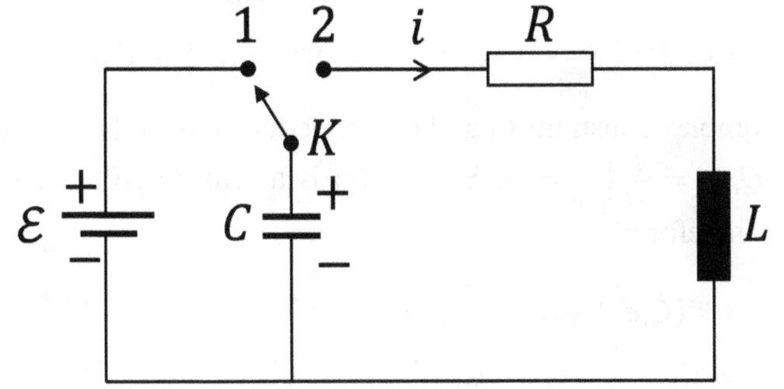

Figure 8.13. Transient RLC circuit.

the current intensity in the circuit has the sense shown in the figure. Kirchhoff's second theorem leads to the equation:

$$L\frac{di}{dt} + Ri + \frac{Q}{C} = 0 \tag{8.59}$$

Rearranging the terms in equation (8.59) yields:

$$\frac{d^2Q}{dt^2} + 2\frac{R}{2L}\frac{dQ}{dt} + \frac{Q}{LC} = 0 \tag{8.60}$$

a second-order differential equation that is linear and homogeneous and has constant coefficients. Using the notations: $\frac{R}{2L} = \alpha$ and $\omega_0^2 = \frac{1}{LC}$, equation (8.60) is rewritten as:

$$\frac{d^2Q}{dt^2} + 2\alpha\frac{dQ}{dt} + \omega_0^2 Q = 0 \tag{8.61}$$

The characteristic equation associated with this differential equation is:

$$\lambda^2 + 2\alpha\lambda + \omega_0^2 = 0 \tag{8.62}$$

and its roots are:

$$\lambda_{1,\,2} = -\alpha \pm \sqrt{\alpha^2 - \omega_0^2} = -\alpha \pm \beta \tag{8.63}$$

(1) At low resistance values, where the attenuation of the oscillations is relatively small, i.e. $\alpha^2 < \omega_0^2$, the expression below the square root is negative. In this case, we note that $\omega^2 = \omega_0^2 - \alpha^2$, which is a positive quantity, and therefore:

$$\omega = \sqrt{\omega_0^2 - \alpha^2} = \sqrt{\frac{1}{LC} - \frac{R^2}{4L^2}} \tag{8.64}$$

The roots of this characteristic equation take the shape:

$$\lambda_{1,\,2} = -\alpha \pm i\omega \tag{8.65}$$

According to the theory of linear and homogeneous differential equations with constant coefficients, the most general solution of equation (8.61) is:

$$Q(t) = C_1 e^{(-\alpha+i\omega)t} + C_2 e^{(-\alpha-i\omega)t} = e^{-\alpha t}\left(C_1 e^{i\omega t} + C_2 e^{-i\omega t}\right) \tag{8.66}$$

The generally complex constants C_1 and C_2 are deduced from the initial conditions at $t = 0$: $Q(0) = Q_0$, $i = \left.\frac{dQ}{dt}\right|_{t=0} = 0$. Since $Q(t)$ is a real magnitude, we must have: $Q(t) = Q^*(t)$. Therefore:

$$e^{-\alpha t}\left(C_1 e^{i\omega t} + C_2 e^{-i\omega t}\right) = e^{-\alpha t}\left(C_1^* e^{-i\omega t} + C_2^* e^{i\omega t}\right)$$
$$\Rightarrow C_1 = C_2^*,\ C_2 = C_1^*$$

If $C_1 = \frac{A}{2}e^{i\delta}$ and $C_2 = \frac{A}{2}e^{-i\delta}$, then:

$$Q(t) = Ae^{-\alpha t}\left[\frac{e^{(i\omega t+\delta)} + e^{-(\delta+i\omega t)}}{2}\right] = Ae^{-\alpha t}\cos(\omega t + \delta)$$

$$\begin{cases} Q(0) = Q_0 \Rightarrow A\cos\delta = Q_0 \\ \left.\dfrac{dQ}{dt}\right|_{t=0} = 0 \Rightarrow -A\alpha\cos\delta - A\omega\sin\delta = 0 \Rightarrow \tan\delta = -\dfrac{\alpha}{\omega} \end{cases}$$

However:

$$\cos\delta = \frac{1}{\sqrt{1+\tan^2\delta}} = \frac{1}{\sqrt{1+\frac{\alpha^2}{\omega^2}}} = \frac{\omega}{\sqrt{\omega^2+\alpha^2}} = \frac{\omega}{\omega_0}, \quad A = \frac{Q_0}{\cos\delta}$$

Finally, we obtain the following expression, which is the solution of the differential equation:

$$Q(t) = \frac{Q_0}{\cos\delta}e^{-\alpha t}\cos(\omega t + \delta) \tag{8.67}$$

This solution differs from that of the LC series circuit in the following ways:

(a) The amplitude of the electrical oscillations $Q_0e^{-\alpha t}$ does not remain constant over time but decreases exponentially. This decrease is more drastic for greater values of the damping constant $\alpha = \frac{R}{2L}$, i.e. at large R and small L. *The damping time constant is* $\tau = \frac{1}{\alpha} = \frac{2L}{R}$, *which represents the time after which the amplitude of oscillations decreases by a factor of e.*

(b) The frequency of the damped electrical oscillations ω is less than the frequency ω_0 of the harmonic oscillations, as shown by (8.64).

(c) The electrical energy originally stored in the circuit in the capacitor's electric field does not remain constant over time but decreases; over time, it is entirely converted to heat due to the Joule effect. Figure 8.14 shows the variation of the electric charge of the capacitor over time.

Knowing the time variation of the electric charge of the capacitor, we can obtain the time variation of the voltage $u_C(t)$ at the terminals of the capacitor and the intensity of the current in the circuit. The rate of energy dissipation in electrical circuits can be characterized, as in the case of damped mechanical oscillations, by the logarithmic decrement Λ of the damping. The logarithmic decrement is:

$$\Lambda = \ln\frac{Q(t)}{Q(t+T)} = \ln\frac{Q_0e^{-\alpha t}\cos(\omega t+\delta)}{Q_0e^{-\alpha(t+T)}\cos[\omega(t+T)+\delta]} = \alpha T$$

Thus,

$$\Lambda = \frac{R}{2L}\frac{2\pi}{\omega} = \frac{\pi}{Q}$$

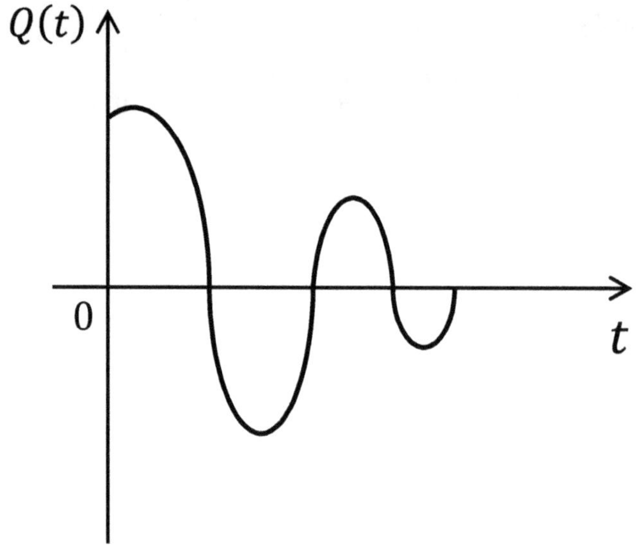

Figure 8.14. Time dependence of the electric charge of the capacitor.

where $\mathcal{Q}$ is the circuit quality factor:

$$\mathcal{Q} = \frac{L\omega}{R} \tag{8.68}$$

It can be shown that the quality factor of a circuit is numerically equal to the ratio of the energy present in the circuit at time t to the electrical energy dissipated in the circuit in the subsequent period, multiplied by a factor of 2π. Indeed, if we choose the moment t such that the charge on the capacitor has its peak value of $Q(t) = Q_0 e^{-\alpha t}$, then the current intensity in the circuit is zero at that time, and all the circuit's energy, i.e. $\frac{1}{2}\frac{Q^2}{C} = \frac{1}{2}\frac{Q_0^2}{C}e^{-2\alpha t}$, is stored in the capacitor. The energy dissipated in the circuit in the following period is:

$$\Delta W(T) = W(t) - W(t+T) = \frac{1}{2}\frac{Q_0^2}{C}e^{-2\alpha t}[1 - e^{-2\alpha T}]$$

Therefore, the required result $\frac{W(t)}{\Delta W(T)}$ is given by:

$$\frac{W(t)}{\Delta W(T)} = (1 - e^{-2\alpha T})^{-1} = [1 - (1 - 2\alpha T)]^{-1} = \frac{1}{2\alpha T} = \frac{1}{2\Lambda} = \frac{\mathcal{Q}}{2\pi}$$

which leads to:

$$\mathcal{Q} = 2\pi \frac{W(t)}{\Delta W(T)} \tag{8.69}$$

(2) If the resistance has higher values, such that $\frac{R^2}{4L^2} = \frac{1}{LC}$, then $\omega = 0$. In this case, $Q(t) = \frac{Q_0}{\cos\delta}e^{-\alpha t}\cos\delta = Q_0 e^{-\alpha t}$, which represents a transition from the periodic

regime to an aperiodic regime. The resistance at which this transition occurs is called the critical resistance, which has the value $R_{cr} = 2\sqrt{\frac{L}{C}}$.

(3) In the case where R is very large (strong damping) such that $\alpha^2 > \omega_0^2 \Rightarrow \alpha^2 - \omega_0^2 > 0$, the roots of the characteristic equation are real:

$$\lambda_{1,2} = -\alpha \pm \beta,\ \beta = \sqrt{\alpha^2 - \omega_0^2}$$

In this case, the solution of the differential equation is:

$$Q(t) = e^{-\alpha t}(C_1 e^{\beta t} + C_2 e^{-\beta t}) \tag{8.70}$$

which contains the real constants C_1 and C_2. The term $e^{-\alpha t}$ shows that there is an exponential decrease in the electric charge over time. If $\alpha^2 \gg \omega_0^2$, so that we can consider $\alpha \cong \beta$, the result is:

$$Q(t) = C_1 e^{-(\alpha-\beta)t} + C_2 e^{-(\alpha+\beta)t} \cong C_1 + C_2 e^{-2\alpha t} \tag{8.71}$$

From the boundary conditions of the problem ($Q(0) = Q_0$ and $Q(\infty) = 0$), it follows that $C_1 = 0$ and $C_2 = Q_0$, such that:

$$Q(t) = Q_0 e^{-2\alpha t} \tag{8.72}$$

A graph of this function is illustrated in figure 8.15. Such a variation in charge is said to be superdamped.

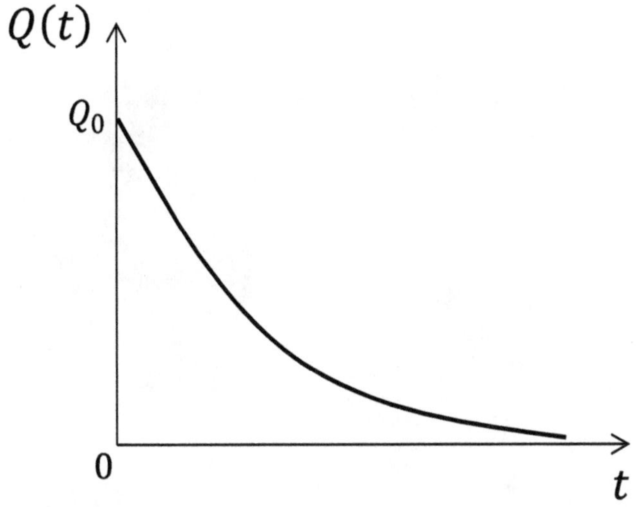

Figure 8.15. Time dependence of the electric charge of a capacitor in the superdamped regime.

Further reading

[1] Antohe Ş and Antohe V-A 2023 *Electrostatics. Formalism of the Electrostatic Field in Vacuum and Matter* 1st edn (Bristol: IOP Publishing) 266 pages

[2] Purcell E M and Morin D J 2013 *Electricity And Magnetism* 3rd edn (Cambridge: Cambridge University Press) 853 pages
[3] Friesen D J 2023 *Electromagnetics for Practicing Engineers: Electrostatics and Magnetostatics* **vol 1** (Norwood, MA: Artech House) 234 pages
[4] Panofsky W K H and Phillips M 2012 *Classical Electricity and Magnetism* 2nd edn (North Chelmsford, MA: Courier Corporation) 512 pages
[5] Jones E T 2007 *Induction Coil: Theory and Applications* (Newberg, OR: Barclay Press) 276 pages
[6] Feynman R P, Leighton R B and Sands M 2011 *Feynman Lectures on Physics, Vol. II: The New Millennium Edition: Mainly Electromagnetism and Matter (Feynman Lectures on Physics)* (New York: Basic Books) 589 pages
[7] Secăreanu I, Ruxandra V, Gherbanovschi N, Logofătu M, Cazan-Corbasca M and Antohe Ş 1984 *Problems of Electricity and Magnetism (Culegere de probleme de Electricitate Şi Magnetism)* (Bucharest: University of Bucharest Publishing House)
[8] Dorf R C and Svoboda J A 2018 *Dorf's Introduction to Electric Circuits* 9th edn (Hoboken, NJ: Wiley Global) 912 pages
[9] Alexander C K and Sadiku M 2020 *ISE Fundamentals of Electric Circuits (ISE HED IRWIN Elec&Computer Enginering)* 7th edn (New York: McGraw-Hill Education)
[10] Zeng G L and Zeng M 2021 *Electric Circuits: A Concise, Conceptual Tutorial* 1st edn (Berlin: Springer) 315 pages

IOP Publishing

Electromagnetism and Special Methods for Electric Circuits Analysis

Ştefan Antohe and Vlad-Andrei Antohe

Chapter 9

Electrical circuits in the alternating current regime

This chapter describes the behavior of electrical circuits in the presence of periodic time-dependent voltages and currents, referred to as the alternating current (AC) regime. We first introduce the physical quantities characterizing this regime, such as the instantaneous values of the current and voltage, their amplitude, and their effective values. We then present the main laws characterizing the electric conduction process in AC circuits using different methods of representation for the time-variable electrical quantities, such as the analytical method, the vector method, and the method of representation through complex numbers. The special methods for the analysis of DC networks introduced in chapter 3 are presented here using the same approach; however, the difference is that all the equations and systems of equations are written in complex forms.

9.1 Physical quantities describing alternating current

9.1.1 Production of single-phase sinusoidal electromotive voltage

Let us consider a magnetic field that has a magnetic induction $\vec{B}$, in which a rectangular circuit rotates. Two conductive rings are fixed at the ends of the circuit, permanently in contact with two brushes providing an electrical connection between the rotating circuit and an external circuit containing a load resistor R (see figure 9.1). When the plane of the rectangular conducting wire (circuit) is perpendicular to the magnetic field lines, its surface is crossed by the maximum flux Φ_m. At a given moment, when a vector normal to the circuit's surface makes an angle α with the magnetic field induction vector $\vec{B}$, the flux passing through its surface is:

$$\Phi = \vec{B} \cdot \vec{S} = BS\cos\alpha = \Phi_m \cos\alpha \tag{9.1}$$

doi:10.1088/978-0-7503-5854-5ch9

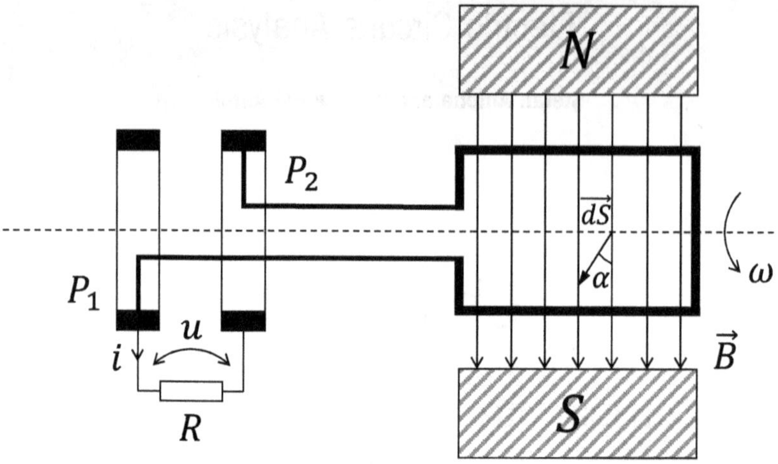

Figure 9.1. Schematic representation of an AC production device.

Considering that the speed of rotation of the circuit is constant over time and is given by:

$$\omega = \frac{\alpha}{t}, \quad \alpha = \omega t \tag{9.2}$$

a voltage is generated between the rings; its value is given by the expression:

$$u = -\frac{d\Phi}{dt} \tag{9.3}$$

In the case of a rectangular frame containing N conducting turns, the induced voltage is:

$$u = -N\frac{d\Phi}{dt} \tag{9.4}$$

Taking into account the expression for the magnetic flux passing through the coil frame, we obtain:

$$u = +N\omega\, \Phi_m \sin \omega t = U_m \sin \omega t \tag{9.5}$$

This instantaneous voltage value is usually expressed as a function of the effective voltage U_{eff} which will be defined below, as follows:

$$u = \sqrt{2}\, U_{\text{eff}} \sin \omega t \tag{9.6}$$

Equations (9.5) and (9.6) yield:

$$\sqrt{2}\, U_{\text{eff}} = N\omega\, \Phi_m \tag{9.7}$$

$$U_{\text{eff}} = \frac{N2\pi\nu\Phi_m}{\sqrt{2}} = 4\frac{\pi}{2\sqrt{2}}N\nu\,\Phi_m = 4.44N\nu\,\Phi_m = 4kN\nu\,\Phi_m \tag{9.8}$$

where k is the so-called form factor, which has a value of 1.11. In reality, in the case of AC generators installed in power plants, the N–S poles (inductor coil) that create the magnetic field constitute the rotor, while the coil with N turns in which the induced current is generated (induction coil) is found on the stator.

9.1.2 Instantaneous, average, and effective values of alternating current intensity and voltage

(1) **Instantaneous values**

Alternating electrical quantities are time-dependent functions; therefore, they are expressed as instantaneous values by periodic functions of the type $u = f(\omega t \pm k\omega T)$, in which T is the period, $\nu = \frac{1}{T}$ is the frequency measured in s^{-1}, and $\omega = \frac{2\pi}{T}$ is the angular frequency (or rotational speed) measured in rad s^{-1}. In the case of sinusoidal AC, the current intensity is represented analytically by the relation:

$$i = I_m \sin(\omega t \pm \varphi) \tag{9.9}$$

where $i(t)$ is the instantaneous value of the current intensity at moment t, I_m is the maximum value, ω is the angular frequency, φ is the initial phase, and $\omega t \pm \varphi$ is the total phase. Correspondingly, the instantaneous value of the alternating sinusoidal voltage is represented by:

$$u = U_m \sin(\omega t \pm \varphi) \tag{9.10}$$

(2) **Average values**

In the case of time-varying physical quantities, it is often necessary to operate with their average values, defined, in the case of AC, based on the electrochemical effect. When current passes through an electrolyte, a charge q is transported within a well-defined time interval. In the case of AC, the charge transported during a period T is:

$$q = \int_0^T i dt \tag{9.11}$$

where i is the instantaneous current intensity. A direct (constant) current would provide transport of the same charge q in the time interval T if the current intensity were:

$$I_{\text{med}} = \frac{q}{T} = \frac{1}{T}\int_0^T i dt \tag{9.12}$$

Relation (9.12) gives an expression for the average AC intensity. When the current intensity varies over time according to a sinusoidal law, then:

$$I_{med} = 0 \tag{9.13}$$

Therefore, in the case of sinusoidal AC, the *average value over a half period* is used, which is:

$$I_{med\frac{1}{2}} = \frac{2}{T}\int_0^{\frac{T}{2}} I_m \sin \omega t dt = -\frac{2I_m \cos \omega t \Big|_0^{\frac{T}{2}}}{T\omega}$$

$$= -\frac{2I_m}{T\omega}(\cos \pi - \cos 0) = \frac{2I_m}{\pi} = 0.636I_m \tag{9.14}$$

By analogy with the reasoning used to define the average current value, the mean voltage value is given by:

$$U_{med} = \frac{1}{T}\int_0^T udt \tag{9.15}$$

Correspondingly, in the case of sinusoidal AC:

$$U_{med} = 0; \ U_{med\frac{1}{2}} = \frac{2U_m}{\pi} = 0.636U_m \tag{9.16}$$

The importance of the average current and voltage values lies in the fact that these values are measured using magnetoelectric-type electrical devices with rectifiers.

(3) **Effective values**

In the case of AC, the instantaneous values of electrical quantities cannot be indicated by mechanical measuring instruments due to their inertia. Such instruments only indicate the so-called *effective values* of the electrical quantities in the circuit. These values are defined on the basis of the electrocaloric effect of alternating electric current. When a voltage u is applied to a circuit, a current of intensity i is established in the circuit, and the instantaneously absorbed power is:

$$P_i = ui \tag{9.17}$$

where u and i are the instantaneous values of voltage and current intensity, respectively. Over a period of time, the AC circuit absorbs the energy:

$$W = \int_0^T uidt \tag{9.18}$$

The ratio W/T gives the average power absorbed over a period:

$$P = \frac{W}{T} = \frac{1}{T}\int_0^T uidt \tag{9.19}$$

If the circuit consists of a resistance R, then $u = Ri$, and the average power is given by the expression:

$$P = \frac{1}{T}\int_0^T Ri^2dt \tag{9.20}$$

This power is dissipated as heat through the Joule effect. The same thermal effect could be produced by a constant current whose intensity I_{eff} satisfies the relationship:

$$P = RI_{\text{eff}}^2 \tag{9.21}$$

From relations (9.20) and (9.21), we can obtain the effective value of the AC intensity:

$$I_{\text{eff}} = \sqrt{\frac{1}{T}\int_0^T i^2dt} \tag{9.22}$$

This relationship defines the effective AC intensity value as the intensity of direct current (DC) which, if it were to flow through the same resistor R, would give off the same heat per unit of time as the AC. Analogously, the effective voltage value has the expression:

$$U_{\text{eff}} = \sqrt{\frac{1}{T}\int_0^T u^2dt} \tag{9.23}$$

9.1.3 Effective values of sinusoidal alternating current intensity and voltage

The intensity of sinusoidal AC depends on time according to the relationship:

$$i = I_m \cos(\omega t - \varphi) \tag{9.24}$$

where I_m is the maximum value of intensity, ω is angular frequency, and φ is its initial phase. Introducing (9.24) into the relation that defines the effective current value (9.22), one obtains:

$$I_{\text{eff}} = \sqrt{\frac{1}{T}\int_0^T I_m^2\cos^2(\omega t - \varphi)dt} = \frac{I_m}{\sqrt{2}} = 0.707I_m \tag{9.25}$$

Analogously, for the effective voltage value, one obtains:

$$U_{\text{eff}} = \frac{U_m}{\sqrt{2}} = 0.707U_m \tag{9.26}$$

Using the above expressions, the instantaneous values of intensity and voltage respectively can be written as:

$$\begin{cases} i = I_{\text{eff}}\sqrt{2}\cos(\omega t - \varphi) \\ u = U_{\text{eff}}\sqrt{2}\cos(\omega t - \varphi) \end{cases} \tag{9.27}$$

Note that the ratio of the effective value to the average value over a half period of sinusoidal AC is precisely the form factor k:

$$\frac{I_{\text{eff}}}{I_{\text{med}\frac{1}{2}}} = \frac{I_m}{\frac{2}{\pi}\sqrt{2}I_m} = \frac{\pi}{2\sqrt{2}} = 1.11 = k \tag{9.28}$$

9.2 Analysis of AC electrical circuits

9.2.1 Behavior of the elements R, L, and C in alternating current

(A) **Circuit consisting of a resistance (*R*) carrying AC**

At the terminals of the circuit formed by resistance R (see figure 9.2), a voltage $u = U_m \cos \omega t = U_{\text{eff}}\sqrt{2}\cos \omega t$ is applied. The current intensity passing through the resistance follows Ohm's law:

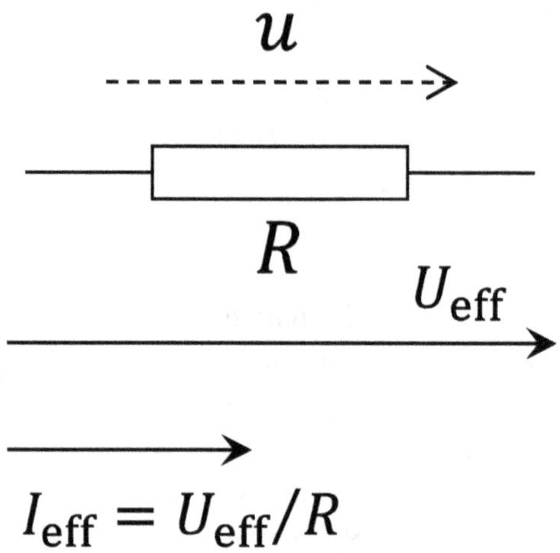

Figure 9.2. Behavior of a resistance R carrying AC.

$$i = \frac{u}{R} = \frac{U_m \cos \omega t}{R} = \frac{U_{\text{eff}}\sqrt{2}}{R}\cos \omega t \tag{9.29}$$

Note that the intensity depends on the time period, and it can be rewritten using the same trigonometric function ($\cos \omega t$) as that used for voltage:

$$i = I_m \cos \omega t = I_{\text{eff}} \sqrt{2} \cos \omega t \tag{9.30}$$

In the case of resistance, the current intensity is in phase with the voltage. In the phase diagram, the vectors of effective intensity and effective voltage are shown as having the same direction and sense (see figure 9.2).

(B) **Circuit consisting of an inductance (L) carrying AC**

At the terminals of an inductance circuit L (see figure 9.3), an alternating voltage $u = U_m \cos \omega t = U_{\text{eff}} \sqrt{2} \cos \omega t$ is applied. The current intensity passing through the inductance can be calculated as follows:

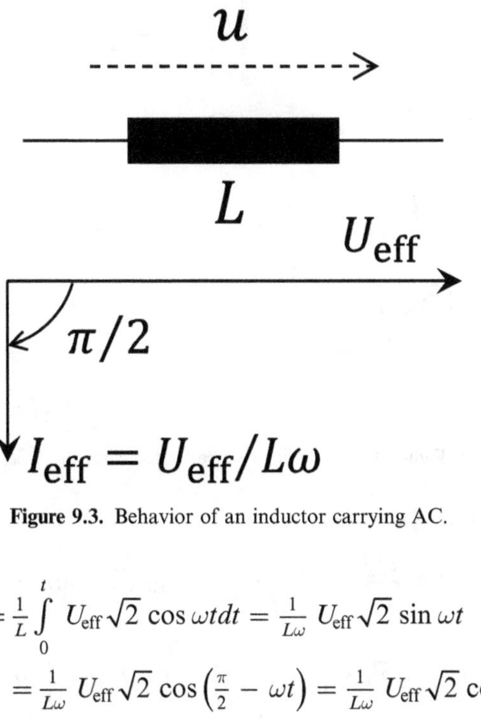

Figure 9.3. Behavior of an inductor carrying AC.

$$i = \frac{1}{L} \int_0^t u dt = \frac{1}{L} \int_0^t U_{\text{eff}} \sqrt{2} \cos \omega t dt = \frac{1}{L\omega} U_{\text{eff}} \sqrt{2} \sin \omega t$$
$$= \frac{1}{L\omega} U_{\text{eff}} \sqrt{2} \cos\left(\frac{\pi}{2} - \omega t\right) = \frac{1}{L\omega} U_{\text{eff}} \sqrt{2} \cos\left(\omega t - \frac{\pi}{2}\right)$$

which results in:

$$i = I_{\text{eff}} \sqrt{2} \cos\left(\omega t - \frac{\pi}{2}\right) \tag{9.31}$$

where $I_{\text{eff}} = \frac{U_{\text{eff}}}{L\omega}$. *Thus, in the case of inductance, the current intensity is phase shifted by* $\frac{\pi}{2}$ *behind the voltage* (see figure 9.3). In the phase diagram, the phasors of the effective intensity of the current flowing through the inductance and the voltage across it form an angle of $\frac{\pi}{2}$ between them. The behavior of an inductor carrying AC is like that of an apparent resistance, called *inductive reactance*, having the value:

$$X_L = L\omega \tag{9.32}$$

(C) **Circuit consisting of a capacitance (*C*) carrying AC**

An alternating voltage $u = U_{\text{eff}}\sqrt{2}\cos\omega t$ is applied to the terminals of a circuit consisting of a capacitance C (see figure 9.4). The current intensity in the circuit is calculated as follows:

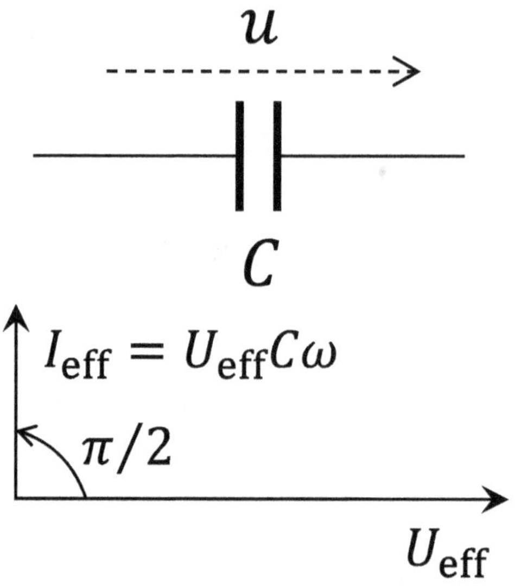

Figure 9.4. Behavior of a capacitor carrying AC.

$$i = \frac{dQ}{dt} = C\frac{du}{dt} = -CU_{\text{eff}}\sqrt{2}\omega\sin\omega t$$

$$= C\omega\, U_{\text{eff}}\sqrt{2}\cos\left(\frac{\pi}{2} + \omega t\right) = I_{\text{eff}}\sqrt{2}\cos\left(\omega t + \frac{\pi}{2}\right)$$

Thus, the final current value is:

$$i = I_{\text{eff}}\sqrt{2}\cos\left(\omega t + \frac{\pi}{2}\right) \tag{9.33}$$

where $I_{\text{eff}} = \frac{U_{\text{eff}}}{X_C} = C\omega\, U_{\text{eff}}$ is *the effective current intensity passing through the capacitance, which is phase shifted by* $\frac{\pi}{2}$ *in front of the voltage* (see figure 9.4). The physical quantity given by the relationship:

$$X_c = \frac{1}{C\omega} \tag{9.34}$$

is called *capacitive reactance*; it plays the role of an apparent resistance introduced into an AC circuit by capacitance.

9.2.2 Circuit consisting of a resistance, an inductance, and a capacitance connected in series (i.e. series RLC) carrying alternating current

At the ends of an RLC series circuit (see figure 9.5), a voltage $u = U_{eff}\sqrt{2}\cos\omega t$ is applied. The voltage is distributed across the three elements of the circuit. The

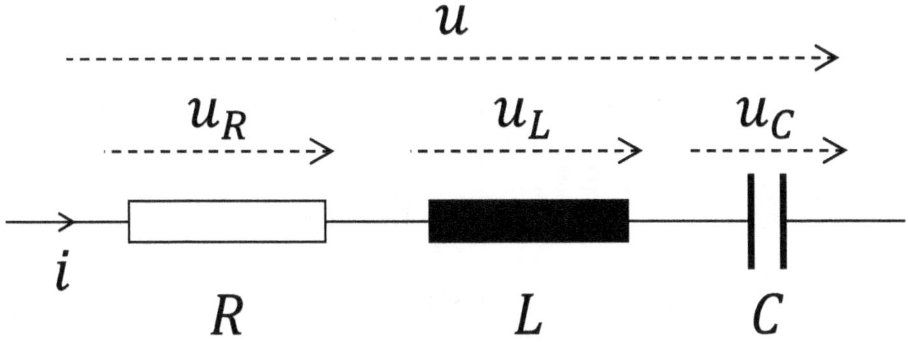

Figure 9.5. Series RLC circuit carrying AC.

voltages present across the circuit elements satisfy the relation (established on the basis of Kirchhoff's second theorem):

$$u_R + u_L + u_C = u \tag{9.35}$$

where u_R, u_L, and u_C are the instantaneous values of the voltages across the resistance, inductance, and capacitance, respectively, obtained using the expressions previously established:

$$u_R = Ri, \quad u_L = L\frac{di}{dt}, \quad u_C = \frac{1}{C}\int idt$$

Using these expressions, relation (9.35) takes the form of a differential–integral equation:

$$Ri + L\frac{di}{dt} + \frac{1}{C}\int idt = U_{eff}\sqrt{2}\cos\omega t \tag{9.36}$$

We therefore seek a solution of the type $i = I_{eff}\sqrt{2}\cos(\omega t - \varphi)$ that satisfies equation (9.36), thus leading to:

$$RI_{eff}\cos(\omega t - \varphi) - L\omega\, I_{eff}\sin(\omega t - \varphi) + \frac{1}{C\omega}I_{eff}\sin(\omega t - \varphi) = U_{eff}\cos\omega t \tag{9.37}$$

By expending $\cos(\omega t - \varphi)$ and $\sin(\omega t - \varphi)$ and equating the terms in $\sin\omega t$ and $\cos\omega t$, from the two members of the above equation, one obtains:

$$RI_{\text{eff}}(\cos\omega t\cos\varphi + \sin\omega t\sin\varphi) + \left(\frac{1}{C\omega} - L\omega\right)I_{\text{eff}}\times$$

$$(\sin\omega t\cos\varphi - \cos\omega t\sin\varphi) = U_{\text{eff}}\cos\omega t$$

resulting in:

$$\begin{cases} RI_{\text{eff}}\cos\varphi + \left(L\omega - \dfrac{1}{C\omega}\right)I_{\text{eff}}\sin\varphi = U_{\text{eff}} \\ RI_{\text{eff}}\sin\varphi - \left(L\omega - \dfrac{1}{C\omega}\right)I_{\text{eff}}\cos\varphi = 0 \end{cases} \tag{9.38}$$

These equations are simultaneously valid at any moment in time. Taking them to the power of two and summing them gives:

$$I_{\text{eff}}^2\left[R^2 + \left(L\omega - \frac{1}{C\omega}\right)^2\right] = U_{\text{eff}}^2 \tag{9.39}$$

and from the second equation, we obtain the tangent of the phase shift angle φ that occurs between current intensity and voltage:

$$\tan\varphi = \frac{L\omega - \frac{1}{C\omega}}{R} \tag{9.40}$$

Relation (9.39) expresses Ohm's law in an AC–RLC circuit that is series connected and carrying AC.

$$I_{\text{eff}} = \frac{U_{\text{eff}}}{\sqrt{R^2 + \left(L\omega - \frac{1}{C\omega}\right)^2}} \tag{9.41}$$

The denominator, which plays the role of an equivalent resistance and depends on the frequency of the source voltage u, is called the *impedance of the circuit* and is denoted by Z, having the expression:

$$Z = \sqrt{R^2 + \left(L\omega - \frac{1}{C\omega}\right)^2} = \sqrt{R^2 + X^2} \tag{9.42}$$

The magnitude $X = L\omega - \frac{1}{C\omega}$ is called *the total reactance of the circuit.* Using these expressions, the form of Ohm's law applicable in an inhomogeneous closed circuit with an alternating voltage source is written as:

$$I_{\text{eff}} = \frac{U_{\text{eff}}}{Z} \text{ or } U_{\text{eff}} = ZI_{\text{eff}} \tag{9.43}$$

The phase diagram, taking I_{eff} as the reference phasor (since it is the constant parameter in a series circuit), looks like the one depicted in figure 9.6. U_{eff_R} is in phase with I_{eff}, U_{eff_L} is $\frac{\pi}{2}$ in advance of I_{eff}, and U_{eff_C} is $\frac{\pi}{2}$ behind I_{eff}. U_{eff} is obtained from the

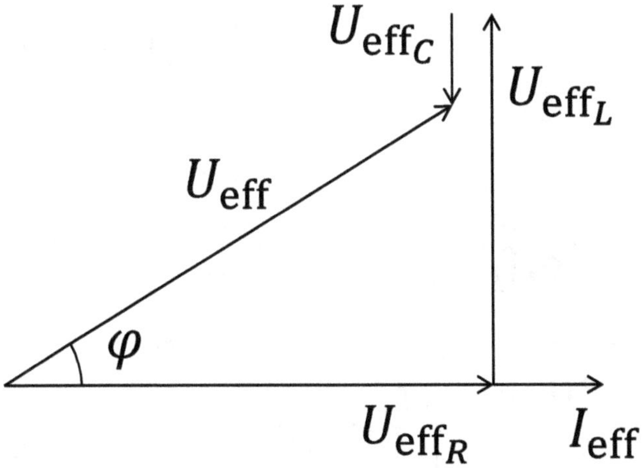

Figure 9.6. Phase diagram of a series RLC circuit carrying AC.

vector sum of U_{eff_R}, U_{eff_L}, and U_{eff_C}. It also results in a phase shift of φ between U_{eff} and I_{eff}, which is actually the phase shift between the current intensity and the voltage.

9.2.3 Circuit consisting of more than two reactances

Consider an electrical circuit consisting of two ideal inductances L_1 and L_2. It is inductively coupled through the mutual inductance M. Two resistors, R_1 and R_2, and a capacitance C are also connected in series (see figure 9.7). The circuit is supplied

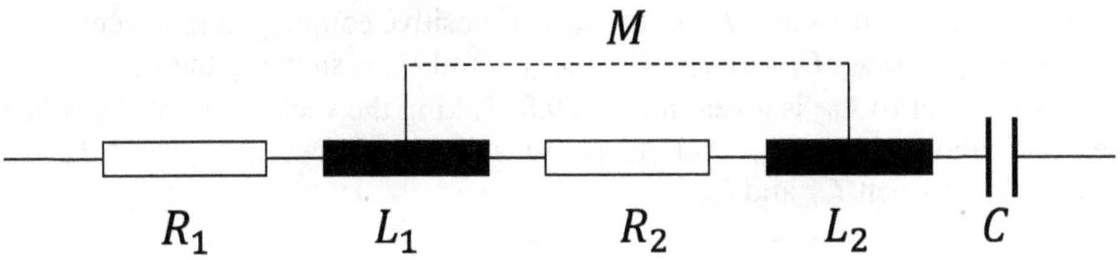

Figure 9.7. Circuit with more than two AC reactances.

with an alternating voltage $u = U_{\mathrm{eff}}\sqrt{2}\cos\omega t$. It is expected that the current intensity will be phase shifted with respect to the voltage and will have the form $i = I_{\mathrm{eff}}\sqrt{2}\cos(\omega t - \varphi)$. Both the phase ($\varphi$) and the relationship between voltage and intensity can be established based on calculations made in the previous case (a series RLC circuit carrying AC). However, this approach can be avoided by using a phase diagram. For all vectors represented in this diagram (see figure 9.8), the vector I_{eff} will again be taken as a reference. The effective voltage across the mutual inductance is represented by $U_{\mathrm{eff}_M} = M\omega\, I_{\mathrm{eff}}$ if it flows in the same direction as in the other circuit elements, or by $U_{\mathrm{eff}_{L_1}} = L_1\omega I_{\mathrm{eff}}$ if it flows in the opposite direction,

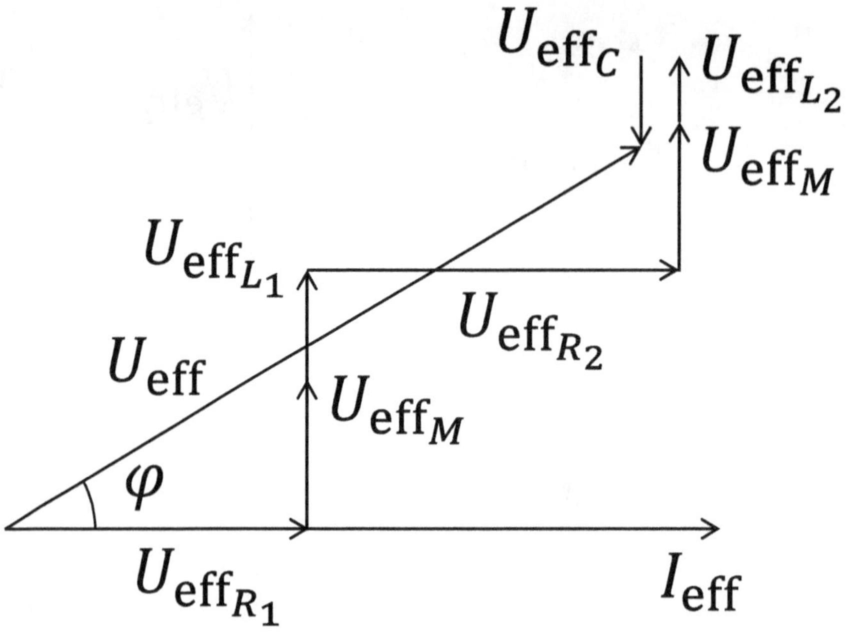

Figure 9.8. Phase diagram of voltages in an AC circuit containing more than two AC reactances.

which depends on the type of coupling between L_1 and L_2 (i.e. whether it is positive or negative). The same U_{eff_M} is again represented by $U_{\text{eff}_{L_2}} = L_2\omega I_{\text{eff}}$ if it flows in the same direction as the latter, or in the opposite direction, based on the sign of the coupling between L_2 and L_1. In the case of positive coupling, a representation of the vectors $U_{\text{eff}_{R_1}}$, $U_{\text{eff}_{L_1}}$, U_{eff_M}, $U_{\text{eff}_{R_2}}$, $U_{\text{eff}_{L_2}}$, U_{eff_M}, and U_{eff_C}, showing the phase shift of each with respect to I_{eff}, is given in figure 9.8. Taking the vector sum of the voltages gives the total voltage U_{eff} and its phase shift φ_{eff} between U_{eff} and I_{eff}. The relationship between I_{eff} and U_{eff} is:

$$U_{\text{eff}} = I_{\text{eff}}\sqrt{(R_1 + R_2)^2 + \left(L_1\omega + L_2\omega \pm 2M\omega - \frac{1}{C\omega}\right)} \tag{9.44}$$

and:

$$\varphi = \text{atan}\frac{L_1\omega + L_2\omega \pm 2M\omega - \frac{1}{C\omega}}{R_1 + R_2} \tag{9.45}$$

In the representations shown in figure 9.8, the knowledge gained in the previous paragraphs regarding the phase shifts of voltages across resistors, inductances, and capacitances relative to the current intensity in a series circuit is taken into account. The qualitative results obtained from phase diagrams can sometimes be sufficient to analyze any series RLC circuit.

9.2.4 Circuit consisting of a resistance, an inductance, and a capacitance connected in parallel (i.e. parallel RLC) carrying alternating current

Let us consider a circuit consisting of a resistance R, an inductance L, and a capacitance C connected in parallel (see figure 9.9(a)). A voltage $u = U_{eff}\sqrt{2}\cos\omega t$

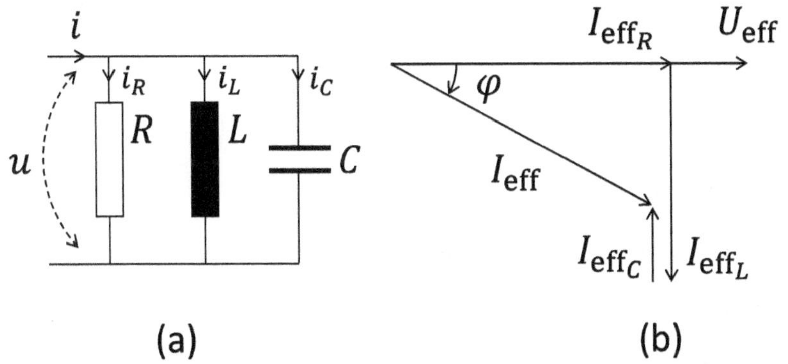

Figure 9.9. AC parallel RLC circuit (a) and its corresponding phase diagram (b).

is applied to its terminals. Kirchhoff's first theorem applied to this circuit leads to:

$$i = i_R + i_L + i_C \tag{9.46}$$

where i_R, i_L, and i_C are the intensities of currents passing through the resistance, inductance, and capacitance, respectively. Using Ohm's law, the effective values of these currents are:

$$i_R = \frac{u}{R} = Gu \tag{9.47}$$

$$i_L = \frac{1}{L}\int u dt \tag{9.48}$$

$$i_C = \frac{dQ}{dt} = C\frac{du}{dt} \tag{9.49}$$

With these expressions, equation (9.46) becomes:

$$Gu + \frac{1}{L}\int u dt + C\frac{du}{dt} = i \tag{9.50}$$

Considering that i and u are of the form $i = I_{eff}\sqrt{2}\cos(\omega t - \varphi)$ and $u = U_{eff}\sqrt{2}\cos\omega t$, and using them to verify equation (9.50), it follows that:

$$\begin{aligned} &GU_{eff}\sqrt{2}\cos\omega t \quad + \frac{1}{L\omega}U_{eff}\sqrt{2}\sin\omega t - C\omega\, U_{eff}\sqrt{2}\sin\omega t \\ &= I_{eff}\sqrt{2}(\cos\omega t\cos\varphi + \sin\omega t\sin\varphi) \end{aligned} \tag{9.51}$$

Hence:

$$GU_{\text{eff}} = I_{\text{eff}} \cos\varphi;\ U_{\text{eff}}\left(\frac{1}{L\omega} - C\omega\right) = I_{\text{eff}} \sin\varphi \tag{9.52}$$

$$U_{\text{eff}}^2\left[G^2 + \left(\frac{1}{L\omega} - C\omega\right)^2\right] = I_{\text{eff}}^2 \tag{9.53}$$

$$\tan\varphi = \frac{\frac{1}{L\omega} - C\omega}{G} \tag{9.54}$$

From (9.53), it follows that:

$$I_{\text{eff}} = U_{\text{eff}}\sqrt{G^2 + \left(\frac{1}{L\omega} - C\omega\right)^2} = U_{\text{eff}}\,Y \tag{9.55}$$

The quantity:

$$Y = \sqrt{\frac{1}{R^2} + \left(\frac{1}{L\omega} - C\omega\right)^2} \tag{9.56}$$

is called the *electric circuit admittance*. Its unit of measurement is Ω^{-1} or S (i.e. *Siemens*). The quantities $B_L = \frac{1}{L\omega}$ and $B_C = -\frac{1}{X_C} = -C\omega$ represent *inductive* and *capacitive electrical susceptance*, respectively, while:

$$B = B_L + B_C \tag{9.57}$$

represents the *total susceptance*. Ohm's law in a parallel RLC circuit is thus:

$$I_{\text{eff}} = U_{\text{eff}}\,Y \tag{9.58}$$

In the phase diagram of the parallel RLC circuit, the phasor U_{eff} is the constant parameter and is thus taken as the reference. The phasors I_{eff_R}, I_{eff_L}, I_{eff_C} as well as their phase shift φ are represented in figure 9.9(b). By vector summation of I_{eff_L} and I_{eff_C}, we obtain:

$$I_{\text{eff}_B} = \left(\frac{1}{L\omega} - C\omega\right)U_{\text{eff}} = (B_L + B_C)U_{\text{eff}} = BU_{\text{eff}} \tag{9.59}$$

$$I_{\text{eff}_R} = I_{\text{eff}_G} = GU_{\text{eff}} \tag{9.60}$$

$$I_{\text{eff}} = \sqrt{I_{\text{eff}_R}^2 + I_{\text{eff}_B}^2} = U_{\text{eff}}\sqrt{G^2 + B^2} = U_{\text{eff}}\,Y \tag{9.61}$$

From the phase diagram, it follows that:

$$I_{\text{eff}_G} = I_{\text{eff}} \cos\varphi = I_{\text{eff}_a}$$

$$I_{\text{eff}_B} = I_{\text{eff}} \sin\varphi = I_{\text{eff}_r} \tag{9.62}$$

which are the active and reactive components of the effective current. We note that we can write:

$$I_{\text{eff}} = \sqrt{I^2_{\text{eff}_a} + I^2_{\text{eff}_r}} \tag{9.63}$$

or Ohm's law in a parallel RLC circuit can be formulated as:

$$I_{\text{eff}} = \frac{U_{\text{eff}}}{\frac{1}{\sqrt{\frac{1}{R^2} + \left(\frac{1}{L\omega} - C\omega\right)^2}}} = \frac{U_{\text{eff}}}{Z_p} \tag{9.64}$$

where Z_p is the *circuit impedance*. The phase shift between intensity and voltage is:

$$\varphi = \text{atan}\frac{\frac{1}{L\omega} - C\omega}{\frac{1}{R}} = \text{atan}\left[R\left(\frac{1}{L\omega} - C\omega\right)\right] \tag{9.65}$$

9.2.5 Parallel circuit with any number of branches

At the terminals of a circuit consisting of n branches connected in parallel (see figure 9.10(a)), a voltage $u = U_{\text{eff}}\sqrt{2}\cos\omega t$ is applied. In the branches of the circuit,

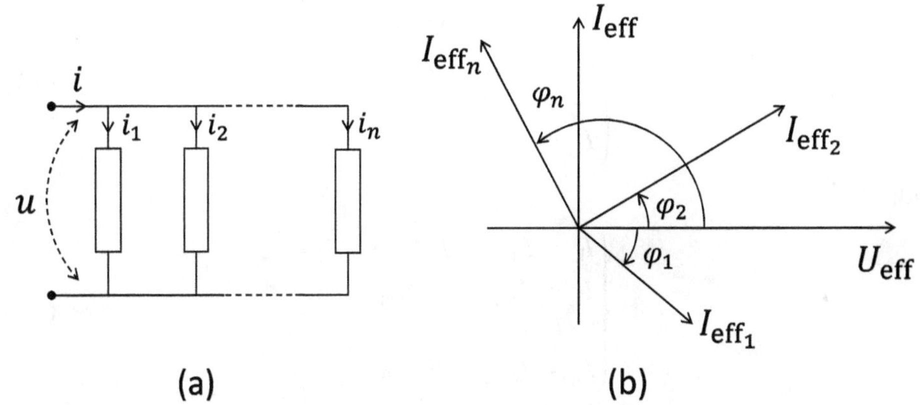

Figure 9.10. Parallel circuit with any number of branches (a) and its corresponding phase diagram (b).

the effective current intensities are: $I_{\text{eff}_1}, I_{\text{eff}_2}, I_{\text{eff}_3}, \cdots, I_{\text{eff}_n}$. The sum of these intensities gives the total intensity I_{eff}. A qualitative analysis of this circuit can be made using the phase diagram shown in figure 9.10(b). The active component of the total intensity is:

$$I_{\text{eff}_a} = I_{\text{eff}_{1_a}} + I_{\text{eff}_{2_a}} + \ldots + I_{\text{eff}_{n_a}} = I_{\text{eff}_1}\cos\varphi_1 + I_{\text{eff}_2}\cos\varphi_2 + \ldots + I_{\text{eff}_n}\cos\varphi_n$$

$$I_{\text{eff}_a} = \sum_{i=1}^{n} I_{\text{eff}_i}\cos\varphi_i \tag{9.66}$$

The reactive component of the total intensity is:

$$I_{\mathrm{eff}_r} = I_{\mathrm{eff}_{1_r}} + I_{\mathrm{eff}_{2_r}} + \ldots + I_{\mathrm{eff}_{n_r}} = I_{\mathrm{eff}_1} \sin \varphi_1 + I_{\mathrm{eff}_2} \sin \varphi_2 + \ldots + I_{\mathrm{eff}_n} \sin \varphi_n$$

$$I_{\mathrm{eff}_r} = \sum_{i=1}^{n} I_{\mathrm{eff}_i} \sin \varphi_i \tag{9.67}$$

Using these expressions, we can obtain the effective current intensity:

$$I_{\mathrm{eff}} = \sqrt{I_{\mathrm{eff}_a}^2 + I_{\mathrm{eff}_r}^2} \tag{9.68}$$

while the phase shift between total intensity and effective voltage is:

$$\tan \varphi = \frac{\sum_{i=1}^{n} I_{\mathrm{eff}_i} \sin \varphi_i}{\sum_{i=1}^{n} I_{\mathrm{eff}_i} \cos \varphi_i} \tag{9.69}$$

On the other hand, as indicated in the previous paragraph, if the conductance and susceptance of each branch are known, we can write:

$$\begin{cases} I_{1_a} = I_{\mathrm{eff}_{G_1}} = G_1 U_{\mathrm{eff}} \\ I_{2_a} = I_{\mathrm{eff}_{G_2}} = G_2 U_{\mathrm{eff}} \\ \vdots \\ I_{n_a} = I_{\mathrm{eff}_{G_n}} = G_n U_{\mathrm{eff}} \end{cases} \tag{9.70}$$

and:

$$\begin{cases} I_{1_r} = I_{\mathrm{eff}_{B_1}} = B_1 U_{\mathrm{eff}} \\ I_{2_r} = I_{\mathrm{eff}_{B_2}} = B_2 U_{\mathrm{eff}} \\ \vdots \\ I_{n_r} = I_{\mathrm{eff}_{B_n}} = B_n U_{\mathrm{eff}} \end{cases} \tag{9.71}$$

Using these, we can obtain the active and reactive components of the total intensity:

$$I_a = U_{\mathrm{eff}} \sum_{i=1}^{n} G_i = U_{\mathrm{eff}} G$$

$$I_r = U_{\mathrm{eff}} \sum_{i=1}^{n} B_i = U_{\mathrm{eff}} B \tag{9.72}$$

and the effective value of the total current intensity is:

$$I_{\mathrm{eff}} = U_{\mathrm{eff}} \sqrt{G^2 + B^2} = U_{\mathrm{eff}} \sqrt{\left(\sum_{i=1}^{n} G_i\right)^2 + \left(\sum_{i=1}^{n} B_i\right)^2} \tag{9.73}$$

The admittance of the circuit as a whole is:

$$Y = \sqrt{G^2 + B^2} \tag{9.74}$$

and so:

$$I_{\text{eff}} = YU_{\text{eff}} \tag{9.75}$$

Using the values I_{eff_a}, I_{eff_r}, and I_{eff} of the total intensity, the phase shift φ between them and the voltage can be obtained using the relations:

$$\begin{cases} \sin\varphi = I_{\text{eff}_r}/I_{\text{eff}} \\ \cos\varphi = I_{\text{eff}_a}/I_{\text{eff}} \\ \tan\varphi = I_{\text{eff}_r}/I_{\text{eff}_a} \end{cases} \tag{9.76}$$

9.2.6 Equivalence relationships when switching from a series RLC circuit to a parallel RLC circuit and vice versa

If the impedance Z, resistance R, and reactance X of a series circuit are known, the equivalent admittance Y, conductance G, and susceptance B of the circuit can be calculated. For the series circuit, the relationship between the effective current intensity value and the voltage value is:

$$I_{\text{eff}} = \frac{U_{\text{eff}}}{Z} \tag{9.77}$$

For the same circuit, I_{eff} can be written according to the equivalent admittance as follows:

$$I_{\text{eff}} = YU_{\text{eff}} \tag{9.78}$$

The following relation therefore holds for the impedance and equivalent electrical admittance of the circuit:

$$Y = \frac{1}{Z} \tag{9.79}$$

Expressing Z as a function of resistance R and reactance X, we can rewrite the expression for Y as follows:

$$Y = \frac{1}{\sqrt{R^2 + X^2}} \tag{9.80}$$

or:

$$Y^2 = \frac{1}{R^2 + X^2} = \frac{R^2 + X^2}{(R^2 + X^2)^2} = \frac{R^2}{(R^2 + X^2)^2} + \frac{X^2}{(R^2 + X^2)^2} \tag{9.81}$$

On the other hand, the admittance can be expressed in terms of an equivalent susceptance B and an equivalent conductance G as follows:

$$Y^2 = G^2 + B^2 \tag{9.82}$$

Comparing the last two relationships, we obtain:

$$G = \frac{R}{R^2 + X^2} = \frac{R}{Z^2} \tag{9.83}$$

$$B = \frac{X}{R^2 + X^2} = \frac{X}{Z^2} \tag{9.84}$$

Conversely, when a parallel circuit is considered, whose admittance Y, conductance G, and susceptance B are known, one can calculate the equivalent impedance Z, resistance R, and reactance X of the circuit by following the above procedure. This time, we write:

$$Z = \frac{1}{Y} = \frac{1}{\sqrt{G^2 + B^2}} \tag{9.85}$$

or:

$$Z^2 = \frac{1}{G^2 + B^2} = \frac{G^2 + B^2}{(G^2 + B^2)^2} = \frac{G^2}{(G^2 + B^2)^2} + \frac{B^2}{(G^2 + B^2)^2} \tag{9.86}$$

However, taking into account (9.42), we obtain $Z^2 = R^2 + X^2$, and then:

- the equivalent circuit resistance:

$$R = \frac{G}{G^2 + B^2} = \frac{G}{Y^2} \tag{9.87}$$

- the equivalent circuit reactance:

$$X = \frac{B}{G^2 + B^2} = \frac{B}{Y^2} \tag{9.88}$$

9.3 Power in alternating current circuits

Let us consider a circuit consisting of a resistance R, an inductance L, and a capacitance C. These are connected in series (see figure 9.11), and an AC is applied at the circuit terminals. The alternating voltage is given by:

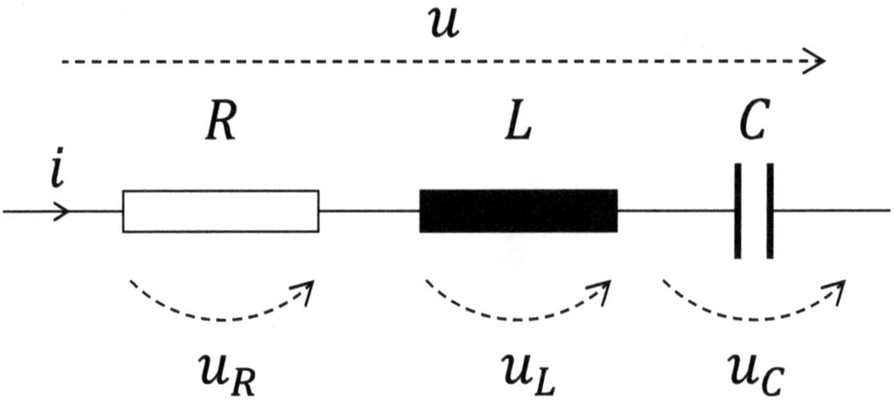

Figure 9.11. Series RLC circuit carrying AC.

$$u = U_{eff}\sqrt{2}\cos\omega t \tag{9.89}$$

The intensity of the current in the circuit is phase shifted by φ in respect to the phase of the voltage u and is given by the expression:

$$i = I_{eff}\sqrt{2}\cos(\omega t - \varphi) \tag{9.90}$$

The following relationship holds between the instantaneous voltage u at the ends of the circuit and the instantaneous current intensity i passing through the circuit:

$$u = Ri + L\frac{di}{dt} + \frac{1}{C}\int idt \tag{9.91}$$

The instantaneous power absorbed by the circuit is given by the relationship:

$$P_i = iu \tag{9.92}$$

Taking into account equation (9.91), the instantaneous power becomes:

$$P_i = Ri^2 + Li\frac{di}{dt} + \frac{i}{C}\int idt \tag{9.93}$$

The three terms on the right-hand side of equation (9.93) have the following meanings:

$P_R = Ri^2$—*instantaneous power absorbed by the resistance*;

$P_L = Li\frac{di}{dt}$—*instantaneous power absorbed by the inductance;*

$P_C = \frac{i}{C}\int idt$—*instantaneous power absorbed by the capacitance.*

Using these notations, equation (9.93) becomes:

$$P_i = P_R + P_L + P_C \tag{9.94}$$

Analyzing each term on the right-hand side of equation (9.94) yields the following:

- *The instantaneous power absorbed by the resistance* is given by the expression:

$$P_R = Ri^2 = 2RI_{eff}^2\cos^2(\omega t - \varphi) = RI_{eff}^2[1 + \cos 2(\omega t - \varphi)] \tag{9.95}$$

According to relation (9.95), and taking into account that the function $\cos 2(\omega t - \varphi)$ takes values in the interval $[-1, 1]$, it follows that P_R is a positive quantity (see figure 9.12). P_R takes values between zero and $2RI_{eff}^2$; it varies periodically over time with an angular frequency of 2ω and a period of $\frac{T}{2}$. Note that the angular frequency ω and the period T are the angular frequency and period of both the voltage and current intensity. If the average value is taken over a period T, the following value is obtained:

$$P_{R_{med}} = \frac{1}{T}\int_0^T P_R dt = \frac{1}{T}\int_0^T RI_{eff}^2[1 + \cos 2(\omega t - \varphi)]dt = RI_{eff}^2 \tag{9.96}$$

The power absorbed by resistance is converted into heat through the electrocaloric effect.

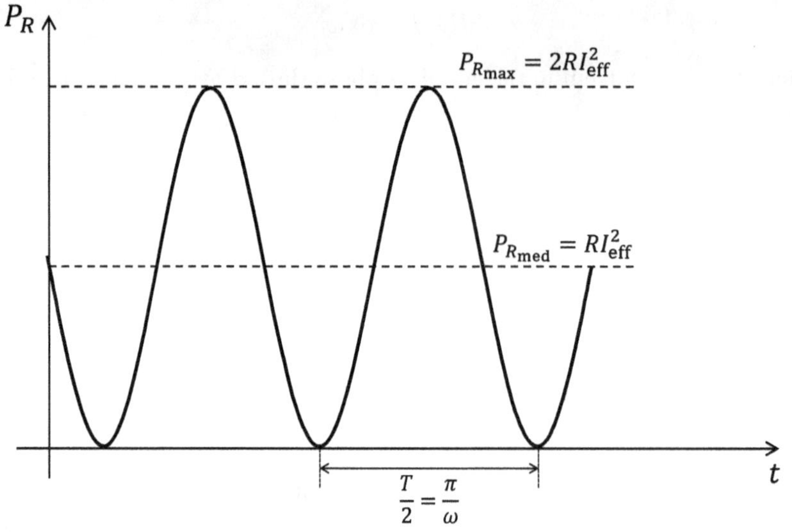

Figure 9.12. Time dependence of instantaneous power absorbed by resistance.

- After replacing the current intensity, the *instantaneous power absorbed by inductance* becomes:

$$P_L = Li\frac{di}{dt} = -2L\omega\, I_{eff}^2 \cos(\omega t - \varphi)\sin(\omega t - \varphi)$$
$$= -L\omega\, I_{eff}^2 \sin 2(\omega t - \varphi) \tag{9.97}$$

- *The instantaneous power absorbed by capacitance* is:

$$P_C = \frac{i}{C}\int idt = 2\frac{I_{eff}^2}{C\omega}\cos(\omega t - \varphi)\sin(\omega t - \varphi)$$
$$= \frac{I_{eff}^2}{C\omega}\sin 2(\omega t - \varphi) \tag{9.98}$$

Summing the instantaneous powers absorbed by the inductance and the capacitance yields the instantaneous power transferred to the circuit's reactance:

$$P_L + P_C = -\left(L\omega - \frac{1}{C\omega}\right)I_{eff}^2 \sin 2(\omega t - \varphi) \tag{9.99}$$

or:

$$P_X = -XI_{eff}^2 \sin 2(\omega t - \varphi) \tag{9.100}$$

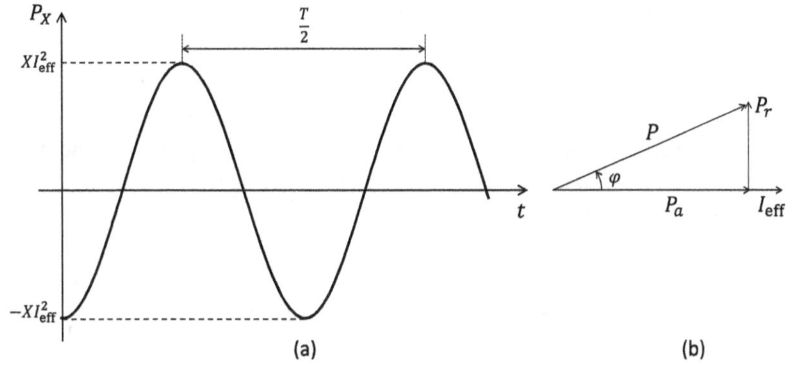

Figure 9.13. Time dependence of instantaneous power exchanged on the circuit reactance.

Plotting the dependence *of* P_X over time (see figure 9.13(a)), it can be seen that positive and negative alternation occurs. In the positive part of the alternation, the reactance of the circuit receives power from the source, and in the negative part of the alternation, it gives this power to the source. Circuit reactance does not consume power. The maximum value of the power received in the positive part of the alternation is: $P_{X_{\max}} = XI_{\text{eff}}^2$. It should be noted that because the average value of a sinusoidal function over a period is zero, then $P_{X_{\text{med}}} = -\frac{1}{T}\int_0^T XI_{\text{eff}}^2 \sin 2(\omega t - \varphi)dt = 0$ as well. The average value of the power absorbed by the resistance is given by:

$$P_{R_{\text{med}}} = P_a \tag{9.101}$$

which it is called the *active power*. Its unit of measurement is the *watt* (W). The maximum value of the power received by the reactance is:

$$P_{X_{\max}} = P_r \tag{9.102}$$

which is called the *reactive power*, measured in *volt-ampere reactive* (VAR). Summing the instantaneous power absorbed by the resistance P_R and that exchanged by the reactance P_X of the circuit, the total *instantaneous power* is given by:

$$P_i = RI_{\text{eff}}^2[1 + \cos 2(\omega t - \varphi)] - XI_{\text{eff}}^2 \sin 2(\omega t - \varphi) \tag{9.103}$$

The average value of the total power over a period is:

$$P_{\text{med}} = \frac{1}{T}\int_0^T P_i dt = \frac{1}{T}\int_0^T P_R dt + \frac{1}{T}\int_0^T P_X dt = RI_{\text{eff}}^2 = P_{R_{\text{med}}} \tag{9.104}$$

It follows that the average value of the total power over a period is equal to the average power absorbed by the resistance, that is, the active power:

$$P_{\text{med}} = P_a = P_{R_{\text{med}}} = RI_{\text{eff}}^2 \tag{9.105}$$

The mathematical treatment presented above highlights the circuit element that consumes electricity, namely the resistance. At the same time, we should emphasize that reactance, without consuming electricity, exchanges power with the electricity source. The energy that the elements receive from the source is stored in their magnetic and electric fields and then given back to the source. This energy exchange has a periodic nature. The problem of total instantaneous power can be treated mathematically and in another way that does not directly reveal the role of the circuit elements, i.e. resistance and reactance. Through the previous treatment, however, knowledge has been obtained that allows easy observation of when resistance and reactance come into play, thus completing the mathematical treatment presented below.

Substituting the expressions for the voltage and intensity into the relation defining instantaneous power, one obtains:

$$P_i = ui = 2U_{\text{eff}} I_{\text{eff}} \cos \omega t \cos (\omega t - \varphi) \tag{9.106}$$

If we plot the time dependencies of voltage u, current i, and power P_i (see figure 9.14), we find that when u and i are positive, P_i is also positive, and when one of the

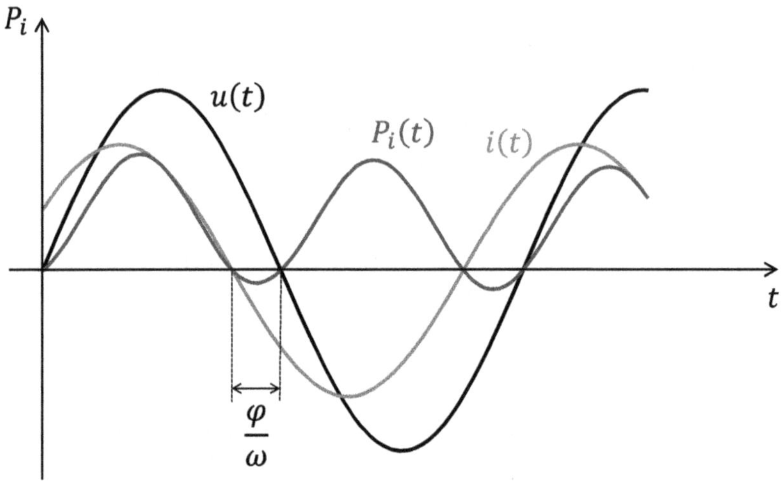

Figure 9.14. Time dependence of the instantaneous values of voltage u, current i, and power P_i in a series RLC AC circuit.

quantities u or i is negative, P_i is negative, too. The time interval $\frac{\varphi}{\omega}$ specific to the negative part of the alternation is the time interval determined by the phase difference between current intensity and voltage. From figure 9.14, it can easily be seen why power has twice the angular frequency of voltage or intensity. It can also be observed that positive portions of the alternation have greater amplitude than negative portions. In the positive parts of the alternation, the source transfers energy to both the resistance and the reactance of the circuit, while in negative parts, energy is only returned by the reactance. Starting from the expression of instantaneous power, a new expression can be found for the average power and implicitly for the active power. Starting from expression (9.106), a few simple transformations lead to the form:

$$P_i = I_{\mathrm{eff}} U_{\mathrm{eff}} [\cos(2\omega t - \varphi) + \cos\varphi] \tag{9.107}$$

The minimum instantaneous power value is:

$$P_{i_{\min}} = I_{\mathrm{eff}} U_{\mathrm{eff}} (-1 + \cos\varphi) \tag{9.108}$$

and the maximum value is:

$$P_{i_{\max}} = I_{\mathrm{eff}} U_{\mathrm{eff}} (1 + \cos\varphi) \tag{9.109}$$

The average value of the total power is:

$$P_{\mathrm{med}} = \frac{1}{T}\int_0^T P_i\,dt = I_{\mathrm{eff}} U_{\mathrm{eff}} \cos\varphi \tag{9.110}$$

The quantity denoted by $P_A = U_{\mathrm{eff}} I_{\mathrm{eff}}$ is called the *apparent power* and has the unit of measurement of the *volt-ampere* (VA). Using this quantity, the active power becomes:

$$P_{\mathrm{med}} = P_{R_{\mathrm{med}}} = P_a = I_{\mathrm{eff}} U_{\mathrm{eff}} \cos\varphi = P_A \cos\varphi \tag{9.111}$$

where $\cos\varphi$ is called the *power factor*. On the vector diagram of voltages in the RLC electric circuit (figure 9.6), it can easily be ascertained that:

$$U_{\mathrm{eff}} \cos\varphi = U_{\mathrm{eff}_R} = RI_{\mathrm{eff}} \tag{9.112}$$

Using this expression, the active power can be written as:

$$P_a = RI_{\mathrm{eff}}^2 = U_{\mathrm{eff}} I_{\mathrm{eff}} \cos\varphi \tag{9.113}$$

From the same vector diagram of voltages, it follows that:

$$XI_{\mathrm{eff}} = U_{\mathrm{eff}} \sin\varphi \tag{9.114}$$

$$P_r = XI_{\mathrm{eff}}^2 = U_{\mathrm{eff}} I_{\mathrm{eff}} \sin\varphi = P_A \sin\varphi \tag{9.115}$$

With these equations, the relationship between apparent power, active power, and reactive power is given by:

$$P_A^2 = P_a^2 + P_r^2 \tag{9.116}$$

This relationship is vectorially transposed through the triangle of powers (see figure 9.13(b)), as the circuit is characterized either by impedance, resistance, and capacitance, or by admittance, conductance, and susceptance. The three powers, i.e. active, reactive, and apparent, have the expressions:

$$P_a = RI_{\mathrm{eff}}^2,\ P_r = XI_{\mathrm{eff}}^2,\ P_A = ZI_{\mathrm{eff}}^2$$

$$P_a = GU_{\mathrm{eff}}^2,\ P_r = BU_{\mathrm{eff}}^2,\ P_A = YU_{\mathrm{eff}}^2$$

9.4 Resonance in AC circuits

9.4.1 Energy conditions for the occurrence of resonant phenomena

Consider a given frequency of a voltage applied across a series RLC electric circuit in the AC regime. In this scenario, the values of the R, L, and C network parameters can be adjusted so that reactive power exchange does not occur between the power source and the network. Energy can be stored in the electric fields of the capacitances and in the magnetic fields of the inductances in the circuit. This energy can be transferred from electric fields to magnetic fields and vice versa but is not transferred back to the source. Under these conditions, *the phenomenon of resonance* is established in the circuit. *Only the active power changes between the source and the circuit.* This means that the power factor is $\cos\varphi = 1$. For any circuit component, we can determine either the equivalent resistance R and the equivalent reactance X, or the equivalent conductance G and equivalent susceptance B. From the condition that at resonance the reactive power is equal to zero ($P_r = XI_{\text{eff}}^2 = 0 = UI\sin\varphi$ or $P_r = BU_{\text{eff}}^2 = 0$), it follows that:

$$X = 0 \text{ and } \sin\varphi = 0 \quad \Rightarrow \varphi = 0 \text{ or } B = 0 \text{ and } \cos\varphi = 1 \tag{9.117}$$

because I_{eff} and U_{eff} are generally nonzero. Starting from relations (9.117), it can be said that for any circuit, one can find a function of the parameters R, L, and C of the network and the angular frequency of the current ω that is equal to zero when the phenomenon of resonance is established in the circuit:

$$f(R, L, C, \omega) = 0 \tag{9.118}$$

When ω is held constant, the parameters R, L, and C can be varied until the resonance phenomenon is reached. Conversely, when L and C have fixed values, one can vary the frequency ω until the function f reaches a value of zero, i.e. until resonance is reached. The resonant angular frequency value is denoted by ω_r.

9.4.2 Resonance in a series RLC circuit

In the case of a series RLC circuit, using the first resonant condition, i.e. considering fixed values of L and C, it follows that the resonant angular frequency:

$$X = L\omega - \frac{1}{C\omega} = X_L - X_C = 0 \Rightarrow L\omega = \frac{1}{C\omega} \Rightarrow \omega_r = \frac{1}{\sqrt{LC}} \tag{9.119}$$

A complete study of the phenomenon of resonance involves analyzing the angular frequency dependence of the electrical quantities in the circuit, namely current intensity, effective voltages across inductors and capacitors, and the phase difference between intensity and voltage. Applying an AC voltage $u = U_{\text{eff}}\sqrt{2}\cos\omega t$ to a series RLC circuit gives rise to a current in the circuit described by the equation $i = I_{\text{eff}}\sqrt{2}\cos(\omega t - \varphi)$ and whose intensity is phase shifted by φ in respect to the voltage:

(a) The effective value of the current intensity in the series RLC AC circuit is given by Ohm's law:

$$I_{\text{eff}} = \frac{U_{\text{eff}}}{\sqrt{R^2 + \left(L\omega - \frac{1}{C\omega}\right)^2}} \tag{9.120}$$

Plotting $I_{\text{eff}}(\omega)$ shows that at $\omega = \omega_r$ the current reaches a maximum, and at $\omega = 0$ and $\omega \to \infty$, $I_{\text{eff}} \to 0$.

(b) The effective value of the voltage drop across the inductance has the expression:

$$U_{\text{eff}_L} = L\omega\, I_{\text{eff}} = \frac{L\omega\, U_{\text{eff}}}{\sqrt{R^2 + \left(L\omega - \frac{1}{C\omega}\right)^2}} \tag{9.121}$$

Note that at $\omega = 0, U_{\text{eff}_L} = 0$; when $\omega \to \infty, U_{\text{eff}_L} = U_{\text{eff}}$; and at $\omega = \omega_r, U_{\text{eff}_L}$ becomes:

$$U_{\text{eff}_L} = \frac{L\omega_r U_{\text{eff}}}{R} = \frac{U_{\text{eff}}}{R}\frac{L}{\sqrt{LC}} = U_{\text{eff}}\frac{1}{R}\sqrt{\frac{L}{C}} \tag{9.122}$$

The quantity $R_C = \sqrt{\frac{L}{C}}$ is called the characteristic resistance or characteristic impedance of the circuit, and the ratio $d = \frac{R}{R_C}$ is called the attenuation factor. The value $\frac{U_{\text{eff}_L}|_{\omega_r}}{U_{\text{eff}}} = Q = \frac{R_C}{R} = \frac{1}{d}$ is called the quality factor of the circuit, and it shows how many times the voltage per inductance or capacitance (considered ideal circuit elements) at resonance is greater than the applied voltage. The voltage across an inductance reaches a maximum value at a given angular frequency ω_L which is determined from the condition $\frac{dU_{\text{eff}_L}}{d\omega} = 0$:

$$\omega_L = \frac{1}{\sqrt{LC - \frac{R^2C^2}{2}}} = \frac{1}{\sqrt{LC}\sqrt{1 - \frac{R^2}{2}\frac{C}{L}}} = \frac{\omega_r}{\sqrt{1 - \frac{d^2}{2}}} > \omega_r \tag{9.123}$$

It is worth noting that $\omega_L > \omega_r$ and $U_{\text{eff}_{L\max}} = \frac{U_{\text{eff}}}{d\sqrt{1 - \frac{d^2}{4}}} > \frac{U_{\text{eff}}}{d} = U_{\text{eff}_{L\text{resonance}}}$.

When $\omega \to \infty \Rightarrow U_{\text{eff}_L} = U_{\text{eff}}$. A graph of $U_{\text{eff}_L} = f(\omega)$ is shown in figure 9.15.

(c) The effective value of a voltage applied to a capacitor is given by the expression:

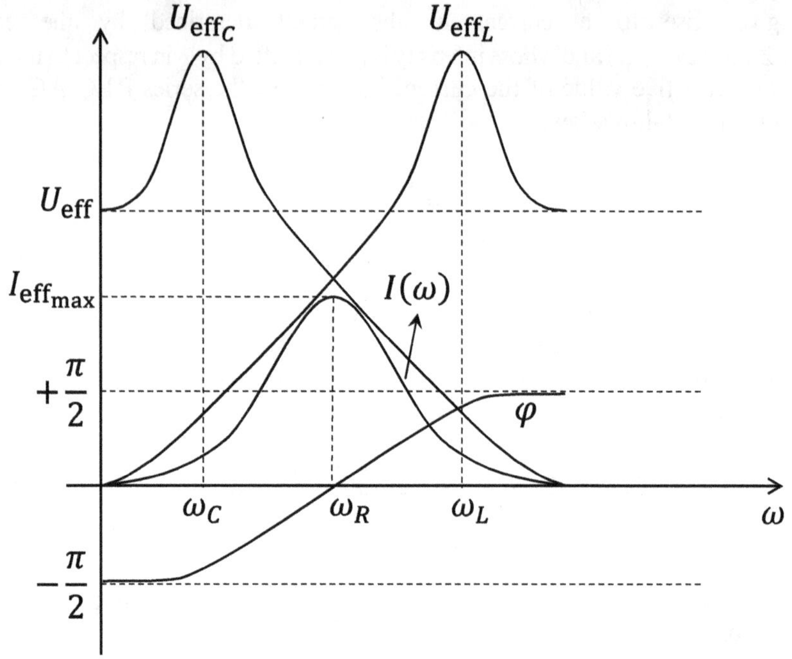

Figure 9.15. Frequency dependence of various quantities in a series RLC circuit carrying AC: current intensity, the voltage across the inductance, the voltage across the capacitance, and the current–voltage phase shift.

$$\begin{cases} U_{\text{eff}_C} = X_C I = \dfrac{1}{C\omega} \dfrac{U_{\text{eff}}}{\sqrt{R^2 + \left(L\omega - \frac{1}{C\omega}\right)^2}} \\ U_{\text{eff}_C}\Big|_{\omega=0} = \dfrac{U_{\text{eff}}}{\sqrt{R^2C^2\omega^2 + (LC\omega^2 - 1)^2}}\Bigg|_{\omega=0} = U_{\text{eff}} \\ U_{\text{eff}_C}\Big|_{\omega=\omega_r} = \dfrac{U_{\text{eff}}}{RC\omega_r} = \dfrac{U_{\text{eff}}}{RC\frac{1}{\sqrt{LC}}} = U_{\text{eff}}\dfrac{1}{R}\sqrt{\dfrac{L}{C}} = \dfrac{U_{\text{eff}}}{d} \end{cases} \tag{9.124}$$

When $\omega \to \infty$, $U_{\text{eff}_C} \to 0$. U_{eff_C} has a maximum value at an angular frequency of ω_C, which results from the condition $\frac{dU_{\text{eff}_C}}{d\omega} = 0$ at a value of:

$$\omega_C = \frac{1}{\sqrt{LC}}\sqrt{1 - \frac{R^2C}{2L}} = \omega_r\sqrt{1 - \frac{d^2}{2}} < \omega_r \tag{9.125}$$

Thus, U_{eff_C} reaches its maximum value at a frequency of $\omega_C < \omega_r$, and in this case, the maximum value of the voltage drop across the capacitance is given by:

$$U_{\text{eff}C_{\max}} = \frac{U_{\text{eff}}}{\sqrt{R^2C^2\omega_C{}^2 + (LC\omega_C{}^2 - 1)^2}}$$
$$= \frac{U_{\text{eff}}}{\sqrt{R^2\frac{C}{L} - \frac{R^4}{2}\frac{C^2}{L^2} + \left(1 - \frac{R^2}{2}\frac{C}{L} - 1\right)^2}} \tag{9.126}$$
$$= \frac{U_{\text{eff}}}{\sqrt{R^2\frac{C}{L}}\sqrt{1 - \frac{R^2}{4}\frac{C}{L}}} = \frac{U_{\text{eff}}}{d\sqrt{1 - \frac{d^2}{4}}}$$

It can be observed that this is higher than the value of the effective voltage per capacitance at resonance. A graph of $U_{\text{eff}_C} = f(\omega)$ is also shown in figure 9.15. Comparing the two strings of effective values for the voltage drop across an inductance and across a capacitance, it can be observed that they are equal in absolute value at $\omega = \omega_r$. It can also be seen that U_{eff_C} starts from the value U_{eff} (at $\omega = 0$), the same value towards which U_{eff_L} tends when $\omega \to \infty$. At $\omega = \omega_r$, the values of U_{eff_C} and U_{eff_L} can be higher than U_{eff}, so the phenomenon studied here is also called voltage resonance.

(d) The tangent of the phase difference φ between the intensity of the current and the voltage is given by the expression:

$$\tan\varphi = \frac{L\omega - \frac{1}{C\omega}}{R} \tag{9.127}$$

By varying the value of the angular frequency, as above, it is found that:

$$\begin{cases} \omega = 0, & \tan\varphi = -\infty \Longrightarrow \varphi = -\frac{\pi}{2} \\ \omega = \omega_r, & \tan\varphi = 0 \Longrightarrow \varphi = 0,\ \cos\varphi = 1 \\ \omega = \infty, & \tan\varphi = \infty \Longrightarrow \varphi = \frac{\pi}{2} \end{cases} \tag{9.128}$$

The dependence of φ on ω is also shown in figure 9.15. At $\omega = \omega_r$, $\tan\varphi = 0 \Longrightarrow \varphi = 0$, and $\cos\varphi = 1$. This means that the circuit only exchanges active power with the source of electricity, which was established from the beginning as a condition of the resonance phenomenon. Returning to the dependence of the physical quantities on the angular frequency, it can be observed that the effective value of the current intensity is at a maximum at $\omega = \omega_r$, having the value $I_{\text{eff}_{\max}} = \frac{U_{\text{eff}}}{R}$. Thus, the maximum value of the intensity depends on the resistance in the circuit and therefore on the attenuation factor (or damping factor) $d = \frac{R}{R_C}$. In addition, the shape of the resonance curve $I = f(\omega)$ depends on the value of the damping factor; the higher it is, the flatter the curve, with the maximum being less pronounced

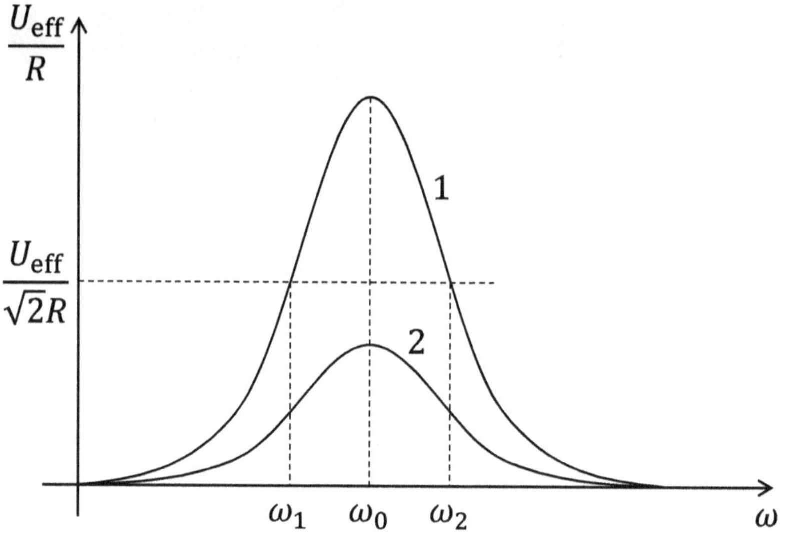

Figure 9.16. Dependence of current on frequency for two values of the resistance in a circuit that satisfies the relationship: $R_1 < R_2$.

(curve 2 in figure 9.16). The reverse is also true: the lower the damping factor, the sharper the curve, with the resonance being more pronounced (curve 1 in figure 9.16). In order to assess the rapidity of the variation of the current near the point of resonance $\omega = \omega_r$, we use the values of the angular frequencies ω_1 and ω_2 obtained using the condition that the current intensity must become $\sqrt{2}$ times smaller than its value at resonance, i.e. the values obtained by solving the equation:

$$I_{\text{eff}} = \frac{U_{\text{eff}}}{Z} = \frac{U_{\text{eff}}}{\sqrt{2}R} \tag{9.129}$$

According to equation (9.129), the values ω_1 and ω_2 of the angular frequencies for which the intensity becomes $\sqrt{2}$ less than that at resonance can be obtained by solving the equation $Z = \sqrt{2}R$, as follows:

$$\begin{cases} \sqrt{R^2 + \left(L\omega - \dfrac{1}{C\omega}\right)^2} = \sqrt{2}R \\ (LC\omega^2 - 1)^2 = R^2C^2\omega^2;\ LC\omega^2 - 1 = \pm RC\omega \end{cases}$$

Dividing the last equation by $LC\omega_r^2$ gives:

$$\left(\frac{\omega}{\omega_r}\right)^2 - \frac{1}{LC\omega_r^2} = \pm\frac{RC\omega}{LC\omega_r^2} \text{ or } \left(\frac{\omega}{\omega_r}\right)^2 \mp \frac{R}{L}\frac{\omega}{\omega_r^2} - 1 = 0$$

which has the solutions:

$$\frac{\omega}{\omega_r} = \frac{1}{2}\left(\mp\frac{R}{L\omega_r} \pm \sqrt{\frac{R^2}{L^2\omega_r^2} + 4}\right) \tag{9.130}$$

Keeping only physically convenient solutions for $\frac{\omega}{\omega_r}$ gives:

$$\begin{cases} \dfrac{\omega_1}{\omega_r} = \dfrac{1}{2}\left(-\dfrac{R}{L\omega_r} + \sqrt{\dfrac{R^2}{L^2\omega_r^2} + 4}\right) \\ \dfrac{\omega_2}{\omega_r} = \dfrac{1}{2}\left(\dfrac{R}{L\omega_r} + \sqrt{\dfrac{R^2}{L^2\omega_r^2} + 4}\right) \end{cases} \tag{9.131}$$

Taking the difference between the two solutions, we obtain:

$$\frac{\Delta\omega}{\omega_r} = \frac{\omega_2}{\omega_r} - \frac{\omega_1}{\omega_r} = \frac{R}{L\omega_r} = \frac{R}{L\sqrt{\frac{1}{LC}}} = \frac{1}{\frac{1}{R}\sqrt{\frac{L}{C}}} = \frac{1}{Q} = d \tag{9.132}$$

Thus, the quantity $\frac{\Delta\omega}{\omega_r} = \frac{1}{Q} = d$ is related to the quality factor Q and the damping factor d. It can be seen that in this way, the quality factor and the damping factor can be determined experimentally. The quantity $\frac{\Delta\omega}{\omega_r}$ shows that the resonance is more pronounced if the ratio $\frac{\Delta\omega}{\omega_r}$ is lower or the quality factor $Q = \frac{\omega_r}{\Delta\omega}$ is higher.

9.4.3 Resonance in a parallel RLC circuit

In the case of a parallel circuit, starting from one of the resonance conditions, it follows that:

$$B = \frac{1}{L\omega} - C\omega = B_L - B_C = 0\,;\; C\omega = \frac{1}{L\omega} \tag{9.133}$$

When L and C have known values, the last relation gives the expression for the resonant angular frequency:

$$\omega_r = \frac{1}{\sqrt{LC}} \tag{9.134}$$

The study of the phenomenon of resonance in a parallel RLC circuit involves tracking the frequency dependence of the actual values of the total current intensity, the current intensity through the capacitance on one hand, and through the inductance on the other, as well as the phase difference between the total intensity and the voltage. If the voltage $u = U_{\text{eff}}\sqrt{2}\cos\omega t$ is applied to a parallel circuit, a current arises in the circuit, the intensity of which is phase shifted by φ in respect to the voltage and which is described by the equation $i = I_{\text{eff}}\sqrt{2}\cos(\omega t - \varphi)$.

(a) In the case of a parallel circuit, the following relationship describes the effective values of the current and the voltage:

$$I_{\text{eff}} = GU_{\text{eff}} = U_{\text{eff}}\sqrt{G^2 + \left(\frac{1}{L\omega} - C\omega\right)^2} \tag{9.135}$$

Plotting $I_{\text{eff}}(\omega)$ (see figure 9.17), we obtain a curve that reaches a minimum at $\omega = \omega_r = \frac{1}{\sqrt{LC}}$ and tends to infinity when $\omega \to 0$ and $\omega \to \infty$. The minimum effective intensity is $I_{\text{eff}_{\text{min}}} = U_{\text{eff}}\, G = \frac{U_{\text{eff}}}{R}$.

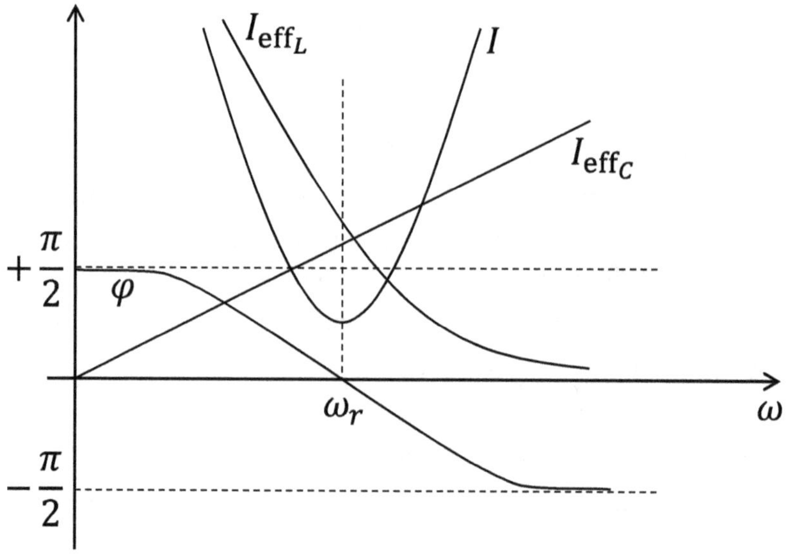

Figure 9.17. Frequency dependence of the currents passing through the branches of a parallel RLC circuit carrying AC.

(b) The effective intensity of the current in the capacitor has a value given by the expression:

$$I_{\text{eff}_C} = \frac{U_{\text{eff}}}{X_C} = U_{\text{eff}}\, C\omega \tag{9.136}$$

Note that the dependence of $I_{\text{eff}_C}(\omega)$ is linear (see figure 9.17). The following notations will be used: $G_C = \sqrt{\frac{C}{L}}$—*the characteristic conductance* or *characteristic admittance* and $\delta = \frac{G}{G_C}$—*the damping factor (attenuation)*:

$$\begin{cases} \omega = 0, & I_{\text{eff}_C} = U_{\text{eff}}\,C\omega = 0 \\ \omega = \omega_r, \; I_{\text{eff}_C} = U_{\text{eff}}\,C\omega_r = U_{\text{eff}}\sqrt{\dfrac{C}{L}} = \dfrac{U_{\text{eff}}}{R}\dfrac{G_C}{G} = \dfrac{I_{\text{eff}_R}}{\delta} \\ \omega = \infty, & I_{\text{eff}_C} = U_{\text{eff}}\,C\omega = \infty \end{cases} \tag{9.137}$$

(c) The effective intensity of the current passing through the inductance has a value given by the expression:

$$I_{\text{eff}_L} = \frac{U_{\text{eff}}}{X_L} = \frac{U_{\text{eff}}}{L\omega} \tag{9.138}$$

Thus, it decreases hyperbolically as the angular frequency increases, as shown in figure 9.17:

$$\begin{cases} \omega = 0, & I_{\text{eff}_L} = \dfrac{U_{\text{eff}}}{L\omega} = \infty \\ \omega = \omega_r, \; I_{\text{eff}_L} = \dfrac{U_{\text{eff}}}{L\omega_r} = U_{\text{eff}}\sqrt{\dfrac{C}{L}} = \dfrac{U_{\text{eff}}}{R}\dfrac{G_C}{G} = \dfrac{I_{\text{eff}_R}}{\delta} \\ \omega = \infty, & I_{\text{eff}_L} = \dfrac{U_{\text{eff}}}{L\omega} = 0 \end{cases} \tag{9.139}$$

Comparing the two strings of effective values of the current intensity in capacitance and inductance, it can be seen that they are equal in absolute value, i.e. $\omega = \omega_r$, at resonance and that they can be greater than I_{eff_R} at $\omega = \omega_r$. This is why it is also said that we are dealing with the *resonance of currents*. Values equal to ∞ are not realized in reality, because we never find ideal capacitances or inductances in electrical circuits.

(d) In the case of a parallel circuit, the tangent of the phase difference φ between the intensity of the current and the voltage has a value given by the expression:

$$\tan\varphi = \frac{\frac{1}{L\omega} - C\omega}{R} \tag{9.140}$$

By varying the value of ω, we obtain:

$$\begin{cases} \omega = 0, \; \tan\varphi = \infty \Longrightarrow \varphi = \dfrac{\pi}{2} \\ \omega = \omega_r, \; \tan\varphi = 0 \Longrightarrow \varphi = 0, \; \cos\varphi = 1 \\ \omega = \infty, \; \tan\varphi = -\infty \Longrightarrow \varphi = -\dfrac{\pi}{2} \end{cases}$$

Graphs of I_{eff_R}, I_{eff_L}, I_{eff_C}, and φ are shown in figure 9.17 at different scales.

9.4.4 Mixed circuits carrying alternating current

In the case of a more complex circuit, the resonant frequency is determined by one of the conditions $X = 0$ or $B = 0$. For example, in the case of the circuit in figure 9.18 (a), the resonant condition is: $B = B_1 + B_2 = 0$, where: $B_1 = \frac{X_1}{Z_1^2}$, $B_2 = \frac{X_2}{Z_2^2}$ or $B_1 = \frac{L\omega}{R_1^2 + L^2\omega^2}$, $B_2 = -\frac{1}{C\omega\,(R_2^2 + \frac{1}{C^2\omega^2})}$.

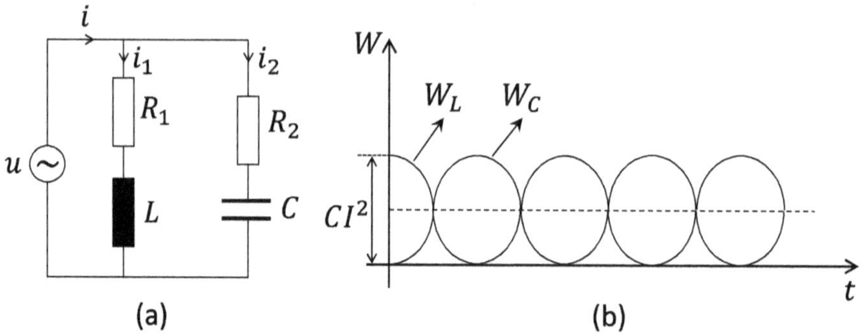

Figure 9.18. (a) Mixed circuit carrying AC. (b) Conservation of energy stored in the electric and magnetic field during resonance.

The resonant condition is: $\frac{L\omega}{R_1^2 + L^2\omega^2} - \frac{1}{C\omega\,(R_2^2 + \frac{1}{C^2\omega^2})} = 0$, from which we obtain the following expression for the resonant frequency:

$$\omega_r = \frac{1}{\sqrt{LC}}\sqrt{\frac{\frac{L}{C} - R_1^2}{\frac{L}{C} - R_2^2}} \tag{9.141}$$

Thus, the resonant frequency also depends on R_1 and R_2. If $R_1 = R_2$ (*characteristic impedance*), then the resonant frequency becomes $\omega_r = \frac{1}{\sqrt{LC}}$; i.e. the same as in the case of series and parallel RLC circuits. For $R_1 = R_2 = Z_0$, the resonant frequency is indeterminate because the circuit is resonant for any value of the angular frequency. For $R_1 > Z_0$ and $R_2 < Z_0$, as for $R_1 < Z_0$ and $R_2 > Z_0$, the result is an imaginary value for ω_r; i.e. resonance is not possible. Finally, for $R_2 = 0$ and $R_1 < Z_0$, we obtain the general result:

$$\omega = \frac{1}{\sqrt{LC}}\sqrt{\frac{\frac{L}{C} - R_1^2}{\frac{L}{C}}} = \frac{1}{\sqrt{LC}}\sqrt{1 - \frac{R_1^2 C}{L}}$$

It follows that:

$$\omega^2 LC = 1 - \frac{R_1^2 C}{L} \Rightarrow 1 - \omega^2 LC = \frac{R_1^2 C}{L} \Rightarrow C = \frac{L}{\omega^2 L^2 + R_1^2} \tag{9.142}$$

This relationship gives the capacitance that must be inserted in parallel with an inductive circuit in order for the assembly to have a phase difference φ of zero between the voltage and the current ($\varphi = 0, \;\cos\varphi = 1$).

9.4.5 Conservation of total energy stored in the magnetic and electric fields of inductances and capacitances during resonance

When the phenomenon of resonance is established in a circuit, there is no exchange of reactive power. However, between the inductive and capacitive elements of the circuit, energy is exchanged in an oscillatory manner, even though the total energy stored in these fields is constant. Let us exemplify the above using a series RLC circuit carrying AC. Taking, in this case, the following expression for the intensity of the AC:

$$i = \sqrt{2}\,I_{\text{eff}}\cos\omega t \tag{9.143}$$

the total energy at a given time, in the magnetic field of inductance, has the expression:

$$W_L = \frac{1}{2}Li_L^2 = LI_{\text{eff}_L}^2\cos^2\omega t \tag{9.144}$$

However, in a series circuit, $I_{\text{eff}_L} = I_{\text{eff}}$, so:

$$W_L = \frac{1}{2}Li_L^2 = LI_{\text{eff}}^2\cos^2\omega t \tag{9.145}$$

The energy stored in the electric field of the capacitance at a given time is given by the expression:

$$W_C = \frac{1}{2}Cu_C^2 \tag{9.146}$$

However, the voltage drop across the capacitance is phase shifted by $-\frac{\pi}{2}$ relative to the current intensity:

$$u_C = \sqrt{2}\,U_{\text{eff}_C}\cos\left(\omega t - \frac{\pi}{2}\right) = \sqrt{2}\,U_{\text{eff}_C}\sin\omega t \tag{9.147}$$

Using this expression for u_C, it follows that:

$$W_C = \frac{1}{2}Cu_C^2 = CU_{\text{eff}_C}^2\sin^2\omega t \tag{9.148}$$

Summing the energies given by (9.145)and (9.148) yields the total energy stored in the magnetic and electric fields:

$$W = W_L + W_C = LI_{\text{eff}}^2\cos^2\omega t + CU_{\text{eff}_C}^2\sin^2\omega t \tag{9.149}$$

At resonance, $U_{\text{eff}_C} = U_{\text{eff}_L} = L\omega_r I_{\text{eff}_L} = L\omega_r I_{\text{eff}}$, so that:

$$CU_{\text{eff}_C}^2 = CL^2\omega_r^2 I_{\text{eff}}^2 = LI_{\text{eff}}^2$$

Taking this into account, the total energy in the electric and magnetic fields is:

$$W = W_L + W_C = LI_{\text{eff}}^2(\cos^2\omega t + \sin^2\omega t) = LI_{\text{eff}}^2 \tag{9.150}$$

According to relationship (9.150), the total energy stored in the magnetic and electric fields of the inductances and capacitances of the circuit is constant at resonance. It can be seen that while the energy W_L stored in the magnetic field decreases, the energy stored in the electric field W_C increases; however, their sum remains constant (see figure 9.18(b)). Only the active energy is absorbed from the source, being consumed by the resistors of the resonant circuit and converted into thermal energy, $W_a = RI_{\text{eff}}^2 t$.

9.5 Complex representation for alternating current

If a time-varying voltage $u(t)$ is applied to a circuit, a current flows through it with an intensity which is a time-dependent function, denoted by $i(t)$. The two electric quantities, i.e. the voltage and the current, must satisfy a relation like Ohm's law:

$$i(t) = \frac{u(t)}{Z} \tag{9.151}$$

where Z is a function dependent on the circuit parameters and frequency but independent of time, current i, and voltage u. Z characterizes the circuit and plays a role similar to the equivalent resistance in DC circuits.

9.5.1 Series RLC circuit

When electric current quantities vary over time, the instantaneous voltage present across the ends of each component in a series RLC electric circuit is expressed as a voltage drop, as follows:

$$Ri + L\frac{di}{dt} + \frac{1}{C}\int idt = u \tag{9.152}$$

By taking the derivative with respect to time of both sides of this equation, we obtain a differential equation of the second order:

$$R\frac{di}{dt} + L\frac{d^2i}{dt^2} + \frac{i}{C} = \frac{du}{dt} \tag{9.153}$$

or:

$$L\frac{d^2i}{dt^2} + R\frac{di}{dt} + \frac{i}{C} - \frac{du}{dt} = 0. \tag{9.154}$$

However,

$$\text{if } i(t) = \frac{u(t)}{Z}, \text{ then } \frac{di}{dt} = \frac{1}{Z}\frac{du}{dt}; \frac{d^2i}{dt^2} = \frac{1}{Z}\frac{d^2u}{dt^2} \tag{9.155}$$

As a result, equation (9.154) becomes:

$$\frac{L}{Z}\frac{d^2u}{dt^2} + \frac{R}{Z}\frac{du}{dt} - \frac{du}{dt} + \frac{1}{ZC}u = 0 \tag{9.156}$$

or:

$$\frac{L}{Z}\frac{d^2u}{dt^2} + \left(\frac{R}{Z} - 1\right)\frac{du}{dt} + \frac{1}{ZC}u = 0 \tag{9.157}$$

This is a second-order differential equation that is homogeneous and has constant coefficients. Suppose that the solution of equation (9.157) has the form:

$$u(t) = U_0 e^{\mathcal{E}t} \tag{9.158}$$

where U_0 is a positive and constant real quantity and $\mathcal{E}$ is a complex quantity. This results from a solution of the characteristic equation associated with equation (9.157):

$$\frac{L}{Z}\mathcal{E}^2 + \left(\frac{R}{Z} - 1\right)\mathcal{E} + \frac{1}{ZC} = 0 \text{ or } \mathcal{E}^2 + \left(\frac{R}{L} - \frac{Z}{L}\right)\mathcal{E} + \frac{1}{LC} = 0 \tag{9.159}$$

The solutions of this equation are:

$$\mathcal{E}_{1,2} = \frac{1}{2}\left[-\left(\frac{R}{L} - \frac{Z}{L}\right) \pm \sqrt{\left(\frac{R}{L} - \frac{Z}{L}\right)^2 - \frac{4}{LC}}\right] = \delta \pm i\omega \tag{9.160}$$

where δ is a real quantity and ω can be a real or imaginary quantity. Using the characteristic equation (9.159), if $\mathcal{E}$ is known, then an expression for the impedance can be deduced, as follows:

$$LC\mathcal{E}^2 + (R - Z)C\mathcal{E} + 1 = 0 \text{ or } LC\mathcal{E}^2 + RC\mathcal{E} - ZC\mathcal{E} + 1 = 0$$

This results in:

$$Z = \frac{1}{C\mathcal{E}}(LC\mathcal{E}^2 + RC\mathcal{E} + 1) = L\mathcal{E} + \frac{1}{C\mathcal{E}} + R \tag{9.161}$$

Given this result, the intensity of the current is $i(t) = \frac{u(t)}{Z} = \frac{U_0 e^{\mathcal{E}t}}{Z}$. The dependence of $i(t)$ takes particular forms, depending on the expressions for $\mathcal{E}$ and implicitly on those for δ and ω:

(a) If $\delta = 0$, $\omega = 0$ and $\mathcal{E} = 0$

In this situation:

$$u = U_0 e^{0t} = U_0 = \text{constant}; Z = \infty; i = 0 \tag{9.162}$$

This is the case of an RLC electric circuit coupled to a DC voltage source U_0 = constant. The circuit is interrupted (i.e. it is an open circuit) as a result of the presence of the capacitance, and therefore the current is zero. If the circuit has no capacitance, then $Z = R$ and $I_0 = \frac{U_0}{R}$, i.e. a DC is established in the circuit that flows through the ideal coil connected in series with the resistance R.

(b) If $\delta = 0$ and $\omega \neq 0$

When $\delta = 0$ and ω is real and nonzero, then $\mathcal{E} = i\omega$ and the voltage takes the form:

$$u = U_0 e^{i\omega t} = U_0(\cos \omega t + i \sin \omega t) = U_0 \underline{\omega t} = \underline{U} \quad (9.163)$$

The voltage is an electric quantity that varies over time and is complex. In this case, the parameter Z is also a complex electric quantity,

$$\underline{Z} = R + iL\omega + \frac{1}{iC\omega} = R + i\left(L\omega - \frac{1}{C\omega}\right) \quad (9.164)$$

which represents *the complex impedance* of the electric circuit. The intensity of the circuit is also a complex electric quantity:

$$\underline{I} = \frac{\underline{U}}{\underline{Z}} = \frac{U_0 e^{i\omega t}}{|\underline{Z}| e^{i\varphi}} = \frac{U_0}{|\underline{Z}|} e^{i(\omega t - \varphi)} = I_0 e^{i(\omega t - \varphi)} \quad (9.165)$$

which is phase shifted with respect to the voltage by an angle of φ. The intensity modulus I_0 is equal to the ratio of the voltage modulus U_0 to the impedance modulus $|\underline{Z}|$:

$$I_0 = \frac{U_0}{|\underline{Z}|} \quad (9.166)$$

where:

$$|\underline{Z}| = \sqrt{R^2 + X^2} = \sqrt{R^2 + \left(L\omega - \frac{1}{C\omega}\right)^2}$$

i.e. the impedance of the series RLC circuit carrying AC. The phase difference φ between intensity and voltage is given by: $\tan \varphi = \frac{\Im[\underline{Z}]}{\Re[\underline{Z}]} = \frac{X}{R} = \frac{L\omega - \frac{1}{C\omega}}{R}$, i.e. the expression for $\tan \varphi$ derived for a series RLC circuit carrying AC. This treatment makes it possible to find all the properties of the series RLC circuit previously studied using an analytical representation (i.e. trigonometric functions).

(c) If $\delta \neq 0$ and $\omega = 0$

In this case, $\mathcal{E} = \delta$ and the voltage takes the form:

$$u = U_0 e^{\delta t} \quad (9.167)$$

Thus, the voltage is a real quantity. The parameter Z is also real:

$$Z = R + L\delta + \frac{1}{C\delta} \quad (9.168)$$

The intensity of the current is also a real quantity:

$$i = \frac{U_0 e^{\delta t}}{Z} = I_0 e^{\delta t} \quad (9.169)$$

When $\delta < 0$, both the voltage and the intensity of the current decrease exponentially over time. We thus encounter one of the cases studied in the transitory regime.

(d) If, again, $\delta \neq 0$ but ω has only an imaginary part
When $\delta \neq 0$ and $\omega = id$, then:

$$\mathcal{E} = \delta + i(id) = \delta - d = D \tag{9.170}$$

The voltage is a real quantity:

$$u = U_0 e^{Dt} \tag{9.171}$$

The parameter Z is also real:

$$Z = R + LD + \frac{1}{CD} \tag{9.172}$$

as is the intensity of the current:

$$i = \frac{U_0 e^{Dt}}{Z} = I_0 e^{Dt} \tag{9.173}$$

where $I_0 = \frac{U_0}{Z}$. If $D < 0$, both voltage and intensity decrease exponentially over time. So, this case is similar to the one analyzed above.

(e) If $\delta \neq 0$ and $\omega \neq 0$ has only a real part, then:

$$\mathcal{E} = \delta + i\omega \tag{9.174}$$

The voltage takes the form:

$$\underline{U} = U_0 e^{\delta t} e^{i\omega t} = U_0 e^{\delta t}(\cos \omega t + i \sin \omega t) = U_0 e^{\delta t} \underline{\omega t} \tag{9.175}$$

When $\delta < 0$, the voltage periodically depends on the time, and its amplitude decreases exponentially. The parameter $\underline{Z}$ is given by:

$$\underline{Z} = R + L(\delta + i\omega) + \frac{1}{C(\delta + i\omega)} \tag{9.176}$$

The intensity of the current is also complex; its amplitude depends on time and is phase shifted by φ with respect to the voltage u:

$$\underline{I} = \frac{\underline{U}}{\underline{Z}} = \frac{U_0 e^{\delta t} e^{i\omega t}}{|\underline{Z}| e^{i\varphi}} = \frac{U_0}{|\underline{Z}|} e^{\delta t} e^{i(\omega t - \varphi)} = I_0 e^{\delta t} \underline{(\omega t - \varphi)} \tag{9.177}$$

This case is identical to the one encountered in the transitory regime for a series RLC circuit with a time-varying voltage source. This circuit generates its own oscillations, which are damped over time.

9.5.2 Complex impedance and complex admittance of AC circuits

In the previous subsections, it was found that in the case of AC, the impedance of the series RLC circuit can be expressed as a complex electric quantity that has the form:

$$\underline{Z} = R + i\omega\, L + \frac{1}{i\omega C} \tag{9.178}$$

If $\underline{X_L} = i\omega L$ and $\underline{X_C} = \frac{1}{i\omega C}$ denote the complex inductive reactance and the complex capacitive reactance, respectively, then the complex impedance is:

$$\underline{Z} = R + \underline{X_L} + \underline{X_C} \tag{9.179}$$

which can be written as:

$$\underline{Z} = R + i\left(L\omega - \frac{1}{C\omega}\right) = R + iX = R + \underline{X} \tag{9.180}$$

where $X = L\omega - \frac{1}{C\omega}$ is the total reactance. The impedance modulus is:

$$|\underline{Z}| = \sqrt{R^2 + \left(L\omega - \frac{1}{C\omega}\right)^2} \tag{9.181}$$

and the phase difference between the phasors U and I appears as the argument of the complex impedance written trigonometrically:

$$\underline{Z} = |\underline{Z}|e^{i\varphi} = |\underline{Z}|(\cos\varphi + \mathrm{i}\sin\varphi) = R + iX \tag{9.182}$$

Thus:

$$\tan\varphi = \frac{\Im[\underline{Z}]}{\Re[\underline{Z}]} = \frac{X}{R} = \frac{L\omega - \frac{1}{C\omega}}{R} \tag{9.183}$$

Figure 9.19 shows a phase diagram of the impedance.

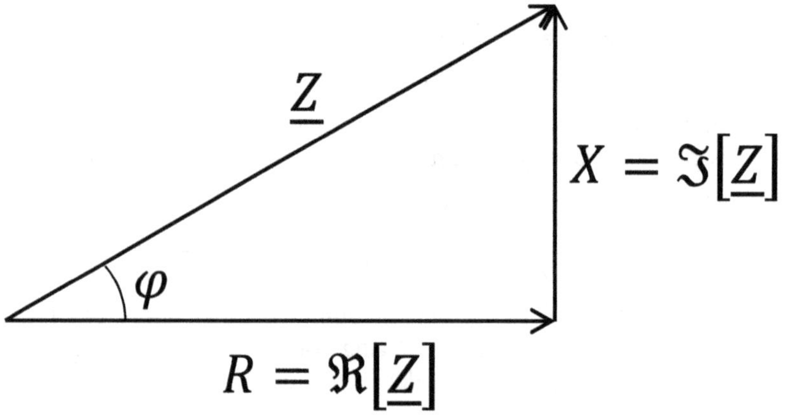

Figure 9.19. The impedance triangle.

9.5.3 Parallel RLC circuit

In the case of the parallel RLC circuit, the admittance can be written in complex form as follows:

$$\underline{Y} = G + \frac{1}{iL\omega} + iC\omega \tag{9.184}$$

or:

$$\underline{Y} = G - i\left(\frac{1}{L\omega} - C\omega\right) \tag{9.185}$$

If $\underline{B_L} = \frac{1}{i\omega L}$ and $\underline{B_C} = i\omega C$ denote the complex inductive susceptance and the complex capacitive susceptance, respectively, and the total complex susceptance is $\underline{B} = \underline{B_L} + \underline{B_C}$, then the admittance takes the form:

$$\underline{Y} = G + \underline{B} = G - iB \tag{9.186}$$

where $B = \frac{1}{L\omega} - C\omega$. For the same circuit, the following relationship exists between the complex quantities $\underline{Z}$ and $\underline{Y}$:

$$\underline{Z}\underline{Y} = 1 \tag{9.187}$$

This relationship allows us to calculate the equivalent complex admittance of a circuit for which the complex impedance is known, and vice versa. If, for any circuit, the complex impedance is given in the form:

$$\underline{Z} = R + iX \tag{9.188}$$

the complex admittance of the same circuit can be expressed as:

$$\underline{Y} = \frac{1}{\underline{Z}} = \frac{1}{R + iX} = \frac{R - iX}{R^2 + X^2} = \frac{R}{R^2 + X^2} - i\frac{X}{R^2 + X^2} \tag{9.189}$$

or, knowing that $\underline{Y} = G - iB$, we obtain:

$$G = \frac{R}{R^2 + X^2}, \quad B = \frac{X}{R^2 + X^2} \tag{9.190}$$

Thus, we have recovered the expressions previously established using the analytic representation. If, for any circuit, the admittance of the circuit $\underline{Y} = G - iB$ is known, then the complex equivalent impedance can be determined from the relation $\underline{Z}\underline{Y} = 1$ as follows:

$$\underline{Z} = \frac{1}{\underline{Y}} = \frac{1}{G - iB} = \frac{G + iB}{G^2 + B^2} = \frac{G}{G^2 + B^2} + i\frac{B}{G^2 + B^2}$$

However, since $\underline{Z} = R + iX$, we obtain:

$$R = \frac{G}{G^2 + B^2}, \quad X = \frac{B}{G^2 + B^2} \tag{9.191}$$

Thus, the expressions previously established using the analytic representation are recovered. The complex electric quantities $\underline{Z}$ and $\underline{Y}$ can be interpreted as mathematical operators. Thus, the complex impedance is the complex operator $\underline{Z}$ which, when applied to the complex intensity $\underline{I}$, transforms it into a complex voltage according to the relation:

$$\underline{U} = \underline{Z}\underline{I} \tag{9.192}$$

The complex admittance is the complex operator $\underline{Y}$ which, when applied to the complex voltage, converts it into a complex intensity according to the relation:

$$\underline{I} = \underline{Y}\,\underline{U} \tag{9.193}$$

The complex quantities $\underline{Z}$ and $\underline{Y}$ can be expressed in different forms. The quantity $\underline{Z}$ can be expressed algebraically as $\underline{Z} = R + iX$, trigonometrically as $\underline{Z} = |\underline{Z}|(\cos\varphi + i\sin\varphi)$, or exponentially as $\underline{Z} = |\underline{Z}|e^{i\varphi}$, where $|\underline{Z}| = \sqrt{R^2 + (L\omega - \frac{1}{C\omega})^2} = \sqrt{R^2 + X^2}$ is the modulus of the complex quantity $\underline{Z}$, and the trigonometric functions that appear in trigonometric form are expressed as functions of R, X, and $|\underline{Z}|$ by $\cos\varphi = \frac{R}{|\underline{Z}|}$, $\sin\varphi = \frac{X}{|\underline{Z}|}$, and $\tan\varphi = \frac{X}{R}$, respectively. If the intensity of the current is expressed in the complex exponential form $\underline{I} = I_0 e^{i(\omega t - \varphi)}$, then, with the help of the complex quantity $\underline{Z} = |\underline{Z}|e^{i\varphi}$ (the impedance operator), we can obtain the complex voltage through the operational equation:

$$\underline{U} = \underline{Z}\underline{I} = |\underline{Z}|e^{i\varphi}\, I_0 e^{i(\omega t - \varphi)} = I_0\, |\underline{Z}|e^{i\omega t} = U_0 e^{i\omega t} \tag{9.194}$$

where:

$$U_0 = I_0\, |\underline{Z}| \text{ or } U_{\text{eff}} = I_{\text{eff}}\, |\underline{Z}| \tag{9.195}$$

Similarly, the quantity $\underline{Y}$ can be expressed algebraically as $\underline{Y} = G - iB$, trigonometrically as $\underline{Y} = |\underline{Y}|(\cos\varphi - i\sin\varphi)$, or exponentially as $\underline{Y} = |\underline{Y}|e^{-i\varphi}$, where $|\underline{Y}| = \sqrt{G^2 + (\frac{1}{L\omega} - C\omega)^2} = \sqrt{G^2 + B^2}$ is the modulus of the complex quantity $\underline{Y}$ and the trigonometric functions that appear in trigonometric form are expressed as functions of G, B, and $|\underline{Y}|$ by $\cos\varphi = \frac{G}{|\underline{Y}|}$, $\sin\varphi = \frac{B}{|\underline{Y}|}$, and $\tan\varphi = \frac{B}{G}$, respectively. If the complex voltage is expressed as $\underline{U} = U_0 e^{i\omega t}$, then, with the help of the complex quantity $\underline{Y} = |\underline{Y}|e^{-i\varphi}$ (the admittance operator), we can obtain the complex intensity $\underline{I}$ through the operational equation:

$$\underline{I} = \underline{Y}\underline{U} = |\underline{Y}|e^{-i\varphi} U_0 e^{i\omega t} = U_0\, |Y|e^{i(\omega t - \varphi)} = I_0 e^{i(\omega t - \varphi)} \tag{9.196}$$

where:

$$I_0 = U_0\, |Y| \text{ or } I_{\text{eff}} = U_{\text{eff}}\, |\underline{Y}| \tag{9.197}$$

The complex representation of AC quantities makes it possible to calculate the exact quantities in a circuit, complementing the phase method, which is an approximative method.

9.5.4 General theorems of linear electric circuits (networks)

1. ***Constituent elements of an electricity grid carrying AC***
   - An *electrical network* is an assembly made up of *active* and *passive* circuit elements, connected to each other in different ways.
   - An element capable of adding energy to the circuit is called an *active circuit element*, while a *passive circuit element* is one that consumes energy.
   - As topography elements, a network consists of nodes, branches (sides), and meshes (loops).
   - In the AC regime, a grid can consist of several subgrids coupled inductively or capacitively.
   - The network is called simple if it consists of a single subnet.
   - The network is complete (or isolated) if it does not have access terminals to the outside.
   - The network is incomplete if it is only a part of the complete network; an incomplete network has access terminals to the outside.
   - When a network has two external access terminals, it is called a dipole; when it has four external access terminals, it is called a quadrupole; and when it has any number $n > 4$ of external terminals, it is generally called multipole.
   - When a network (subnet) has N nodes, the number L of sides that can link the nodes together two by two is:

$$L = \frac{N(N-1)}{2} = \sum_{k=1}^{N-1} k \tag{9.198}$$

   - The number B of independent loops (meshes) is: $B = L - N + S$, where S is the number of subnets. In the case of DC networks, $S = 1$. This means that we only have to deal with a single subgrid, because inductive or capacitive couplings cannot be made between DC subnetworks.
   - In the case of an AC network consisting of S subnetworks, one can write:

$$p = N - S \tag{9.199}$$

an independent equation from Kirchhoff's first theorem. Indeed, if there are S subnets and each subnet has $N_1, N_2, \ldots, N_n$ nodes, then, at each subnet, we can write $N_k - 1$ independent equations. As a result, for S subnets, we can write:

$$p = \sum_{k=1}^{S}(N_k - 1) = (N_1 + N_2 + \ldots + N_S - S) = N - S \tag{9.200}$$

A network is fully characterized if the following are known:

- A schematic of the network, representing the way in which the elements of the network are linked to each other;
- Electricity sources, i.e. their number, the values of the electromotive voltages, and their impedances;
- Energy receptors and their impedances.

2. ***Power generators***

   (a) *The ideal voltage generator.*

   An ideal voltage generator is a source of electricity that produces a time-dependent voltage across its terminals. This voltage remains constant regardless of the load impedance connected to the generator's terminals. An ideal voltage generator is graphically represented in figure 9.20(a), where $\mathcal{E}$ denotes the electromotive force (EMF) of the generator.

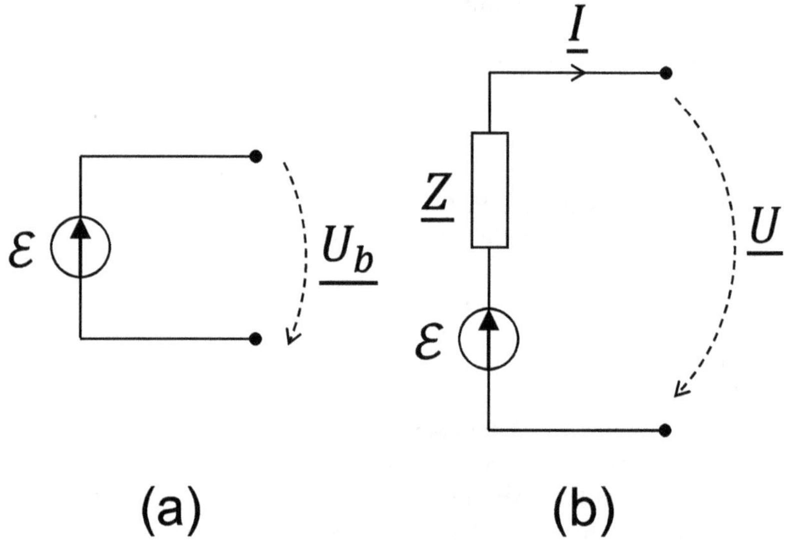

Figure 9.20. Schematic representations of an ideal voltage generator (a) and a voltage generator (b).

   (b) *The real voltage generator.*

   If, in series with the EMF, the generator also has an impedance Z, then it is a *real voltage generator* (see figure 9.20(b)). The voltage U_b at the terminals differs from the EMF $\mathcal{E}$ when a circuit is connected to the terminals. Under open-circuit conditions, the voltage at the terminals U_b becomes equal to the EMF $\mathcal{E}$. This occurs because the current passing through the circuit is zero, resulting in a zero voltage drop across the internal impedance Z. The positive direction of the voltage at the terminals is from a high potential to a low potential, and the positive direction of the electric field induced by the source is from a low potential to a high potential.

(c) *The ideal current generator.*

An ideal current generator is a source of electricity which guarantees that a certain current flows through a circuit connected to its terminals, regardless of the total load impedance of the circuit. This happens when the internal impedance of the generator tends to infinity, which is equivalent to the presence of an internal admittance that tends to zero. We graphically represent an ideal current generator in figure 9.21(a), where the intensity of the current produced by the generator is denoted by J.

(d) *The real current generator.*

If, in parallel with the generator defined above, there is an admittance Y, we are dealing with a real current generator (see figure 9.21(b)). Between terminals, the positive direction of the current is from a higher potential to a lower one, while inside the generator, the positive direction of the current is from a lower potential to a higher one. The intensity of the current passing through the load circuit is:

$$\underline{I} = \underline{J} - \underline{I_Y} \tag{9.201}$$

which is the difference between the short-circuit current and the current absorbed by the internal admittance, $\underline{I_Y}$.

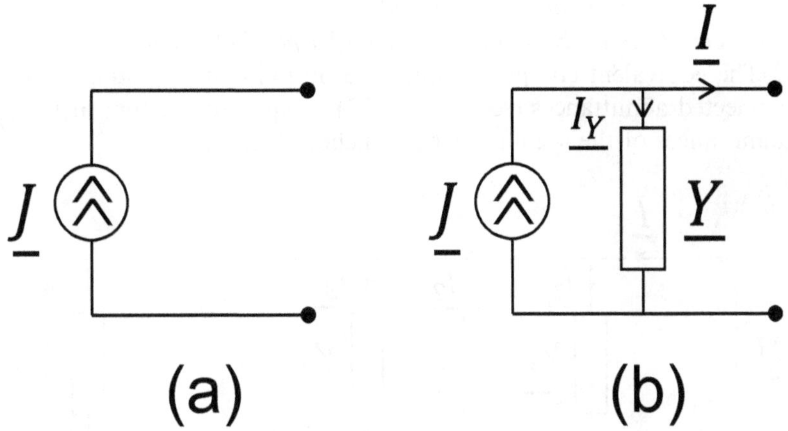

Figure 9.21. Schematic representations of an ideal current generator (a) and a real current generator (b).

3. ***Theorem of equivalent impedance of a series circuit***

Let us consider a circuit consisting of n impedances connected in series (see figure 9.22). The equivalent complex impedance of the circuit is equal to the sum of the complex impedances of the n component parts of the circuit:

$$\underline{Z} = (\underline{Z_1} + \underline{Z_2} + \ldots + \underline{Z_n}) \tag{9.202}$$

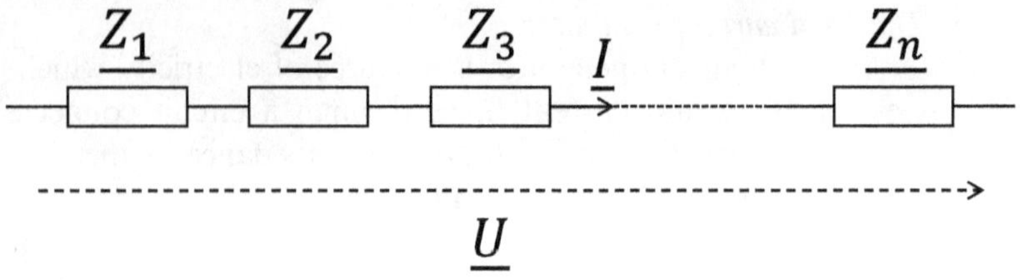

Figure 9.22. Series circuit consisting of n impedances.

Each of the complex impedances in the circuit is composed of a real resistance and an imaginary reactance:

$$\underline{Z_n} = R_n + iX_n \tag{9.203}$$

As a result, equation (9.202) becomes:

$$\underline{Z} = \sum_n R_n + i\sum_n X_n = R + iX \tag{9.204}$$

where:

$$R = \sum_n R_n, \quad X = \sum_n X_n \tag{9.205}$$

Thus, *the equivalent resistance of a series circuit is equal to the sum of its component resistances, and its equivalent reactance is equal to the sum of the reactances of its n component impedances.*

4. ***Theorem of the equivalent admittance of a parallel circuit***

   The equivalent complex admittance of a circuit consisting of n parallel-connected admittances (see figure 9.23) is equal to the sum of the complex admittances of the n component branches of the circuit:

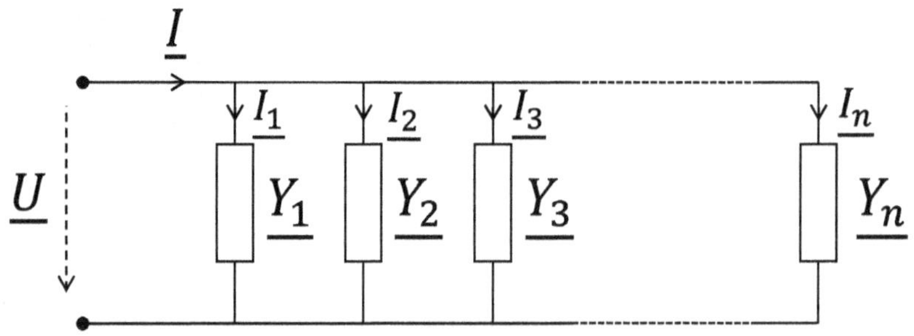

Figure 9.23. Parallel circuit consisting of n branches.

$$\underline{Y} = (\underline{Y_1} + \underline{Y_2} + \ldots + \underline{Y_n}) \tag{9.206}$$

The admittance of any branch of the circuit consists of a real conductance and an imaginary susceptance:

$$\underline{Y_n} = G_n - iB_n \tag{9.207}$$

Taking into account relation (9.206), relation (9.207) becomes:

$$\underline{Y} = \sum_n G_n - i\sum_n B_n = G - iB \tag{9.208}$$

where:

$$G = \sum_n G_n, \quad B = \sum_n B_n \tag{9.209}$$

Therefore, *the equivalent conductance of a parallel circuit is equal to the sum of the conductances of its branches, and its equivalent susceptance is equal to the sum of the susceptances of its component branches.*

5. ***Ohm's law in AC circuits***

For a portion of a circuit that contains voltage sources in addition to passive elements (see figure 9.24), the relationship between the complex voltages takes the form:

$$\underline{U_b} = \underline{\text{I}}\underline{\text{Z}} + \underline{\mathcal{E}} = \underline{U_a} + \underline{\mathcal{E}} \tag{9.210}$$

The electromotive voltage $\mathcal{E}$ has a positive sign '+' if it has the same direction as the current flowing through that branch, and a negative sign '–' if it has the opposite direction to the current in that branch, respectively. Relation (9.210) is the mathematical expression of the complex form of Ohm's theorem.

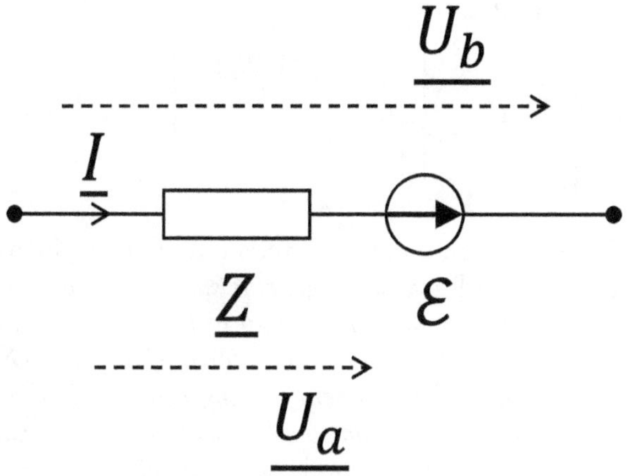

Figure 9.24. Side of an AC circuit containing a voltage generator.

6. ***Kirchhoff's Laws in AC circuits.***

I. Kirchhoff's first theorem refers to a network node by stating that *the algebraic sum of the intensities of the currents competing at a grid node is equal to zero.* For a circuit with $n - 1$ independent nodes, the theorem can be written in the form:

$$\sum_{k=1}^{n} N_{kp}\underline{I_k} = 0 \tag{9.211}$$

where $\underline{I_k}$ represents the intensity in the complex form of the current intensity passing through the k-branch and N_{kp} are the coefficients of node members. The number of lattice sides is equal to l, so $k = 1, 2, \ldots, l$. The index p refers to the number of the node, so $p = 1, 2, 3, \ldots, n - 1$. The N_{kp} coefficient takes the values:

- $+1$ when the current passing through branch k enters node p;
- -1 when the current passing through branch k exits node p;
- 0 when the current passing through branch k does not compete at node p.

Having $n - 1$ equations in the form of relation (9.211), we can move on to the matrix form of Kirchhoff's first law:

$$[N]^t[\underline{I}] = 0 \tag{9.212}$$

where the intensities of the currents competing at the $n - 1$ independent nodes form a column matrix:

$$[\underline{I}] = \begin{Vmatrix} \underline{I_1} \\ \underline{I_2} \\ \underline{I_l} \end{Vmatrix} \tag{9.213}$$

and $[N]^t$ is transposed into the matrix $[N]$ of member coefficients, the latter having the form:

$$[\underline{N}]^t = \begin{bmatrix} N_{11} & N_{12} \cdots N_{1,n-1} \\ & \vdots \ddots \vdots \\ N_{l1} & N_{l2} \cdots N_{l,n-1} \end{bmatrix} \tag{9.214}$$

II. Kirchhoff's second theorem refers to a loop in a network and is formulated as follows: *the algebraic sum of the electromotive voltages in complex form along a loop is equal to the algebraic sum of the products of the total impedance in complex form of one branch and the intensity in complex form of the current flowing through that branch (the voltage drop in complex form) that occur on the network branches of the loop.* In order to write the algebraic sums of the analytic quantities, we use the positive direction of the loop. For a network containing a single loop, Kirchhoff's second theorem can be written in the form:

$$\sum_{k=1}^{l} T_{kq}\underline{\mathcal{E}_k} = \sum_{k=1}^{l} T_{kq}\underline{U_k} \tag{9.215}$$

where l is the number of the sides belonging to one loop of the network (circuit) and q is the number of fundamental loops. For electromotive voltages, as for the voltage drops, complex expressions are used. In equation (9.215), the coefficients T_{kq}, called the coefficients of loop membership, take the following values:

- $+1$ when branch k belongs to loop q and has the chosen positive sense;
- -1 when branch k belongs to loop q and has the opposite direction to that considered above;
- 0 when branch k does not belong to loop q.

With these, Kirchhoff's second theorem can be written using the matrices for the electric quantities in equation (9.215), as follows:

$$[T]^t[\underline{\mathcal{E}}] = [T]^t[\underline{U}] \tag{9.216}$$

where the electromotive voltages and voltage drops are written in the form of column matrices and $[T]^t$ is transposed into the matrix $[T]$ of the member coefficients, the latter being:

$$[T] = \begin{bmatrix} T_{11} & T_{12} \cdots T_{1o} \\ & \vdots \ddots \vdots \\ T_{l1} & T_{l2} \cdots N_{lo} \end{bmatrix} \tag{9.217}$$

9.6 Special methods for the analysis of AC electric circuits

Based on Kirchhoff's theorems, as in the case of DC networks, various methods of solving networks based on the complex representation of known and unknown electrical quantities can also be used for AC circuits. Some methods use theorems to more easily solve circuits in the AC regime, as follows.

(1) ***Superposition theorem***

In a complete electrical network, the complex current intensity in any branch equals the sum of the complex currents that would flow through that branch if each EMF source were activated individually. When each source is considered separately, the other sources are replaced by their complex internal impedances:

$$\underline{I}_j = \sum_{k=1}^{n} \underline{I}_{jk} \tag{9.218}$$

As a method of network analysis, this method is applied to circuits with two or, at most, three sources. If several sources are present, the number of calculations increases considerably compared to other methods.

(2) ***The cyclic current method***

This method reduces the system of complex equations given by Kirchhoff's theorems from a system of L equations with L unknowns to a system that has a number $B = L - N + S$ of equations and B unknowns, which are fictional contour currents in complex form, $\underline{I}_{c1}, \underline{I}_{c2}, \ldots, \underline{I}_{cB}$, supposed to be forced to flow in each fundamental loop of the network. The following system can be written for them:

$$\begin{cases} \underline{I_{c1}}\,\underline{Z_{11}} + \underline{I_{c2}}\,\underline{Z_{12}} + \ldots + \underline{I_{cB}}\,\underline{Z_{1B}} = \underline{\mathcal{E}_{c1}} \\ \underline{I_{c1}}\,\underline{Z_{21}} + \underline{I_{c2}}\,\underline{Z_{22}} + \ldots + \underline{I_{cB}}\,\underline{Z_{2B}} = \underline{\mathcal{E}_{c2}} \\ \underline{I_{c1}}\,\underline{Z_{j1}} + \ldots + \underline{I_{cj}}\,\underline{Z_{jj}} + \underline{I_{ck}}\underline{Z_{jk}} + \ldots + \underline{I_{cB}}\,\underline{Z_{jB}} = \underline{\mathcal{E}_{cj}} \\ \underline{I_{c1}}\,\underline{Z_{B1}} + \underline{I_{c2}}\,\underline{Z_{B2}} + \ldots + \underline{I_{cB}}\,\underline{Z_{BB}} = \underline{\mathcal{E}_{cB}} \end{cases} \tag{9.219}$$

where $\underline{Z_{jj}}$ is the total complex impedance of loop j. $\underline{Z_{jk}}$ is the complex impedance of the side common to loop j and loop k; it is given the sign '+' when the neighboring cyclic currents $\underline{I_{cj}}$ and $\underline{I_{ck}}$ flow through it in the same direction and sense, and given the sign '−' when the neighboring cyclic currents $\underline{I_{cj}}$ and $\underline{I_{ck}}$ flow through it in opposite senses. $\underline{\mathcal{E}_{cj}}$ is the algebraic sum of the complex electromotive forces from the network present along loop j.

(3) ***The theorem of nodes' potentials***

The method is recommended for circuits for which $N - 1 < B$ and it allows the determination of the potentials of the network nodes in relation to the potential of one of the nodes, taken as a reference potential. If, between two nodes α and β of a network, there is an admittance $\underline{Y_{\alpha\beta}}$ and a complex voltage $\underline{\mathcal{E}_{\alpha\beta}}$, then for node α, we can write the equation:

$$\underline{V_\alpha}\sum_{\beta=1}^{N}\underline{Y_{\alpha\beta}} - \sum_{\beta=1}^{N}\underline{V_\beta}\,\underline{Y_{\alpha\beta}} + \sum_{\beta=1}^{N}\underline{\mathcal{E}_{\alpha\beta}}\,\underline{Y_{\alpha\beta}} = 0 \tag{9.220}$$

In a circuit with N nodes, this yields $N - 1$ equations of the form (9.220). One of the nodes can be associated with the value 0 as a reference potential. The system's unknowns are the potentials of the $N - 1$ nodes. After solving the system of $N - 1$ equations with $N - 1$ unknowns and thus obtaining the potentials of the $N - 1$ nodes, we can find the currents on each branch of the network using an equation of the type:

$$\underline{I_{\alpha\beta}} = (\underline{V_\alpha} - \underline{V_\beta} + \underline{\mathcal{E}_{\alpha\beta}})\underline{Y_{\alpha\beta}} \tag{9.221}$$

(4) ***The equivalent voltage generator theorem (Thévenin's theorem)***

The intensity of a current passing through a passive element connected between nodes A and B of a complex active network (see figure 9.25) is equal to the ratio between the voltage that would be established at the ends of the passive element in its absence, $\underline{U_{AB_0}}$ (the open-circuit voltage), and the sum of the complex impedance $\underline{Z}$ of the passive element and the equivalent complex impedance of the passivated active network $\underline{Z_{AB_0}}$:

$$\underline{I_{AB}} = \frac{\underline{U_{AB_0}}}{\underline{Z} + \underline{Z_{AB_0}}} \tag{9.222}$$

By 'the impedance of the passivated active network', we mean the equivalent impedance $\underline{Z_{AB_0}}$ between nodes A and B of the active network, from which

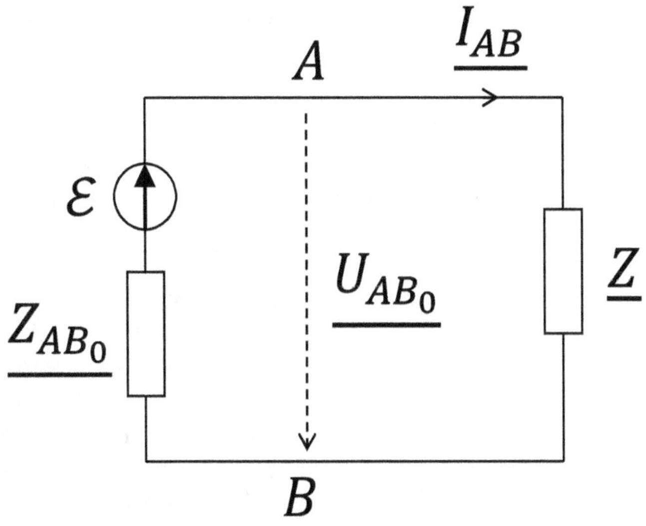

Figure 9.25. Illustration of Thévenin's theorem.

the electromotive voltages of the sources have been removed but their internal complex impedances have been preserved. Based on Thévenin's theorem, an active part of the network can be equated with a voltage source.

(5) ***The equivalent current generator theorem (Norton's theorem)***

The voltage across the terminals A and B of a passive element that is part of a complex active network (see figure 9.26) is equal to the ratio between the intensity $\underline{I_{AB_{sc}}}$ of the current flowing between nodes A and B under short-circuit conditions and the sum of the complex admittance $\underline{Y}$ of the

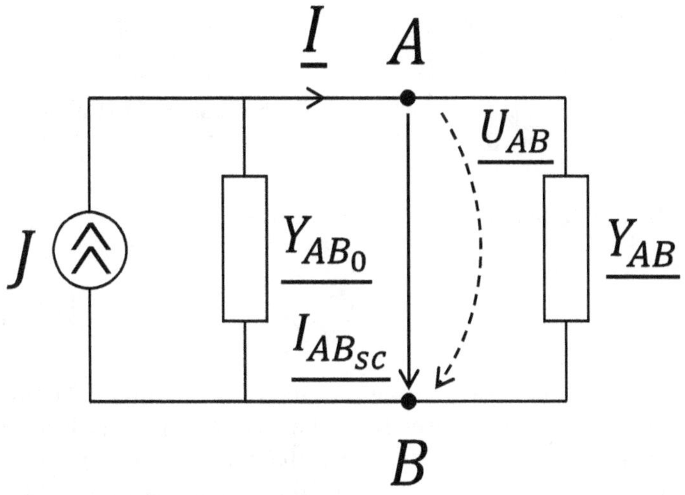

Figure 9.26. Illustration of Norton's theorem.

passive element and the equivalent admittance $\underline{Y}_{AB_0}$ of the passivated active network between the two terminals A and B:

$$\underline{U}_{AB} = \frac{\underline{I}_{AB_{sc}}}{\underline{Y} + \underline{Y}_{AB}} \tag{9.223}$$

Based on Norton's theorem, an active network can be equated with a source of electric current.

(6) ***Theorem of maximum active power transfer in AC networks.***

This theorem establishes the conditions for the transfer of maximum active power from an EMF source (or an active network) to a receiver (consumer). The theorem can be formulated as follows: *a voltage source with an EMF of $\underline{\mathcal{E}}$ and a complex internal impedance of $\underline{Z}_i$ (see figure 9.27) which feeds a load impedance network $\underline{Z}$ will transfer to this the maximum active power when the complex impedance of the passive network is equal to the internal impedance of the source in conjugate complex form*, i.e. $\underline{Z} = \underline{Z}_i^*$.

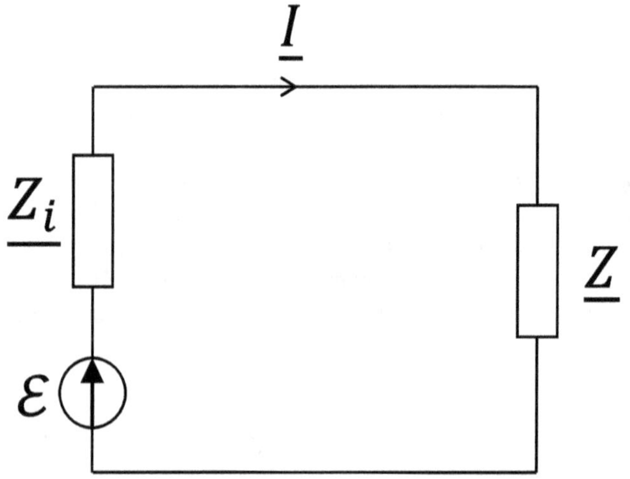

Figure 9.27. Simple AC circuit containing a voltage source with an EMF of $\underline{\mathcal{E}}$ and an internal impedance $\underline{Z}_i$ which feeds an impedance network $\underline{Z}$.

The maximum active power transferred is:

$$P_{a_{\max}} = \frac{\mathcal{E}_{\mathrm{eff}}^2}{4R_i} \tag{9.224}$$

where R_i is the internal resistance of the voltage source. The problem of the transfer of maximum power from the source to load resistance in the DC regime has been dealt with in the chapters on DC circuits. In order to highlight the specific elements of the problem in the case of AC circuits, the proof of this theorem will be presented below. Thus, with the exception of the active power transfer theorem, the other theorems stated above have practically the same formulations as in the case of DC networks, the difference being that the equations and systems of equations are written in terms of complex numbers and complex quantities. The intensity of the current in the circuit shown in figure 9.27 is:

$$\underline{I} = \frac{\underline{\mathcal{E}}}{\underline{Z} + \underline{Z}_i} \tag{9.225}$$

where: $\underline{Z} = R + iX$ and $\underline{Z}_i = R_i + iX_i$. The relationship between the actual values of the current intensity and the electromotive voltage is also valid:

$$I_{\text{eff}} = \frac{\mathcal{E}_{\text{eff}}}{\sqrt{(R + R_i)^2 + (X + X_i)^2}} \tag{9.226}$$

where R_i and X_i are the resistance and reactance of the generator and R and X are the resistance and reactance of the receiver (consumer). The active power transferred from the generator to the receiver has the general expression:

$$P_a = RI_{\text{eff}}^2 = \frac{R\mathcal{E}_{\text{eff}}^2}{(R + R_i)^2 + (X + X_i)^2} \tag{9.227}$$

The expression for the active power P_a has variables for both R and X. Therefore, we can write the condition for the maximum power transfer between the source and the circuit as follows:

$$\begin{aligned} \left.\frac{\partial P_a}{\partial X}\right|_{R,\, R_i,\, X_i} &= 0 \\ \left.\frac{\partial P_a}{\partial R}\right|_{R_i,\, X,\, X_i} &= 0 \end{aligned} \tag{9.228}$$

Keeping R, R_i, and X_i constant and varying X gives us the first condition under which the maximum active power transfer is obtained:

$$\frac{\partial P_a}{\partial X} = -\frac{2R\mathcal{E}_{\text{eff}}^2(X + X_i)}{[(R + R_i)^2 + (X + X_i)^2]^2} = 0 \tag{9.229}$$

which leads to:

$$X = -X_i \tag{9.230}$$

In other words, in order for the maximum active power to be transferred from the source to the receiver, the reactance of the receiver must be equal to that of the generator but with the opposite sign. Keeping R_i, X_i, and X constant and varying R, we find a second condition for the maximum transfer of active power. Starting from:

$$\frac{\partial P_a}{\partial R} = \frac{\mathcal{E}_{\text{eff}}^2\{[(R+R_i)^2 + (X+X_i)^2] - 2R(R+R_i)\}}{[(R+R_i)^2 + (X+X_i)^2]^2} = 0$$

and taking into account that $X = -X_i$, we obtain:

$$R = R_i \tag{9.231}$$

The simultaneous fulfillment of conditions (9.230) and (9.231) leads to the expression for the receiver impedance:

$$\underline{Z} = R_i - iX_i \tag{9.232}$$

That is, in order for the maximum active power to be transferred from the generator to the receiver, the receiver's impedance must equal the conjugate complex of the internal impedance of the generator, which is given by: $\underline{Z_i^*} = R_i - iX_i$.

Application

As an example of the methods of analysis of AC networks in complex representation, a 'Boucherot' circuit will be studied. Let us consider the circuit shown in figure 9.28, to which an alternating voltage is applied, represented in complex form by $\underline{U} = U_0 e^{i\omega t}$. Find out under what conditions the intensity of the current passing through the resistor R does not depend on its value.

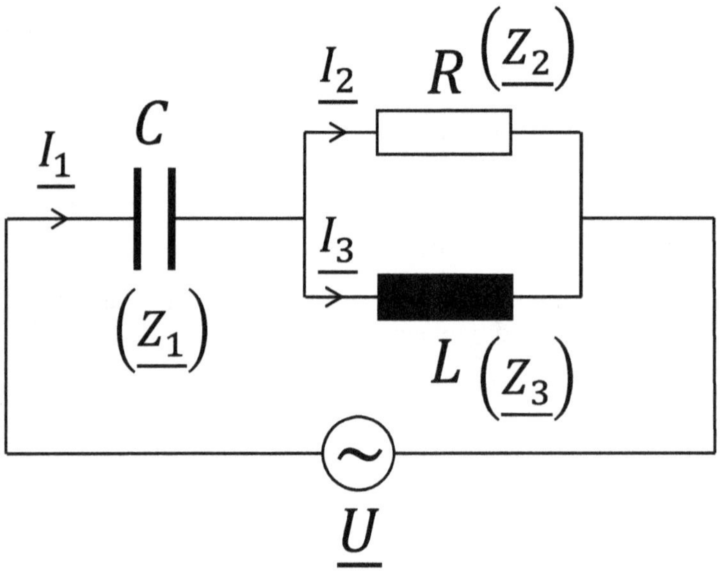

Figure 9.28. *'Boucherot'-type circuit.*

Solution:

To solve the problem, we first calculate the intensity of the current passing through R; then, the condition must be applied that it does not depend on R. Recalling the complex representation, the complex impedances associated with the passive circuit elements are:

$$\underline{Z_1} = \frac{1}{iC\omega}, \quad \underline{Z_2} = R, \quad \underline{Z_3} = iL\omega \tag{9.233}$$

and by using the equivalence theorems of the series and parallel groups for impedances, the total impedance of the circuit is obtained, as follows:

$$\underline{Z} = \underline{Z_1} + \frac{R\underline{Z_3}}{R + \underline{Z_3}} \tag{9.234}$$

Considering the complex voltage $\underline{U}$ and the total complex impedance $\underline{Z}$ of the circuit, according to Ohm's law, the complex intensity of the current can be expressed as:

$$\underline{I_1} = \frac{\underline{U}}{\underline{Z_1} + \frac{R\underline{Z_3}}{R+\underline{Z_3}}} \tag{9.235}$$

or:

$$\underline{I_1} = \frac{\underline{U}(R + \underline{Z_3})}{(R + \underline{Z_3})\underline{Z_1} + R\underline{Z_3}} \tag{9.236}$$

For the loop formed by R and $\underline{Z_3}$, according to Kirchhoff's second theorem, one can write:

$$R\underline{I_2} = \underline{Z_3}\,\underline{I_3} \tag{9.237}$$

and for the node in which R, $\underline{Z_1}$, and $\underline{Z_2}$ compete, Kirchhoff's first law leads to: $\underline{I_1} = \underline{I_2} + \underline{I_3}$. Then, $\underline{I_2}$ is:

$$\underline{I_2} = \underline{I_1} - \underline{I_3} = \underline{I_1} - \frac{R\underline{I_2}}{\underline{Z_3}} \tag{9.238}$$

or:

$$\underline{I_2} = \frac{\underline{I_1}}{1 + \frac{R}{\underline{Z_3}}} = \frac{\frac{(R+\underline{Z_3})\underline{U}}{(R+\underline{Z_3})\underline{Z_1} + R\underline{Z_3}}}{\frac{(R+\underline{Z_3})}{\underline{Z_3}}}$$

$$\Longrightarrow \underline{I_2} = \frac{\underline{U}\;\underline{Z_3}}{(\underline{Z_1} + \underline{Z_3})R + \underline{Z_1}\underline{Z_3}} \tag{9.239}$$

From this equation, we find that $\underline{I_2}$ does not depend on R if the condition $\underline{Z_1} + \underline{Z_3} = 0$ is satisfied or:

$$\underline{Z_1} = -\underline{Z_3} \tag{9.240}$$

Taking into account the complex expressions for $\underline{Z_1}$ and $\underline{Z_3}$, it follows that:

$$\frac{1}{iC\omega} = -\, iL\omega \Longrightarrow \omega = \frac{1}{\sqrt{LC}} \tag{9.241}$$

Thus, $\underline{I_2}$ does not depend on R if the inductance L and the capacitance C have values that satisfy a resonant condition. When this condition is met, the current $\underline{I_2}$ passing through resistance R becomes:

$$\underline{I_2} = \frac{\underline{U}\ \underline{Z_3}}{\underline{Z_1}\underline{Z_3}} = \frac{\underline{U}}{\underline{Z_1}} = \underline{U}iC\omega \tag{9.242}$$

a relation which shows that $\underline{I_2}$ is phase shifted by $\frac{\pi}{2}$ with respect to the applied voltage.

9.7 Complex representation of power in AC circuits

9.7.1 Complex apparent power

By associating complex numbers with the electrical quantities of the current, in an appropriate form, the complex expression for the apparent power can be found. A series RLC circuit carrying AC is considered again. For an applied voltage U, the associated complex electrical quantity is:

$$\underline{U} = U_{\text{eff}}\, e^{i\omega t} \tag{9.243}$$

and for the intensity of the current, the complex electrical quantity is:

$$\underline{I} = I_{\text{eff}}\, e^{i(\omega t-\varphi)} \tag{9.244}$$

where φ is the phase shift of the current with respect to the voltage. The conjugate complex form of the current intensity is:

$$\underline{I^*} = I_{\text{eff}}\, e^{-i(\omega t-\varphi)} \tag{9.245}$$

By taking the product of $\underline{U}$ and $\underline{I^*}$, we obtain:

$$\underline{U}\underline{I^*} = U_{\text{eff}} I_{\text{eff}}\, e^{i\varphi} \tag{9.246}$$

or:

$$\underline{U}\underline{I^*} = U_{\text{eff}} I_{\text{eff}} (\cos\varphi + i\sin\varphi) \tag{9.247}$$

Using the real part of expression (9.247), we find the active power:

$$\Re[\underline{U}\underline{I^*}] = U_{\text{eff}} I_{\text{eff}} \cos\varphi = P_a \tag{9.248}$$

and using the imaginary part of the expression (9.247), we find the reactive power:

$$\Im[\underline{U}\underline{I^*}] = U_{\text{eff}} I_{\text{eff}} \sin\varphi = P_r \tag{9.249}$$

Thus, the product of $\underline{U}$ and $\underline{I^*}$ can be rewritten as:

$$\underline{U}\underline{I^*} = P_a + iP_r \tag{9.250}$$

By taking the product of the complex quantity $P_a + iP_r$ and its conjugate complex $P_a - iP_r$, we obtain the square of the apparent power:

$$(P_a + iP_r)(P_a - iP_r) = P_a^2 + P_r^2 = P_A^2 \qquad (9.251)$$

Thus, a known result has been reached that allows us to equate the apparent power $\underline{P_A}$ in complex form with the product $\underline{U}\underline{I}^*$:

$$\underline{P_A} = \underline{U}\underline{I}^* = P_a + iP_r \qquad (9.252)$$

9.7.2 Oscillating power expression

In order to explain the title of this subsection, we recall the expression for instantaneous power:

$$P_i = UI = 2U_{\text{eff}}I_{\text{eff}} \cos \omega t \cos (\omega t - \varphi) \qquad (9.253)$$

which can be changed into the form:

$$P_i = U_{\text{eff}}I_{\text{eff}} \cos (2\omega t - \varphi) + U_{\text{eff}}I_{\text{eff}} \cos \varphi \qquad (9.254)$$

where the second term is the active power:

$$P_a = U_{\text{eff}}I_{\text{eff}} \cos \varphi \qquad (9.255)$$

This is the average power consumed in the form of Joule heating in the resistance of the circuit, and its first term is called the *oscillating power*:

$$P_o = U_{\text{eff}}I_{\text{eff}} \cos (2\omega t - \varphi) \qquad (9.256)$$

The oscillating power can also be expressed as the real part of the product of the complex quantities $\underline{U}$ and $\underline{I}$ given by relations (9.243) and (9.244), respectively:

$$\underline{U}\,\underline{I} = U_{\text{eff}}I_{\text{eff}}e^{i(2\omega t - \varphi)}$$

$$\Longrightarrow \underline{U}\,\underline{I} = U_{\text{eff}}I_{\text{eff}}[\cos (2\omega t - \varphi) + i \sin (2\omega t - \varphi)] \qquad (9.257)$$

Thus, the real part of the product $\underline{U}\,\underline{I}$ is exactly the oscillating power:

$$P_o = \Re[\underline{U}\,\underline{I}] = U_{\text{eff}}I_{\text{eff}} \cos (2\omega t - \varphi) \qquad (9.258)$$

9.7.3 Diagram of powers in the complex plane

The total instantaneous power and power components, the expressions of which have been determined using both analytic and complex representations, can be rendered using a representation in the complex plane called a *power diagram*. The power diagram is constructed in the complex plane with the help of a circle whose radius, drawn between the center of the circle and the origin of the axes, makes an angle of φ with the axis of the real numbers (see figure 9.29). The radius OC is equal to the product $U_{\text{eff}}I_{\text{eff}}$ and is therefore equal to the apparent power (P_A). The projection of OP on the real axis

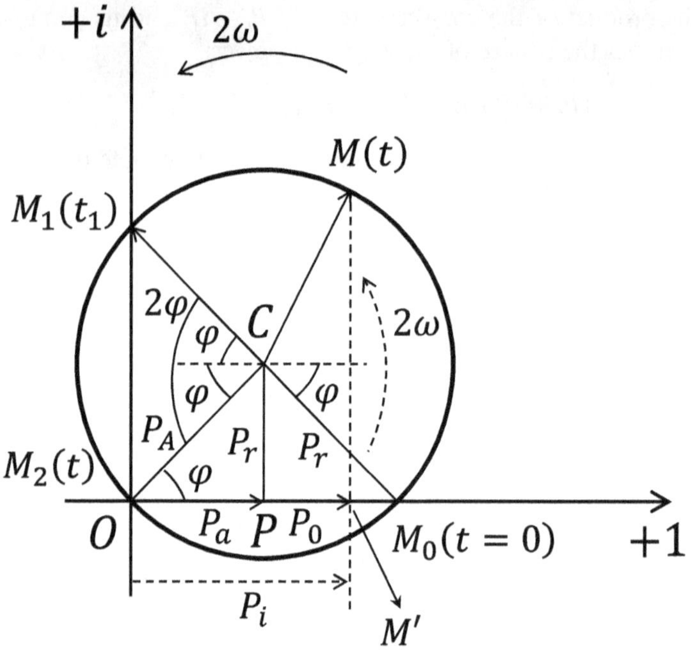

Figure 9.29. Power diagram in the complex plane.

is equal to the active power P_a, and the projection on the imaginary axis is equal to the reactive power P_r. Let us imagine a vector CM that behaves according to $P_A = U_{\text{eff}} I_{\text{eff}}$. It rotates in the trigonometric direction with angular velocity 2ω. Note that its projection on the real axis is the oscillating power, i.e. $P_o = U_{\text{eff}} I_{\text{eff}} \cos(2\omega t - \varphi)$. On the real axis, we now have the sum $P_a + P_o$, which is precisely the instantaneous power $P_i = P_a + P_o = U_{\text{eff}} I_{\text{eff}} [\cos\varphi + \cos(2\omega t - \varphi)] = OM'$. When the projection is on the positive side of the real axis, we have the ability to find the power transferred from the source to the circuit at a given moment. When the projection is on the negative side of the real axis, we have the ability to find the power transferred from the circuit to the source. Between points M_1 and M_2, we have an arc of a circle swept by the extremity of the rotating vector in the time interval $\Delta t = t_2 - t_1 = \frac{\varphi}{\omega} = \frac{2\varphi}{2\omega}$, the interval determined by the phase shift φ between intensity and voltage.

If $\varphi = 0$, the circle's center is on the real axis and tangent to the imaginary axis at the origin (see figure 9.30(a)). In this case, $P_a = P_A = U_{\text{eff}} I_{\text{eff}}$, meaning that the source transfers power to the circuit without receiving anything back. The instantaneous power P_i takes values between 0 and $2U_{\text{eff}} I_{\text{eff}}$. This is true for the case where the circuit consists of a resistor.

If $\varphi = \frac{\pi}{2}$, the circle has its center on the imaginary axis and is tangent at to the real axis at the origin, see figure 9.30(b). We note that $P_r = P_A = U_{\text{eff}} I_{\text{eff}}$, meaning that the power transferred to the circuit is received entirely by the source in an oscillatory manner. The circuit does not consume power (energy). The instantaneous power P_i

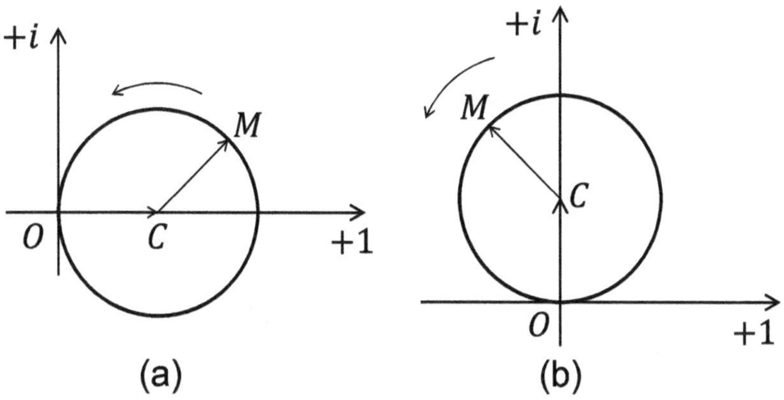

Figure 9.30. Power diagrams in the complex plane for: (a) a purely resistive circuit and (b) a purely reactive circuit.

takes values between $-U_{\text{eff}}I_{\text{eff}}$ and $+U_{\text{eff}}I_{\text{eff}}$. This situation occurs when the circuit consists of a reactance.

Further reading

[1] Purcell E M and Morin D J 2013 *Electricity and Magnetism* 3rd edn (Cambridge: Cambridge University Press) 853 pages

[2] Feynman R P, Leighton R B and Sands M 2011 *Feynman Lectures on Physics, Vol. II: The New Millennium Edition: Mainly Electromagnetism and Matter (Feynman Lectures on Physics)* (New York: Basic Books) 589 pages

[3] Secăreanu I, Ruxandra V, Gherbanovschi N, Logofătu M, Cazan-Corbasca M and Antohe Ş 1984 *Problems of Electricity and Magnetism (Culegere de probleme de Electricitate şi Magnetism)* (Bucharest: University of Bucharest Publishing House)

[4] Dorf R C and Svoboda J A 2018 *Dorf's Introduction to Electric Circuits* 9th edn (Hoboken, NJ: Wiley Global) 912 pages

[5] Alexander C K and Sadiku M 2020 *ISE Fundamentals of Electric Circuits (ISE HED IRWIN ELEC&COMPUTER ENGINERING)* 7th edn (New York: McGraw-Hill Education)

[6] Zeng G L and Zeng M 2021 *Electric Circuits: A Concise, Conceptual Tutorial* 1st edn (Berlin: Springer) 315 pages

[7] Barr J R 2023 *Principles of Direct-Current Electrical Engineering* (London: Palala Press) 592 pages

[8] O'Malley J 2011 *Schaum's Outline of Basic Circuit Analysis* 2nd edn (New York: McGraw-Hill) 432 pages

[9] Nasar S 1988 *3000 Solved Problems in Electrical Circuits* 1st edn (New York: McGraw-Hill) 768 pages

[10] Fiore J M 2020 *DC Electrical Circuit Analysis: A Practical Approach* (New York: Independently published) 374 pages

IOP Publishing

Electromagnetism and Special Methods for Electric Circuits Analysis

Ştefan Antohe and Vlad-Andrei Antohe

Chapter 10

Electromagnetic fields and electromagnetic waves

Based on a phenomenological analysis of the main laws of classical electromagnetism (i.e. the law of electromagnetic induction, the law of magnetic circuits, as well as the law of electric and magnetic flux), it has generally been found that the electromagnetic field can be described with the help of the local forms of these laws. These forms represent the most appropriate expression (corresponding to the concept of localization of all physical actions and properties) of the structure of this physical system called the electromagnetic field.

10.1 Maxwell's equations

Following the path presented above to describe the electromagnetic field, a system of equations with partial derivatives was built over time. This system essentially represents the local forms of the general laws of electromagnetic fields in stationary environments (i.e. where the local velocity is zero) and in the domains of continuity and smoothness of the local physical properties. It is called the *table of Maxwell's equations* and consists of the following equations:

$$\begin{cases} \nabla \times \vec{E} = -\dfrac{\partial \vec{B}}{\partial t} \\ \nabla \times \vec{H} = \vec{j} + \dfrac{\partial \vec{D}}{\partial t} \\ \nabla \vec{D} = \rho \\ \nabla \vec{B} = 0 \end{cases} \tag{10.1}$$

doi:10.1088/978-0-7503-5854-5ch10

Maxwell's system of equations is completed by the relations between $\vec{D}$ and $\vec{E}$, between $\vec{B}$ and $\vec{H}$, and between $\vec{j}$ and $\vec{E}$, which in linear environments represent *the constitutive equations*:

$$\begin{cases} \vec{D} = \varepsilon \vec{E} \\ \vec{B} = \mu \vec{H} \\ \vec{j} = \sigma \vec{E} \end{cases} \tag{10.2}$$

From a mathematical perspective, Maxwell's system of equations is a system of eight equations with partial derivatives, simultaneously satisfied by six unknowns, which are components of the vectors $\vec{E}(\vec{r}, t)$ and $\vec{H}(\vec{r}, t)$:

$$\begin{cases} \frac{\partial E_z}{\partial y} - \frac{\partial E_y}{\partial z} = -\mu \frac{\partial H_x}{\partial t} \\ \frac{\partial E_x}{\partial z} - \frac{\partial E_z}{\partial x} = -\mu \frac{\partial H_y}{\partial t} \\ \frac{\partial E_y}{\partial x} - \frac{\partial E_x}{\partial y} = -\mu \frac{\partial H_z}{\partial t} \\ \frac{\partial H_z}{\partial y} - \frac{\partial H_y}{\partial z} = j_x + \varepsilon \frac{\partial E_x}{\partial t} \\ \frac{\partial H_x}{\partial z} - \frac{\partial H_z}{\partial x} = j_y + \varepsilon \frac{\partial E_y}{\partial t} \\ \frac{\partial H_y}{\partial x} - \frac{\partial H_x}{\partial y} = j_z + \varepsilon \frac{\partial E_z}{\partial t} \\ \frac{\partial E_x}{\partial x} + \frac{\partial E_y}{\partial y} + \frac{\partial E_z}{\partial z} = \frac{\rho}{\varepsilon} \\ \frac{\partial H_x}{\partial x} + \frac{\partial H_y}{\partial y} + \frac{\partial H_z}{\partial z} = 0 \end{cases} \tag{10.3}$$

In the theory of systems of equations with partial derivatives, it is shown that the solutions $\vec{E}(\vec{r}, t)$ and $\vec{H}(\vec{r}, t)$ of such a system are uniquely determined if the following are known: the constants ε and μ, the sources of the fields ρ and j_x, j_y, j_z, as well as the initial and boundary conditions of the components of the electromagnetic field. The lack of symmetry in $\vec{E}(\vec{r}, t)$ and $\vec{B}(\vec{r}, t)$ in Maxwell's equations is due, as observed, to the presence of the electric charge density ρ and the conduction current $\vec{j}$. In vacuum, these terms are zero, and Maxwell's equations become:

$$\begin{cases} \nabla \times \vec{E} = -\frac{\partial \vec{B}}{\partial t} = -\mu_0 \frac{\partial \vec{H}}{\partial t} \\ \nabla \times \vec{H} = \frac{\partial \vec{D}}{\partial t} = \varepsilon_0 \frac{\partial \vec{E}}{\partial t} \\ \nabla \vec{E} = 0 \\ \nabla \vec{H} = 0 \end{cases} \tag{10.4}$$

Here, the term for the displacement current density $\frac{\partial \vec{D}}{\partial t}$ is very important. Together with its analog in the first equation, this term allows the equations to be symmetric. This suggests the possibility of the mutual generation of an electric and magnetic field and its detachment from its sources, thus leading to the notions of the electromagnetic field and the electromagnetic wave. Recognizing this, Maxwell went on to develop the electromagnetic theory of light, i.e. *a disturbance of the electromagnetic field that propagates away from the sources of the field, forming an electromagnetic wave.*

10.2 The electromagnetic wave equation

As shown above, from an analysis of the system of laws of electromagnetic fields and from an analysis of Maxwell's system of equations, we find that in time-varying fields there is a double causal connection between the electric field and the magnetic field. This is due to the law of electromagnetic induction and, respectively, due to the displacement current expressed by the law of magnetic circuits. This connection confirms the existence of electromagnetic fields.

Starting from Maxwell's system of equations in vacuum ($\rho = 0$, $\vec{j} = 0$, $\sigma = 0$, $\varepsilon = \varepsilon_0$, and $\mu = \mu_0$), we will obtain the wave equation for the component $\vec{E}$ or $\vec{H}$ of the electromagnetic field, as follows:

$$\begin{cases} \nabla \times (\nabla \times \vec{E}) = - \nabla \times \left(\mu_0 \dfrac{\partial \vec{H}}{\partial t} \right) \\ \nabla(\nabla \vec{E}) - \Delta \vec{E} = -\mu_0 \dfrac{\partial}{\partial t} (\nabla \times \vec{H}) \\ - \Delta \vec{E} = -\mu_0 \dfrac{\partial}{\partial t} \left(\varepsilon_0 \dfrac{\partial \vec{E}}{\partial t} \right) \\ \Delta \vec{E} = \varepsilon_0 \mu_0 \dfrac{\partial^2 \vec{E}}{\partial t^2} \end{cases} \tag{10.5}$$

The Laplace operator acts on each component of the electric field, so the electromagnetic wave equation for a component of the electric field is:

$$\frac{\partial^2 E_x}{\partial x^2} + \frac{\partial^2 E_x}{\partial y^2} + \frac{\partial^2 E_x}{\partial z^2} - \varepsilon_0 \mu_0 \frac{\partial^2 E_x}{\partial t^2} = 0 \tag{10.6}$$

Similarly, for the magnetic component of the electromagnetic field, the following are obtained:

$$\begin{cases} \nabla \times (\nabla \times \vec{H}) = \nabla \times \left(\varepsilon_0 \dfrac{\partial \vec{E}}{\partial t} \right) \\ \nabla(\nabla \vec{H}) - \Delta \vec{H} = \varepsilon_0 \dfrac{\partial}{\partial t} (\nabla \times \vec{E}) \\ - \Delta \vec{H} = \varepsilon_0 \dfrac{\partial}{\partial t} (\nabla \times \vec{E}) = -\varepsilon_0 \mu_0 \dfrac{\partial^2 \vec{H}}{\partial t^2} \\ \Delta \vec{H} = \varepsilon_0 \mu_0 \dfrac{\partial^2 \vec{H}}{\partial t^2} \end{cases} \tag{10.7}$$

or, for a component of the magnetic field:

$$\frac{\partial^2 H_x}{\partial x^2} + \frac{\partial^2 H_x}{\partial y^2} + \frac{\partial^2 H_x}{\partial z^2} - \varepsilon_0 \mu_0 \frac{\partial^2 H_x}{\partial t^2} = 0 \tag{10.8}$$

Therefore, the two components E_x and H_x satisfy the same equation. If we choose only the electrical component, the solution of equation (10.5) can be written in the form of a complex variable function:

$$\vec{E} = \vec{E_0} e^{i\omega\left(t - \frac{x}{v}\right)} \tag{10.9}$$

or in the form of a trigonometric function:

$$\vec{E} = \vec{E_0} \cos \omega \left(t - \frac{x}{v}\right) \tag{10.10}$$

where $\vec{E_0}$ is the amplitude of the wave, ω its angular frequency, v is the speed of propagation of the monochromatic plane wave, and t and x are the temporal and spatial variables, respectively. In relations (10.9) and (10.10), the direction OX was chosen as the direction of propagation. Assuming that one of the functions (10.9) or (10.10) verifies the wave equation (10.6) for the component E_x of the electric field, we can obtain the speed of propagation of the wave ($\vec{v}$):

$$\begin{aligned}
&\frac{\partial^2 E_x}{\partial x^2} + \frac{\partial^2 E_x}{\partial y^2} + \frac{\partial^2 E_x}{\partial z^2} - \varepsilon_0 \mu_0 \frac{\partial^2 E_x}{\partial t^2} = 0 \\
&\Rightarrow \vec{E_0}\left(-\frac{i\omega}{v}\right)\left(-\frac{i\omega}{v}\right) e^{i\omega\left(t - \frac{x}{v}\right)} - \varepsilon_0 \mu_0 \vec{E_0} (i\omega)^2 e^{i\omega\left(t - \frac{x}{v}\right)} = 0 \\
&\Rightarrow \left(-\frac{\omega^2}{v^2} + \varepsilon_0 \mu_0 \omega^2\right)\vec{E} = 0 \\
&\Rightarrow v = \frac{1}{\sqrt{\varepsilon_0 \mu_0}}
\end{aligned} \tag{10.11}$$

In this case of propagation of an electromagnetic wave in vacuum, the speed of propagation is denoted by $c = \frac{1}{\sqrt{\varepsilon_0 \mu_0}}$, which represents the speed of light. In an environment with ε and μ, the speed of propagation is:

$$v = \frac{1}{\sqrt{\varepsilon \mu}} = \frac{1}{\sqrt{\varepsilon_0 \mu_0}} \frac{1}{\sqrt{\varepsilon_r \mu_r}} = \frac{c}{\sqrt{\varepsilon_r \mu_r}} = \frac{c}{n} \tag{10.12}$$

Thus, the refractive index of the respective medium n, defined as the ratio $\frac{c}{v}$ is:

$$n = \frac{c}{v} = \sqrt{\varepsilon_r \mu_r} \tag{10.13}$$

In the case of an electromagnetic field whose local state quantities have (at a given moment) the same value at all points of a perpendicular plane in a preferred direction (the OX-axis), the solution of the Maxwell's equations is a plane electromagnetic wave, i.e. a transverse wave that has no components in the direction of propagation ($E_x = 0$, $H_x = 0$), as shown in figure 10.1. In such a wave, the vectors $\vec{E}$ and $\vec{H}$ oscillate in planes perpendicular to the direction of propagation.

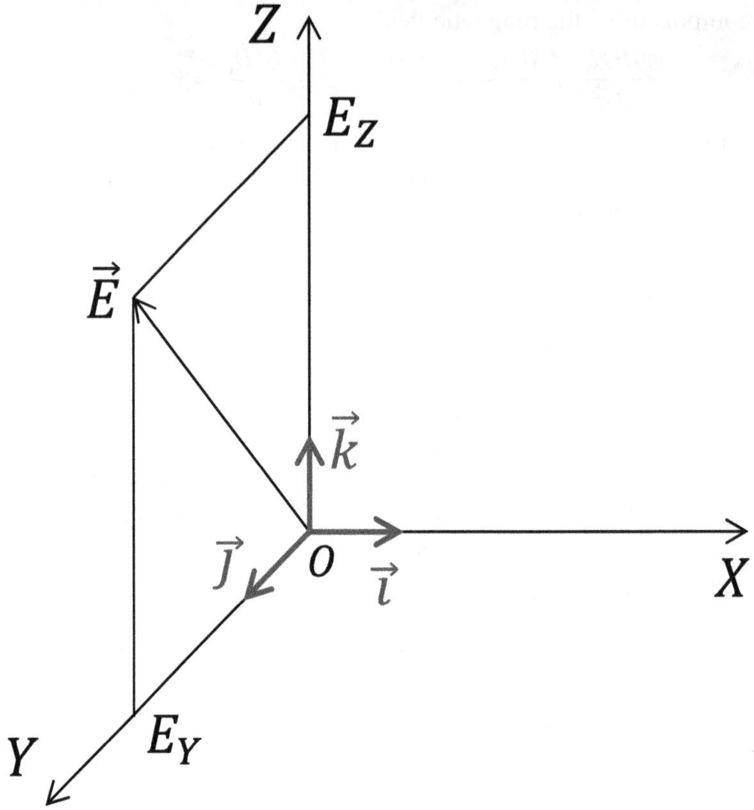

Figure 10.1. Electrical component of an electromagnetic wave.

Indeed, considering the velocity of propagation in the direction of OX and $E_x = 0, H_x = 0$ (and taking into account the fact that the values of the states $\vec{E}$ and $\vec{H}$ do not vary in the directions Y and Z, i.e. ($\frac{\partial}{\partial y} = 0$, $\frac{\partial}{\partial z} = 0$)), Maxwell's equations (10.3) become:

$$\begin{cases} \dfrac{\partial H_x}{\partial t} = 0, \ \dfrac{\partial E_z}{\partial x} = \mu\dfrac{\partial H_y}{\partial t}, \ \dfrac{\partial E_y}{\partial x} = -\mu\dfrac{\partial H_z}{\partial t}, \ \dfrac{\partial E_x}{\partial x} = 0 \\ \dfrac{\partial E_x}{\partial t} = 0, \ -\dfrac{\partial H_z}{\partial x} = \varepsilon\dfrac{\partial E_y}{\partial t}, \ \dfrac{\partial H_y}{\partial x} = \varepsilon\dfrac{\partial E_z}{\partial t}, \ \dfrac{\partial H_x}{\partial x} = 0 \end{cases} \tag{10.14}$$

(a) From these equations it can be seen that E_x and H_x do not vary either in time or space, i.e. they may constitute static fields but not time-varying fields, i.e. waves.
(b) The remaining unused equations of (10.14) may be grouped into two pairs of equations:

- one referring only to the quantities E_z and H_y:

$$\frac{\partial E_z}{\partial x} = \mu \frac{\partial H_y}{\partial t}$$
$$\frac{\partial H_y}{\partial x} = \varepsilon \frac{\partial E_z}{\partial t}$$

- and another only at the sizes E_y and H_z:

$$\frac{\partial E_y}{\partial x} = -\mu \frac{\partial H_z}{\partial t}$$
$$\frac{\partial H_z}{\partial x} = -\varepsilon \frac{\partial E_y}{\partial t}$$

These result in the following:

$$\begin{cases} \dfrac{\partial^2 E_z}{\partial x^2} - \varepsilon\mu \dfrac{\partial^2 E_z}{\partial t^2} = 0 \\ \dfrac{\partial^2 H_y}{\partial x^2} - \varepsilon\mu \dfrac{\partial^2 H_y}{\partial t^2} = 0 \end{cases}$$

$$\begin{cases} \dfrac{\partial^2 E_y}{\partial x^2} - \varepsilon\mu \dfrac{\partial^2 E_y}{\partial t^2} = 0 \\ \dfrac{\partial^2 H_z}{\partial x^2} - \varepsilon\mu \dfrac{\partial^2 H_z}{\partial t^2} = 0 \end{cases} \tag{10.15}$$

The pairs (E_z, H_y) and (E_y, H_z) are not related in any way, being independent of each other. If all the above electrical quantities are nonzero, the preceding equations express the existence of two superimposed waves that do not influence each other. Since each of these waves constitutes a transverse oscillation of unchanging direction (described in the field of optics as linear polarization), it follows that any plane electromagnetic wave comes from the superposition of two waves with linear polarizations in two orthogonal directions that are independent of each other (see figure 10.2). This observation allows the study to be restricted to only one of these waves; for example, the one described by the pair of physical quantities (E_y, H_z)—i.e. it is assumed that ($E_z = 0$, $H_y = 0$). The vectors $\vec{E} = \vec{j}\,E_y$ and $\vec{H} = \vec{k}H_z$ are perpendicular to each other and both perpendicular to the direction of propagation. In the general case of an electric field $\vec{E}$ in the YOZ plane ($\vec{E} = \vec{j}\,E_y + \vec{k}E_z$), applying $\nabla \times \vec{E} = -\frac{\partial \vec{B}}{\partial t}$ leads to:

$$\nabla \times \vec{E} = \begin{vmatrix} \vec{i} & \vec{j} & \vec{k} \\ \dfrac{\partial}{\partial x} & \dfrac{\partial}{\partial y} = 0 & \dfrac{\partial}{\partial z} = 0 \\ 0 & E_y & E_z \end{vmatrix} = -\vec{j}\frac{\partial E_z}{\partial x} + \vec{k}\frac{\partial E_y}{\partial x} = -\frac{\partial \vec{B}}{\partial t}$$

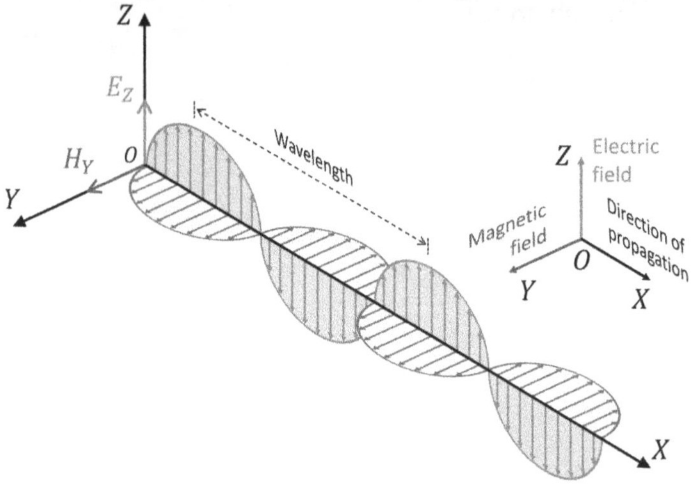

Figure 10.2. Propagation of an electromagnetic wave.

In other words:

$$\vec{i} \times \frac{\partial \vec{E}}{\partial x} = -\frac{\partial \vec{B}}{\partial t}$$

If:

$$\vec{E} = \vec{E_0} \cos \omega\left(t - \frac{x}{v}\right) \Longrightarrow (\vec{i} \times \vec{E_0}) \sin \omega\left(t - \frac{x}{v}\right)\frac{\omega}{v} = -\frac{\partial \vec{B}}{\partial t}$$

$$\Longrightarrow \vec{B} = (\vec{i} \times \vec{E_0}) \cos \omega\left(t - \frac{x}{v}\right) \ \frac{\omega}{v}\frac{1}{\omega}$$

$$\Longrightarrow \vec{H} = (\vec{i} \times \vec{E_0}) \cos \omega\left(t - \frac{x}{v}\right)\frac{1}{v\mu} \Longrightarrow \vec{H} \perp (\vec{i} \times \vec{E_0})$$

From the above calculus, we find that the intensity of the magnetic field $\vec{H}$ is perpendicular to $\vec{i}$ (i.e. to the direction of propagation) but also to $\vec{E_0}$ (i.e. to the electric field). In the particular case of $\vec{E} = \vec{j}\, E_y$, we obtain:

$$\vec{H} = (\vec{i} \times \vec{j})\frac{E_{0y} \cos \omega\left(t - \frac{x}{v}\right)}{v\mu} = \vec{k}\frac{E_{0y} \cos \omega\left(t - \frac{x}{v}\right)}{v\mu} = \vec{k}\frac{E_y}{v\mu} = H_z$$

$$\Longrightarrow \vec{H} \perp \vec{E} \text{ and } E_y = v\ \mu\ H_z = \zeta H_z$$

where the new quantity denoted by ζ represents the ratio of the intensity of the electric field to the intensity of the magnetic field; at a given point, this is called *wave impedance*. It is a physical quantity describing the properties of the material of the medium in which the wave propagates and is related to the constants characterizing the electrical (ε) and magnetic (μ) properties of the medium as follows:

$$\zeta = \frac{E_y}{H_z} = v\,\mu = \frac{\mu}{\sqrt{\varepsilon\mu}} = \sqrt{\frac{\mu}{\varepsilon}} = \sqrt{\frac{\mu_0}{\varepsilon_0}}\sqrt{\frac{\mu_r}{\varepsilon_r}} = \zeta_0\sqrt{\frac{\mu_r}{\varepsilon_r}} \tag{10.16}$$

When the wave propagates in vacuum, this physical quantity is called the *wave's impedance in vacuum*, which has the value:

$$\zeta_0 = \sqrt{\frac{\mu_0}{\varepsilon_0}} = \sqrt{\frac{4\pi \times 10^{-7}}{8.85 \times 10^{-12}}} \cong 377\ \Omega \tag{10.17}$$

Using this relationship, the intensity of the magnetic field can be expressed as a function of the intensity of the electric field:

$$H_z = \frac{E_y}{\zeta} \tag{10.18}$$

That is, at each point in space, the intensity of the magnetic field is proportional to and in phase with the intensity of the electric field. The expression $\zeta = \frac{E_y}{H_z}$ justifies the name of 'impedance' given to this physical quantity ζ, because E_y is measured in $\mathrm{V\,m^{-1}}$ and the intensity H_z is measured in $\mathrm{A\,m^{-1}}$; therefore, $\langle\zeta\rangle = \frac{V}{A} = \Omega$, similar to the impedance introduced for alternating current circuits. At this point, the problem of determining the physical quantities of the electromagnetic field in the plane electromagnetic wave is completely solved, since the study of the pair of quantities H_z and E_y would have led to completely analogous results. The following conditions can therefore be stated regarding plane electromagnetic waves:

(a) In homogeneous, linear, and isotropic (ε and μ are constant), electrically uncharged ($\rho = 0$), insulating ($\vec{j} = 0$), and indefinitely extended media, the solutions of Maxwell's equations that depend on a single spatial variable (corresponding, for example, to the direction OX) are superpositions of monochromatic plane waves propagating at a constant velocity of $v = \frac{1}{\sqrt{\varepsilon\mu}}$ or $c = \frac{1}{\sqrt{\varepsilon_0\mu_0}}$ in the case of vacuum.

(b) In each elementary wave, the vectors $\vec{E}$ and $\vec{H}$ are perpendicular to each other and perpendicular to the direction of propagation, so that the vectors $\vec{E}$, $\vec{H}$, and $\vec{v}$ form a right triorthogonal trihedron, i.e. the vector product $\vec{E} \times \vec{H}$ has the same direction as the direction of propagation.

(c) The variation over time of the quantities $\vec{E}$ and $\vec{H}$ is arbitrary and depends on the conditions of wave generation; at each point and at each moment, the values $\vec{E}$ and $\vec{H}$ are proportional, their ratio being the wave impedance of the respective medium.

The properties of the time-varying electromagnetic field deduced from Maxwell's equations led him to develop the electromagnetic theory of light (in 1865) by interpreting light as a particular form of electromagnetic wave characterized by the frequency spectrum to which the human eye is sensitive. According to this interpretation, the speed of electromagnetic waves in vacuum, i.e. $c = \frac{1}{\sqrt{\varepsilon_0 \mu_0}}$, must be interpreted as the speed of light in vacuum, and the quantity $n = \frac{c}{v} = \sqrt{\varepsilon_r \mu_r}$ must be interpreted as the refractive index of the medium (with $\mu_r = 1$ in non-ferromagnetic media). Additional experiments have shown that the relationship between c (measured optically) and ε_0, μ_0 (measured electrodynamically) is very well verified and that all conclusions regarding the electromagnetic nature of light are in agreement with the experimental data.

10.3 Electromagnetic energy theorem and the Poynting vector

The concept of the electromagnetic field as a form of matter capable of accumulating and transmitting energy allows us to energetically interpret a certain consequence of Maxwell's equations called *the electromagnetic energy theorem*. To prove this theorem, we will consider a volume V_Σ delimited by a surface Σ enclosed in an electromagnetic field that contains immovable bodies ($\vec{v} = 0$) with linear material properties (without hysteresis, i.e. ε and μ are independent of the respective electric field and magnetic field). Considering the density of the energy to be:

$$w = \frac{\vec{E}\vec{D} + \vec{H}\vec{B}}{2} = \frac{\varepsilon E^2}{2} + \frac{\mu H^2}{2} = \frac{D^2}{2\varepsilon} + \frac{B^2}{2\mu} \tag{10.19}$$

the total energy W stored in the electromagnetic field of the volume V_Σ is:

$$W = \int_{V_\Sigma} \frac{\vec{E}\vec{D} + \vec{H}\vec{B}}{2} dv \tag{10.20}$$

From the principle of conservation of energy, it follows that at any moment of time, the variation over time of the energy of the physical system constituted by the electromagnetic field inside the surface Σ (in other words, the rate of decrease of this energy) must be equal to the sum of the powers transferred by this physical system to other physical systems inside and outside the system. The electromagnetic field is in direct contact only with the immovable bodies inside the surface, which receive power only in the form specified by the law of transformation into conductive media (i.e. they receive only heat by the electrocaloric effect, not in the form of mechanical work). The electromagnetic field exchanges energy with the immovable bodies inside the surface at a power denoted by P_i; however, it also exchanges energy with the electromagnetic field outside the surface Σ, which is denoted by P_Σ. So, according to the law of conservation of energy, we can write:

$$-\frac{dW}{dt} = P_i + P_\Sigma \tag{10.21}$$

This relation describes the conservation of energy theorem as applied to the electromagnetic field, in which W is the total energy stored in the electromagnetic field inside the surface Σ. $P_i = \int_{V_\Sigma} \vec{E}\vec{j}\ dv$ is the power transmitted by the field to the immovable bodies inside the surface Σ during the process of charge transport, described by the conduction current density $\vec{j}$, in accordance with the Joule–Lenz law (in the local form $w = \vec{j}\,\vec{E} = \sigma E^2 = \rho j^2$). P_Σ is the flow of electromagnetic energy, i.e. the power transmitted by the electromagnetic field through the closed surface Σ, the expression for which is still unknown and needs to be evaluated. Relationship (10.21) can be written as:

$$-\frac{d}{dt}\int_{V_\Sigma}\left(\frac{D^2}{2\varepsilon}+\frac{B^2}{2\mu}\right)dv = \int_{V_\Sigma}\vec{E}\vec{j}\ dv + P_\Sigma \tag{10.22}$$

Since the environment inside the surface is immovable, on the left-hand side of equation (10.22), one can switch the differentiation operator with the integration operator, resulting in:

$$-\int_{V_\Sigma}\left(\frac{2\vec{D}}{2\varepsilon}\frac{\partial\vec{D}}{\partial t}+\frac{2\vec{B}}{2\mu}\frac{\partial\vec{B}}{\partial t}\right)dv = \int_{V_\Sigma}\vec{E}\vec{j}\ dv + P_\Sigma \tag{10.23}$$

Using Maxwell's equations, it follows that:

$$-\int_{V_\Sigma}\left\{\frac{\vec{D}}{\varepsilon}[(\nabla\times\vec{H})-\vec{j}\,]+\frac{\vec{B}}{\mu}[-(\nabla\times\vec{E})]\right\}dv = \int_{V_\Sigma}\vec{E}\vec{j}\ dv + P_\Sigma$$

or:

$$-\int_{V_\Sigma}[\vec{E}(\nabla\times\vec{H})-\vec{E}\vec{j}-\vec{H}(\nabla\times\vec{E})]dv = \int_{V_\Sigma}\vec{E}\vec{j}\ dv + P_\Sigma \tag{10.24}$$

However, $\nabla(\vec{E}\times\vec{H}) = \vec{H}(\nabla\times\vec{E}) - \vec{E}(\nabla\times\vec{H})$ (a relationship that can also be demonstrated for its components) and therefore:

$$\int_{V_\Sigma}\vec{E}\vec{j}\ dv + \int_{V_\Sigma}[\vec{H}(\nabla\times\vec{E})-\vec{E}(\nabla\times\vec{H})]dv = \int_{V_\Sigma}\vec{E}\vec{j}\ dv + P_\Sigma$$

or:

$$\int_{V_\Sigma}\nabla(\vec{E}\times\vec{H})\ \ dv = \int_{\Sigma}(\vec{E}\times\vec{H})ds = P_\Sigma \tag{10.25}$$

Thus, we have obtained the following expression for the flow of electromagnetic energy (the energy transmitted by the electromagnetic field through the surface Σ):

$$P_\Sigma = \int_{\Sigma}(\vec{E}\times\vec{H})\vec{ds} \tag{10.26}$$

and, respectively, *the electromagnetic energy theorem*, which is written in full as follows:

$$-\frac{d}{dt}\int_{V_\Sigma}\frac{\vec{E}\vec{D}+\vec{H}\vec{B}}{2}dv \quad = \int_{V_\Sigma}\vec{E}\vec{j}\ dv + \int_{\Sigma}(\vec{E}\times\vec{H})\vec{ds} \tag{10.27}$$

being a direct consequence of Maxwell's equations. It states that *the rate of decrease in the energy of an electromagnetic field in a domain limited by the closed surface Σ is equal to the sum of the power transferred to the bodies within the domain (only by the Joule–Lenz effect in the case of immovable bodies) and the flux through the surface Σ of the vector $\vec{Y}$ called the flux density of electromagnetic energy or the Poynting vector:*

$$\vec{Y} = \vec{E}\times\vec{H} \tag{10.28}$$

The Poynting vector is oriented, as we have seen, in the direction of propagation, $\vec{E}\times\vec{H}$. For the plane electromagnetic wave described by E_y and H_z, it follows that:

$$\begin{aligned}\vec{Y} &= \vec{j}\,E_y\times\vec{k}H_z = \vec{i}\,E_yH_z = \vec{i}\,\zeta H^2 = \vec{i}\,\mu v\,\frac{B^2}{\mu^2} = \vec{v}\mu H^2 = \vec{v}w\\ &= \vec{i}\,E_y\frac{E_y}{\zeta} = \vec{i}\frac{\sqrt{\varepsilon}E^2}{\sqrt{\mu}} = \vec{i}\,\varepsilon\,\frac{E^2}{\sqrt{\varepsilon\mu}} = \vec{i}\,v\,\varepsilon\,E^2 = \vec{v}\varepsilon E^2 = \vec{v}w\end{aligned} \tag{10.29}$$

where the energy density of the electromagnetic field is:

$$\begin{aligned}w &= \frac{\varepsilon E^2}{2}+\frac{\mu H^2}{2} = \frac{\varepsilon E^2}{2}+\frac{\mu E^2}{2\zeta^2} = \varepsilon E^2\\ &= \frac{\varepsilon\zeta^2H^2}{2}+\frac{\mu H^2}{2} = \frac{\mu H^2}{2}+\frac{\mu H^2}{2} = \mu H^2\end{aligned} \tag{10.30}$$

So, indeed, the Poynting vector is the product of the propagation velocity and the energy density. This relationship, i.e. $\vec{Y}=\vec{v}w$, shows that the electromagnetic energy flux density or Poynting vector represents the energy carried by the electromagnetic field per unit time through the unit surface from a plane perpendicular to the direction of propagation of the electromagnetic wave.

Further reading

[1] Antohe Ş and Antohe V-A 2023 *Electrostatics. Formalism of the Electrostatic Field in Vacuum and Matter* 1st edn (Bristol: IOP Publishing) 266 pages

[2] Purcell E M and Morin D J 2013 *Electricity and Magnetism* 3rd edn (Cambridge: Cambridge University Press) 853 pages

[3] Feynman R P, Leighton R B and Sands M 2011 *Feynman Lectures on Physics, Vol. II: The New Millennium Edition: Mainly Electromagnetism and Matter (Feynman Lectures on Physics)* (New York: Basic Books) 589 pages

[4] Sadiku M and Nelatury S 2021 *Elements of Electromagnetics (Oxford Series Electric Computer Engineer)* (Oxford: Oxford University Press)

[5] Jackson J D 1998 *Classical Electrodynamics* 3rd edn (New York: Wiley) 832 pages

[6] Bleaney B I and Bleaney B 1965 *Electricity and Magnetism* 3rd edn vol 1 (Oxford: Oxford University Press)

[7] Maxwell J C 2003 *A Treatise on Electricity and Magnetism* **vol 1** (Mineola, NY: Dover) 506 pages

[8] Griffiths D J 2017 *Introduction to Electrodynamics* 4th edn (Cambridge: Cambridge University Press)

[9] Clayton R, Keith W and Nasar S 1997 *Introduction to Electromagnetic Fields* (New York: McGraw-Hill)

[10] Kao M-S and Chang C-F 2021 *Understanding Electromagnetic Waves* 1st edn (Berlin: Springer) 468 pages

www.ingramcontent.com/pod-product-compliance
Lightning Source LLC
Chambersburg PA
CBHW082132210326
41599CB00031B/5953